Lecture Notes in Artificial Intelligence   1476

Subseries of Lecture Notes in Computer Science
Edited by J. G. Carbonell and J. Siekmann

Lecture Notes in Computer Science
Edited by G. Goos, J. Hartmanis and J. van Leeuwen

Springer
*Berlin
Heidelberg
New York
Barcelona
Budapest
Hong Kong
London
Milan
Paris
Singapore
Tokyo*

Jacques Calmet   Jan Plaza (Eds.)

Artificial Intelligence and Symbolic Computation

International Conference AISC'98
Plattsburgh, New York, USA
September 16-18, 1998
Proceedings

 Springer

Series Editors

Jaime G. Carbonell, Carnegie Mellon University, Pittsburgh, PA, USA
Jörg Siekmann, University of Saarland, Saarbrücken, Germany

Volume Editors

Jacques Calmet
Universität Karlsruhe, Fakultät Informatik
Am Fasanengarten 5, Postfach 6980, D-76128 Karlsruhe, Germany
E-mail: calmet@ira.uka.de

Jan Plaza
Plattsburgh State University of New York
101 Broad Street, Plattsburgh, NY 12901, USA
E-mail: plazaja@splava.cc.plattsburgh.edu

Cataloging-in-Publication Data applied for

Die Deutsche Bibliothek - CIP-Einheitsaufnahme

Artificial intelligence and symbolic computation : proceedings /
International Conference AISC '98, Plattsburgh, New York, USA,
September 16 - 18, 1998. Jacques Calmet ; Jan Plaza (ed.). - Berlin ;
Heidelberg ; New York ; Barcelona ; Budapest ; Hong Kong ;
London ; Milan ; Paris ; Singapore ; Tokyo : Springer, 1998
    (Lecture notes in computer science ; Vol. 1476 : Lecture notes in
    artificial intelligence)
    ISBN 3-540-64960-3

CR Subject Classification (1991): I.2.1-4, I.1, G.1-2, F.4.1

ISBN 3-540-64960-3 Springer-Verlag Berlin Heidelberg New York

© Springer-Verlag Berlin Heidelberg 1998
Printed in Germany

Typesetting: Camera ready by author
SPIN 10638716   06/3142 – 5 4 3 2 1 0   Printed on acid-free paper

Foreword

This volume contains invited papers and contributed papers accepted for the Fourth International Conference on Artificial Intelligence and Symbolic Computation (AISC'98) held in USA at the Plattsburgh State University of New York on September 16–18, 1998. The conference belongs to a series that started in Karlsruhe, Germany in 1992 and continued at King's College, Cambridge, UK (1994) and in Steyr, Austria (1996). The proceedings of these earlier meetings appeared in the LNCS series as Volumes 737, 958, and 1138. The first three conferences were called AISMC, where M stands for "mathematical"; the organizers of the current meeting decided to drop the adjective "mathematical" and to emphasize that the conference is concerned with all aspects of symbolic computation in AI: mathematical foundations, implementations, and applications, including applications in industry and academia.

In the opening paper in the proceedings of AISMC-1, the founders of the series, J. Calmet and J. Campbell, gave a brief history of the relationship between artificial intelligence and computer algebra. They also described the scope and goals of the conferences. The goal was to emphasize the interaction of methods and problem solving approaches from AI and symbolic mathematical computation. This year, the conference name has been modified not to exclude mathematics but to illustrate a broadening of the scope. Mathematics with its universally accepted language and methodology remains an ideal environment for constructing precise models of symbolic computation in AI. The scope was in fact already extended in Steyr where the relationship between symbolic mathematical computing and engineering was emphasized. A motivation for extending the scope comes also from the fact that in recent years research on coupling computing and deduction systems has flourished. One of the goals of this series of conferences is to become a forum for such research. This is already exemplified by some of the papers appearing in this volume. Another goal is to move the center of gravity from computer algebra and to set it in artificial intelligence. This is also why the current proceedings are appearing in the LNAI series rather than LNCS.

As in previous conferences, the number of accepted papers has been kept small. This allows for longer than usual presentations by the authors with enough time for discussion, and makes it possible to avoid parallel sessions. Such arrangements provide better conditions for exchange of ideas. Since the scope of the conference is interdisciplinary, this is a mandatory condition to make the meetings successful and lively.

The site for the next conference is already decided. It will be Madrid, Spain, in the year 2000.

We gratefully acknowledge the sponsorship of AISC'98 by the Plattsburgh State University of York. We owe thanks to the members of the Program Committee and Steering Committee for refereeing contributed papers. Other organizations and individuals who contributed to the conference cannot be listed here at the time when these proceedings go to print. We are grateful to them all. Special thanks go to Christoph Zenger who devoted a lot of time to preparing the manuscript for this volume.

June 1998   Jacques Calmet and Jan Plaza

Steering Committee

Jacques Calmet (Karlsruhe, Germany)
John Campbell (London, Great Britain)
Jochen Pfalzgraf (Salzburg, Austria)
Jan Plaza (Plattsburgh, USA), Conference Chairman

Program Committee

Francois Arlabosse (Framatome, France)
Bruno Buchberger (Linz, Austria)
Gregory Butler (Montreal, Canada)
Luigia Carlucci Aiello (Rome, Italy)
James Cunningham (London, Great Britain)
John Debenham (Sydney, Australia)
Ruediger Dillmann (Karlsruhe, Germany)
Fausto Giunchiglia (Trento, Italy)
Stan Klasa (Montreal, Canada)
Alfonso Miola (Rome, Italy)
Lin Padgham (Melbourne, Australia)
Zbigniew W. Ras (Charlotte, USA)
Klaus U. Schulz (Munich, Germany)
Joerg H. Siekmann (Saarbruecken, Germany)
Andrzej Skowron (Warsaw, Poland)
Stanly Steinberg (Albuquerque, USA)
Karel Stokkermans (Salzburg, Austria)
Carolyn Talcott (Stanford, USA)
Peder Thusgaard Ruhoff (Odense, Denmark)
Dongming Wang (Grenoble, France)

Contents

An Inductive Logic Programming Query Language for Database Mining (Extended Abstract)

Luc De Raedt

Department of Computer Science, Katholieke Universiteit Leuven
Celestijnenlaan 200A, B-3001 Heverlee, Belgium
email: Luc.DeRaedt@cs.kuleuven.ac.be
Tel: ++ 32 16 32 76 43 Fax : ++ 32 16 32 79 96

Abstract. First, a short introduction to inductive logic programming and machine learning is presented and then an inductive database mining query language RDM (Relational Database Mining language). RDM integrates concepts from inductive logic programming, constraint logic programming, deductive databases and meta-programming into a flexible environment for relational knowledge discovery in databases. The approach is motivated by the view of data mining as a querying process (see Imielinkski and Mannila, CACM 96). Because the primitives of the presented query language can easily be combined with the Prolog programming language, complex systems and behaviour can be specified declaratively. Integrating a database mining querying language with principles of inductive logic programming has the added benefit that it becomes feasible to search for regularities involving multiple relations in a database. The proposal for an inductive logic programming query language puts inductive logic programming into a new perspective.
Keywords : database mining query language, inductive logic programming, relational learning, inductive query language, data mining.

1 Introduction

The first part of the paper provides a short introduction to the field of inductive logic programming. Inductive logic programming [20, 7] is the study of machine learning and data mining using the first order representations offered by computational logic. Classical approaches to machine learning and data mining use the so-called attribute-value learning representations, which essentially correspond to propositional logics. The use of computational logic is beneficial for a variety of reasons. Firstly, inductive logic programming can rely on the theory of logic programming concerning semantics, inference rules, and execution mechanisms. Secondly, using a richer representation language permits to tackle applications, which cannot be handled by classical techniques (see e.g. [4, 12]) for surveys. Thirdly, the use of computational logic permits to employ background knowledge in the induction process.

The second part of the paper introduces a novel framework for inductive logic programming and data mining in the form of an inductive database mining query

Jacques Calmet and Jan Plaza (Eds.): AISC'98, LNAI 1476, pp. 1–13, 1998.
© Springer-Verlag Berlin Heidelberg 1998

language. This framework is inspired on [13] who view knowledge discovery as a querying process to a database mining system. Querying for knowledge discovery requires an extended query language (w.r.t. database languages) that supports the manipulation, mining and discovery of rules, as well as data. The integration of such rule querying facilities provides new challenges for database technology.

(Inductive) logic programming is particularly well-suited as a basis for such a language. First, a general query language for data mining should not be restricted to handling single relations in a database; rather it should support the handling of multiple relations in a database [11]. Inductive logic programming supports this very naturally. Second, techniques such as meta-programming and query optimization have been studied extensively within the field of logic programming, constraint logic programming and deductive databases. Meta-programming is required to support the manipulation of data and rules, and query optimization is important for efficiency purposes. Third, using logic programs as queries it is easy to provide a semantics for the query language. Fourth, the development of this language relates inductive logic programming to constraint logic programming as the key questions that arise are similar to those addressed in constraint logic programming: semantics, constraint handling, development of solvers, properties of solvers, etc. Finally, though the field of inductive logic programming [15, 20, 7] is sometimes regarded as too inefficient for practical purposes, it provides an excellent conceptual framework for reasoning about data mining. In sum, the embedding of the query language within logic programming results in a very expressive tool for knowledge discovery and in a new view on inductive logic programming and data mining.

This paper is organised as follows: Section 2 contains an intuitive introduction to inductive logic programming and data mining, Section 3 presents the RDM query language, Section 4 shows RDM at work, Section 5 shows how RDM can be implemented, and finally Section 6 concludes.

2   Machine learning and data mining: the ILP view

Current data mining approaches are often distinguished on the basis of their predictive or descriptive nature. In predictive data mining one is given a set of examples or observations that are classified into a finite number of classes. Typically, there are two (or more) classes, one that is called positive, and the other that is negative. The aim then is to induce a hypothesis that correctly classifies all the given (and unseen) examples. Consider Figure 1a where one is given two types of example (+ and -). H is a correct hypothesis as it correctly discriminates the positives from the negatives. The purpose of predictive data mining is thus to generate hypotheses that can be used for classification. Common predictive data mining techniques include decision tree induction (e.g. C4.5 [21]) and rule induction (e.g. CN2 [6, 5] or AQ [17]).

In descriptive data mining one is given a set of unclassified examples and the aim is to find regularities within these examples. Furthermore, it is the aim to characterize as much as possible the given examples. Therefore as many

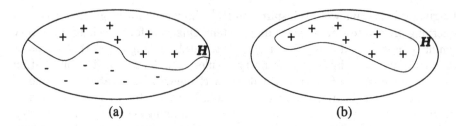

Fig. 1. Predictive versus descriptive induction

properties or regularities as possible are derived. Together they form a kind of most specific hypothesis that covers (explains) all the examples. E.g. in Figure 1b, the hypothesis H characterizes all the given examples. The most popular descriptive data mining technique is that of discovering association rules [1, 2, 26].

Let us illustrate these differences on a simple example from basket analysis in supermarkets. Using an attribute value learning technique the most straightforward representation is given in the following table:

Trans	Customer	Beer	Sausage	...	Mustard	Party
t190	c9	48	48	...	2	yes
t110	c88	12	0	...	1	no
t389	c133	0	6	...	1	no
t444	c9	48	96	...	3	yes

Fig. 2. A table for basket-analysis

There is one tuple for each visit to the supermarket. E.g. the first tuple states that transaction t190 concerns customer c9 who bought 48 units of beer and sausage and 2 units of mustard. Furthermore, his intention was to have a party with this.

Predictive machine learning in this context could be concerned with predicting whether a given transaction has as intention to organise a party or not. E.g. one rule could state that if the quantity of beer purchased is larger than 36 then party=yes; otherwise, party=no. Descriptive data mining on the other hand would be considered with regularity finding. It could e.g. find regularities such as if mustard and beer is purchased, then with probability 80 per cent beer will also be purchased. Or, if customer c9 purchases, then he will organise a party, ...

Another major difference between various machine learning and data mining techniques concerns the representations employed. In the above illustration of attribute-value learning, an example was a *single* tuple from a *single* relation in a database. In contrast, in inductive logic programming, examples can be encoded in multiple tuples of multiple relations. Using inductive logic programming, it would be possible to reason about several transactions of the same customer, or about relations among customers, etc. One might for instance have another table $relative(C1, C2)$ that expresses that $C1$ and $C2$ belong to the same family. Rules that could be induced in such a context are e.g. that if a customer buys Q1 units of product X at time T and there are advertisements for product X after T, then on the next visit the same customer will buy more than Q1 units of product X.

Typically, the relational representations employed within inductive logic programming are needed when tackling structural problems (e.g. in molecular biology), or when reasoning about time or space, or sequences, ...

Finally, using inductive logic programming it is also easy to incorporate background knowledge in the induction process. In the above example this could take the form of taxonomies for the different products. In this way, wine and beer could be grouped under alcohol, knowledge about advertising could be added, etc.

At this point the reader may not (yet?) be convinced of the need for relational representations. I'd like to challenge those readers to represent the Bongard problem in Figure 2 within the attribute value representation, that is each scene (example) should be represented by single tuple in a single relation. using a single tuple in a single relation. Though the Bongard problem is a toy problem its representations are very similar to those needed to represent the structure of e.g. molecules (cf. [25]).

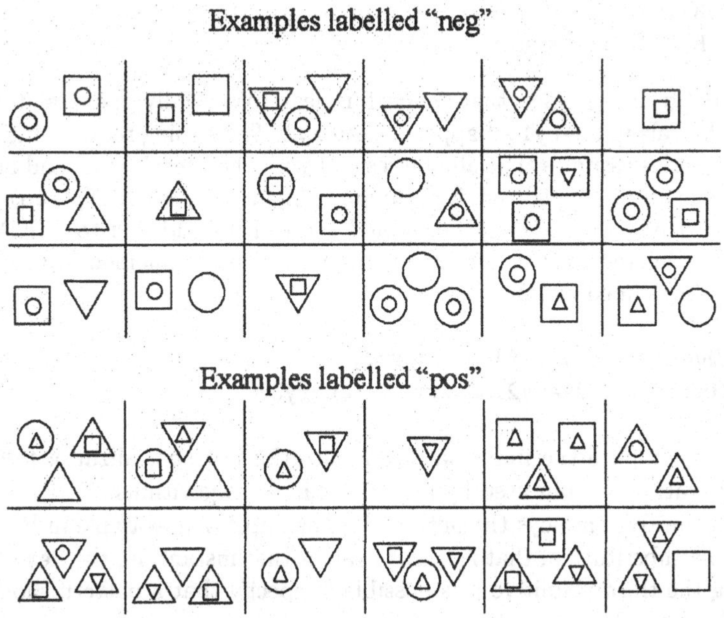

Fig. 3. Bongard problems

3   RDM : a relational database mining query language

In this section, a novel database mining query language is proposed. We do assume some familiarity with Prolog [3] and with machine learning [18].

The key notion in RDM is that of a pattern. A pattern (Query,Key) is a tuple where Query is a Prolog query, and Key is a list of variables that appears in the Query. Data mining with RDM is then the search for patterns that satisfy certain requirements. Three primitives are foreseen to specify requirements in RDM. The first concerns the frequency of patterns, which is defined as follows :

```
frequency(Query,Key,Frequency) :-
        findall(Key,Query,Set),length(Set,Frequency).
```

Thus the frequency of a pattern (Query,Key) is the number of instantiations of the Key for which the Query holds. E.g. the pattern
(purchase(T,C,B,...,yes),[T,C]) has frequency 2 in the above table.

The second primitive concerns the syntactic form of the Query. The primitive $\preceq$ specifies the relation of the Query to specified queries. This corresponds to encoding the 'is more general than'[1] relation. E.g.

```
Query1 ⪯ Query2 :- subset(Query1,Query2).

subset((S,RS),Super) :- occurs(S,Super,NSuper), subset(RS,NSuper).
subset(true,Super).

occurs(Q,(Q,R),R).
occurs(Q,(T,R),NR) :- occurs(Q,R,NR).
```

This Prolog program specifies that Query1 is less (or equally) general than Query2 if the literals in Query1 all appear in Query2. The Prolog representation employed for queries assumes that all queries end with the literal 'true' and also that the non-true literals are sorted. E.g. running (human(X),male(X),true) $\preceq$ (female(X),human(X),male(X),true) would succeed. In addition to $\preceq$ RDM also employs $\prec$. Related to this primitive is the notion of a refinement operator rho which can be defined as follows :

```
rho(Query1,Query2) :- Query1 ≺ Query2,
        not (Query3 ≺ Query2, Query1 ≺ Query3).
```

rho is useful to compute the immediate generalizations or specializations of a query. This operation is often used by machine learning algorithms.

The third primitive employs the notion of an example. An example in RDM corresponds to a substitution, that is a set of values that instantiate the Key of a pattern. Using the third primitive it is possible to specify that a pattern should cover a specific example:

---

[1] Ideally, this should be implemented through the use of Plotkin's θ-subsumption.

```
covers(Query,Key,Example) :- \+ \+ ( Key = Example, Query).
```

If the database would consist of the following facts : `human(luc)`, `human(lieve)`, `male(luc)`, `female(lieve)` then the pattern `((human(X),male(X),true),[X])` would cover the example [luc].

Finally, RDM has a number of extensions of the above primitives in order to provide more flexibility. One extension allows to work with lists of examples. E.g. in the above toy-database one might want to work with a list of examples such as e.g.[[luc],[lieve]]. Using such lists of examples it is possible to refine the above primitives. More specifically, RDM provides the following variants of the above predicates :

- `covers-all(Query,Key,ExampleList)` (resp. `covers-none(Query, Key, ExampleList)` succeeds if all examples in `ExampleList` are covered (resp. are not covered) by the pattern `(Query,Key)`,
- `frequency(Query,Key,ExampleList,Freq)` which succeeds if `Freq` is the number of examples on `ExampleList` that are covered by the pattern `(Query, Key)`.

The second extension concerns the definition of asssociation rules in the typical manner [2].

```
associationrule( (Conclusion ← Condition), Key, Acc, Freq) :-
      Condition ≺ Conclusion, frequency(Condition,Key,Freq1),
      frequency(Conclusion,Key,Freq), Acc is ( Freq1 + 0.0)/ Freq.
```

`Acc` expresses the accuracy of the rule, i.e. the conditional probability that `Conclusion` will succeed given that `Condition` succeeded, and `Freq` expresses the number of times that both `Condition` and `Conclusion` part succeed. This type of rule is very popular in descriptive data mining.

The third extension allows to employ negation in various forms, e.g. to specify that a query is not more specific than a given other query e.g. not `Query` ≼ `(human(X),female(X),true)`.

Observe also that it is easy to enhance RDM by providing further predicate definitions, such as e.g.

- `sample(ExampleList,SampleList,Probability)` generates a sample of the examples (drawn according to `Probability`).

This and other extensions are however quite straightforward, and merely provide further syntactic sugar. Therefore we will not discuss such extensions any further here.

The above predicate definitions provide the semantics of the RDM language. A database mining query is now simply a Prolog query employing the above introduced primitives. So, RDM is embedded within Prolog. It should be mentioned that not all database mining queries are safe, but that it is possible to specify safe queries using mode declarations (as typically done in Prolog manuals). For safe execution, it is e.g. required that for all patterns `(Query,Key)` searched for the data mining query in Prolog must include $Query \preceq (lit_1, ..., lit_n, true)$ before any further reference to $Query$ is made. This is necessary in order to specify the range of queries searched for. Also, without bounding $Query$ in this manner, many of the predicates defined above would not operate correctly. We will not go into further (implementation) details of the semantics at this point. We also wish to stress that the semantical definition is - of course - inefficient. The only purpose of the definitions above is to specify an operational semantics of data mining queries and not to provide an optimal execution mechanism. However, in section 4, we will show how data mining queries can be executed efficiently using techniques from the versionspace [19] and APRIORI algorithms [2].

4   Data mining with RDM

We now basically have all primitives needed to generate database mining queries. In this section, we demonstrate the expressive power of RDM by example.

First, the query language can naturally be used to discover a wide range of frequent patterns. E.g. inducing frequent item-sets goes as follows :

```
?- Query ≼ (Item-1(X), ..., Item-n(X),true), frequency(Query,X,F),
F > 100 .
```

This corresponds to the simplest form of frequent pattern that is currently employed in e.g. marketing applications. The database then merely contains facts which state for all transactions X which items have been bought. The above query then finds frequent combinations of items that are bought together.

Similarly, it is possible to discover first order queries such as those typically employed within inductive logic programming. E.g. in marketing

```
?- (trans(T1,Cust,Time1),trans(T2,Cust,Time2), before(Time1,Time2),
true) ≼ Query, Query ≼ (trans(T1,Cust,Time1),trans(T2,Cust,Time2),
before(Time1,Time2), (Item-1(T1,Q1),Item-1(T2,Q2),Q2 > Q1), ...,
(Item-n(T1,Q1),Item-n(T2,Q2),Q2 > Q1),true), frequency(Query,[T1,T2],F),
F > 100.
```

This database mining query would generate all patterns involving two (ordered) transactions of the same customer where the quantities of bought items in the last transaction is larger. E.g.

```
(trans(T1,Cust,Time1),trans(T2,Cust,Time2), before(Time1,Time2),
(beer(T1,Q1), beer(T2,Q2), T2 > T1), true)
```

would tell something about the frequency of increased sales in beer.

Another first order example of database mining query in a biochemical domain:

```
?-(atom(M,A), true) ⪯ Query1, Query1 ⪯ (atom(M,A), property-1(A),
..., property-n(A),true), frequency(Query1,M,F1), F1 > 20,
(bond(M,A,B), atom(B), true) ⪯ Next, Next ⪯ (bond(M,A,B), atom(B),
property-1(B), ... , property-n(B),true), concatenate(Query1, Next,
Query2), frequency(Query2, M, F2), F2 > 15 .
```

This database mining query will generate two patterns the first of which involves a single atom (and its properties) and the second pattern which is an extension of the first pattern involving two atoms linked with a certain type of bond. The results of this query could be used to generate association rules where the first pattern serves as the condition part and the second one as the conclusion part.

Further details of the use of such first order patterns and the relation to existing frameworks for frequent pattern discovery are provided in a recent paper by Dehaspe and Toivonen [10]. They show that nearly any type of frequent pattern considered in the data mining literature can be expressed in the presented first order framework. Hence, it is easy to simulate these in RDM.

A second type of query is to generate association rules. E.g.

```
?- Query2 ⪯ (Item-1(X), ... , Item-n(X),true),
associationrule((Query1 ← Query2),X,Freq,Acc),
Freq > 100, Acc > 0.5.
```

would generate all association rules (over item-sets) with minimum accuracy of 50 per cent and frequency of 100. This type of query is similar to the queries one can express within Imielinski and Virmani 's M-SQL [14]. Using this same type of queries it is possible to emulate the essence of Dehaspe and De Raedt's WARMR [9] and De Raedt and Dehaspe's Claudien [8].

The third type of queries is aimed at predictive data mining instead of descriptive datamining. It therefore employs both positive and negative examples. E.g.

```
?- Query ⪯ (Item-1(X), ... , Item-n(X),true),
frequency(Query,X,'Positives',P), frequency(Query,X,'Negatives',N),
N < 10, Acc is P / (P+N), Acc > 0.9.
```

This query would generate condition parts of rules that are at least 90 per cent correct, and that cover at most 10 negative examples. Once such rules are found, they can (partly) define a predicate with the Key as argument(s). This is akin to what many classification systems search for. Also, rather than involving a minimum frequency, this type of rule querying imposes a maximum frequency (on the negative examples).

Fourthly, sometimes predictive data mining approaches employ seed examples. These seed examples are used to constrain the rules searched for. E.g. when positive examples are used as seeds, the induced rule should cover the positive. This can easily be expressed with **covers**. In typical database mining situations, these different constructs will be combined in order to find relevant (sets of) patterns.

Of course, the most expressive way of using the query language is to embed the primitives in simple Prolog programs. Using this facility it is easy to emulate inductive logic programming systems (such as e.g. Quinlan's FOIL [22]) in a one page meta-program. This requires the operator rho to be used.

5   Answering database mining queries

Obviously, plainly using the primitives as they have been defined in Prolog will be desastrous from the efficiency point of view. This is because patterns would simply be enumerated using the $\preceq$ primitive, and tested afterwards. The question then is : how can we efficiently produce answers to the queries formulated above?

Let us here concentrate on simple database mining queries. Simple queries search for a single pattern only.

The key insight that leads to efficient algorithms for answering database mining queries is that 1) the space of patterns $\mathcal{L}$ is partially ordered by the $\preceq$ relation, and 2) that all primitives provided in RDM behave monotonically with regard to $\preceq$. Therefore each constraint on a pattern allows to prune the search. Indeed consider the constraint that $Query \preceq q$, which states that the $Query$ searched for should be more general than a specific query q. Given e.g. a candidate query c that is not more general than q, all specializations of c can be pruned as well (as they also won't be a generalization of q). Similar properties hold for the other primitives supported by RDM.

Furthermore, a set of constraints defines a versionspace of acceptable patterns. This versionspace is completely determined by its S and G sets (as defined by [19]).

$S = \{s \in \mathcal{L} \mid$ s is maximally specific in $\mathcal{L}$ and consistent with all constraints$\}$
$G = \{g \in \mathcal{L} \mid$ g is maximally specific in $\mathcal{L}$ and consistent with all constraints$\}$

These sets determine the boundaries of consistent queries. All solutions c then have the property that there is an element $s \in S$ and an element $g \in G$ such that $g \preceq c \preceq s$. We have the following kind of constraints or information elements :

- $Query \preceq q$ (resp. $q \preceq Query$) which states that the target $Query$ is more general (resp. more specific than a given query q; similar elements can be defined wrt. $\prec$).
- $frequency(Query, Key, Data, Freq), Freq > t$ (resp. $\geq$) : which specifies a constraint on the frequency of the pattern (on the specified $Data$)
- $covers(Query, Key, Example)$ which expresses that the $Query$ should cover the $Example$
- also, each of the above primitives can be negated

Information elements of the first type (i.e. concerning $\preceq$) are directly handled by Mellish's description identification algorithm [16, 23], which extends Mitchell's versionspace algorithm [19]. Fact is that the statements of the form $Query \preceq q$ (resp. *not* $Query \preceq q$) can be interpreted as $Query$ covers the 'positive example' q (resp. 'negative example'). In this view the query q is regarded an example (in a different example space than that of RDM). Mellish's description identification algorithm extends Mitchell's well-known versionspace algorithm by handling statements of the form $q \preceq Query$ and its negation. These additional information elements are dual in the original ones, and so is its algorithmic treatment. Now it is easy to process the elements concerning $\preceq$ with the versionspace algorithm. First define the greatest lower bound glb, the least upper bound lub, the maximally general specialisations mgs, and the maximally specific generalizations msg (max and min find the maximal respectively minimal elements w.r.t. $\preceq$):

$$glb(a, b) = \max \{d \in \mathcal{L} \mid a \preceq d \text{ and } b \preceq d\}$$
$$lub(a, b) = \min \{d \in \mathcal{L} \mid d \preceq a \text{ and } d \preceq b\}$$
$$mgs(a, b) = \max \{d \in \mathcal{L} \mid a \preceq d \text{ and } not(d \preceq b)\}$$
$$msg(a, b) = \min \{d \in \mathcal{L} \mid d \preceq a \text{ and } not(b \preceq d)\}$$

Then the algorithm in Figure 4 can be used to compute S and G, which completely determine the set of solutions.

When analysing the above algorithm, the glb, lub, mgs and msg operators can be straightforwardly and efficiently implemented. However, the key implementation issue is how to compute the S set for the case of $frequency(Q, K, D, F), F > f$. Similar questions arise for $F < f$ and the use of covers. In the data mining literature, a lot of attention has been devoted to this, resulting in algorithms such as APRIORI [2].

We can find the S set for this case in two different manners, that correspond to traversing the space in two dual methods. When working general to specific, one can use an algorithm similar to APRIORI.

The algorithm for computing answers to simple queries should be upgraded towards complex queries. Complex queries are answered by handling the literals in the query from left to right (as in Prolog) and mixing that with the versionspace model. This is however the subject of further work.

6   Conclusions

A declarative relational database mining query language has been proposed as well as a procedure to efficiently answer simple queries.

It was argued that the proposed framework for query language puts inductive logic programming into a new perspective and also that it raises several questions of interest to the field of computational logic and data mining. In the author's view, the current proposal and the resulting line of research puts inductive logic programming on the same methodological basis as constraint logic

$S := \{\top\}; G := \{\bot\};$
for all constraints i **do**
    case i of $q \preceq Query$:
        $S := \{s \in S \mid q \preceq s\}$
        $G := \max \{m \mid \exists g \in G : m \in glb(q, g) and \exists s \in S : m \preceq s\}$
    case i of $Query \preceq q$:
        $G := \{g \in G \mid g \preceq q\}$
        $S := \min \{m \mid \exists s \in S : m \in lub(q, s) and \exists g \in G : g \preceq m\}$
    case i of $not\ Query \preceq q$:
        $S := \{s \in S \mid not(s \preceq q)\}$
        $G := \max \{m \mid \exists g \in G : m \in mgs(g, q) and \exists s \in S : m \preceq s\}$
    case i of $not\ q \preceq Query$:
        $G := \{g \in G \mid not(q \preceq g)\}$
        $S := \min \{m \mid \exists s \in S : m \in msg(s, q) and \exists g \in G : g \preceq m\}$
    case i of $covers(Query, e)$:
        $G := \{g \in G \mid covers(g, e)\}$
        $S := \min \{s' \mid covers(s', e) and \exists s \in S : s' \preceq s and \exists g \in G : g \preceq s'\}$
    case i of $not\ covers(Query, e)$:
        $S := \{s \in S \mid not\ covers(s, e)\}$
        $G := \max \{g' \mid not\ covers(g', e) and \exists g \in G : g \preceq g' and \exists s \in S : g' \preceq s\}$
    case i of $frequency(Query, Key, D, F), F > f$:
        $G := \{g \in G \mid frequency(g, Key, D, F), F > f\}$
        $S := \min \{s' \mid frequency(s', Key, D, F), F > f$
            $and \exists s \in S : s' \preceq s and \exists g \in G : g \preceq s'\}$
    case i of $frequency(Query, Key, D, F), F < f$:
        $S := \{s \in S \mid frequency(s, Key, D, F), F < f\}$
        $G := \max \{g' \mid frequency(g', Key, D, F), F < f$
            $and \exists g \in G : g \preceq g' and \exists s \in S : g' \preceq s\}$

Fig. 4. Computing S and G

programming. It also provides one possible foundation for database mining with query languages.

For what concerns related work, our query language is motivated and inspired by Imielinski and Mannila's view put forward in the recent issue of Communications of the ACM. The presented query language seems also a generalization of the M-SQL queries by [14], the MetaQueries by [24] and our own Claudien system [8]. It differs from M-SQL in that it supports the manipulation of clauses (instead of association rules), and from the other two approaches in that it allows to manipulate rules explicitly using a variety of logical and statistical quality criteria.

Acknowledgements

The author is supported by the Fund for Scientific Research, Flanders, and by the ESPRIT IV project no 20237 on Inductive Logic Programming II. He is grateful to Maurice Bruynooghe and Luc Dehaspe for interesting discussions on this, to Hendrik Blockeel for comments on this paper, to the ML team for their interest and figures and especially to Imielinski and Mannila for inspiring this work through their CACM paper.

References

1. R. Agrawal, T. Imielinski, and A. Swami. Mining association rules between sets of items in large databases. In *Proceedings of the 1993 International Conference on Management of Data (SIGMOD 93)*, pages 207–216, May 1993.
2. R. Agrawal, H. Mannila, R. Srikant, H. Toivonen, and A.I. Verkamo. Fast discovery of association rules. In U. Fayyad, G. Piatetsky-Shapiro, P. Smyth, and R. Uthurusamy, editors, *Advances in Knowledge Discovery and Data Mining*, pages 307–328. The MIT Press, 1996.
3. I. Bratko. *Prolog Programming for Artificial Intelligence.* Addison-Wesley, 1990. 2nd Edition.
4. I. Bratko and S. Muggleton. Applications of inductive logic programming. *Communications of the ACM*, 38(11):65–70, 1995.
5. P. Clark and R. Boswell. Rule induction with CN2: Some recent improvements. In Yves Kodratoff, editor, *Proceedings of the 5th European Working Session on Learning*, volume 482 of *Lecture Notes in Artificial Intelligence*, pages 151–163. Springer-Verlag, 1991.
6. P. Clark and T. Niblett. The CN2 algorithm. *Machine Learning*, 3(4):261–284, 1989.
7. L. De Raedt, editor. *Advances in Inductive Logic Programming*, volume 32 of *Frontiers in Artificial Intelligence and Applications*. IOS Press, 1996.
8. L. De Raedt and L. Dehaspe. Clausal discovery. *Machine Learning*, 26:99–146, 1997.
9. L. Dehaspe and L. De Raedt. Mining association rules in multiple relations. In *Proceedings of the 7th International Workshop on Inductive Logic Programming*, volume 1297 of *Lecture Notes in Artificial Intelligence*, pages 125–132. Springer-Verlag, 1997.
10. L. Dehaspe and H. Toivonen. Frequent query discovery: a unifying ILP approach to association rule mining. Technical Report CW-258, Department of Computer Science, Katholieke Universiteit Leuven, March 1998. http://www.cs.kuleuven.ac.be/publicaties/rapporten/CW1998.html.
11. S. Džeroski. Inductive logic programming and knowledge discovery in databases. In U. Fayyad, G. Piatetsky-Shapiro, P. Smyth, and R. Uthurusamy, editors, *Advances in Knowledge Discovery and Data Mining*, pages 118–152. The MIT Press, 1996.
12. S. Džeroski and I. Bratko. Applications of inductive logic programming. In L. De Raedt, editor, *Advances in inductive logic programming*, volume 32 of *Frontiers in Artificial Intelligence and Applications*, pages 65–81. IOS Press, 1996.

13. T. Imielinski and H. Mannila. A database perspectivce on knowledge discovery. *Communications of the ACM*, 39(11):58–64, 1996.
14. T. Imielinski, A. Virmani, and A. Abdulghani. A discovery board application programming interface and query language for database mining. In *Proceedings of KDD 96*. AAAI Press, 1996.
15. N. Lavrač and S. Džeroski. *Inductive Logic Programming: Techniques and Applications*. Ellis Horwood, 1994.
16. C. Mellish. The description identification problem. *Artificial Intelligence*, 52:151 – 167, 1991.
17. R.S. Michalski. A theory and methodology of inductive learning. In R.S Michalski, J.G. Carbonell, and T.M. Mitchell, editors, *Machine Learning: an artificial intelligence approach*, volume 1. Morgan Kaufmann, 1983.
18. T. Mitchell. *Machine Learning*. McGraw-Hill, 1997.
19. T.M. Mitchell. Generalization as search. *Artificial Intelligence*, 18:203–226, 1982.
20. S. Muggleton and L. De Raedt. Inductive logic programming : Theory and methods. *Journal of Logic Programming*, 19,20:629–679, 1994.
21. J. Ross Quinlan. *C4.5: Programs for Machine Learning*. Morgan Kaufmann series in machine learning. Morgan Kaufmann, 1993.
22. J.R. Quinlan. Learning logical definitions from relations. *Machine Learning*, 5:239–266, 1990.
23. G. Sablon, L. De Raedt, and M. Bruynooghe. Iterative versionspaces. *Artificial Intelligence*, 69:393–409, 1994.
24. W. Shen, K. Ong, B. Mitbander, and C. Zaniolo. Metaqueries for data mining. In U. Fayyad, G. Piatetsky-Shapiro, P. Smyth, and R. Uthurusamy, editors, *Advances in Knowledge Discovery and Data Mining*, pages 375–398. The MIT Press, 1996.
25. A. Srinivasan, S.H. Muggleton, M.J.E. Sternberg, and R.D. King. Theories for mutagenicity: A study in first-order and feature-based induction. *Artificial Intelligence*, 85, 1996.
26. H. Toivonen, M. Klemettinen, P. Ronkainen, K. Hätönen, and H. Mannila. Pruning and grouping discovered association rules. In Y. Kodratoff, G. Nakhaeizadeh, and G. Taylor, editors, *Proceedings of the MLnet Familiarization Workshop on Statistics, Machine Learning and Knowledge Discovery in Databases*, pages 47–52, Heraklion, Crete, Greece, 1995.

Bertrand Russell, Herbrand's Theorem, and the Assignment Statement

Melvin Fitting

Dept. Mathematics and Computer Science
Lehman College (CUNY), Bronx, NY 10468
fitting@alpha.lehman.cuny.edu
http://math240.lehman.cuny.edu/fitting

Abstract. While propositional modal logic is a standard tool, first-order modal logic is not. Indeed, it is not generally understood that conventional first-order syntax is insufficiently expressible. In this paper we sketch a natural syntax and semantics for first-order modal logic, and show how it easily copes with well-known problems. And we provide formal tableau proof rules to go with the semantics, rules that are, at least in principle, automatable.

1 Introduction

Propositional modal logic has become a fairly standard item in certain areas of artificial intelligence and computer science. Computational states are often thought of as possible worlds, while knowledge is frequently modeled using Hintikka's idea that a multiplicity of possible worlds reflects our ignorance about the actual one. But *first-order* modal logic is less familiar. It is often seen as a labyrinth full of twists and problems (see [5], for instance). Indeed, standard first-order syntax is actually of inadequate expressive power for modal logic. This is behind many of the well-known "paradoxes" of modal logic. But, a solution to the expressiveness problems of first-order modal logic exists, though it is still not as well-known as it deserves to be. It was explicitly introduced to modal logic in [8,9], though as we will see, the underlying ideas are much earlier.

It is always interesting to find that problems in disparate areas have a common solution. In our case the areas we wish to look at include: the theory of definite descriptions of Bertrand Russell; certain classical philosophical problems; inadequate expressibility of logics of knowledge; and the treatment of the assignment statement in dynamic logic. The source of difficulty in all these cases is the same: classical symbolic logic is taken as the standard. In classical logic assumptions are made that preclude certain kinds of problems—they cannot arise. Since these problems do not arise in classical logic, machinery for a solution is missing as well. Since the problems are significant, and the solution is trivial, the issues should be better known.

The material in this paper is a drastically abbreviated version of a book-length presentation forthcoming soon [2]. While our book is primarily aimed at philosophers, it contains much discussion of formal semantics and tableau-based proof procedures. Automation of these proof procedures still awaits.

Jacques Calmet and Jan Plaza (Eds.): AISC'98, LNAI 1476, pp. 14–28, 1998.
© Springer-Verlag Berlin Heidelberg 1998

2   Frege's Puzzle

The philosopher Frege introduced a famous distinction between *sense* (sinn) and *denotation* (bedeutung), [3]. In part, this was to deal with problems like the following. The morning star is the evening star. Let us symbolize this by $m = e$, a statement of identity. When the ancients came to know the truth of this statement it was a discovery of astronomy. But why could it not have been established by pure logic as follows. Certainly the ancients knew objects were self-identical; in particular, $K(m = m)$—we are using K as a knowledge modality. Since $m = e$ in fact, using substitutivity of equals for equals, occurrences of m can be replaced by occurrences of e. Doing so with the second occurrence in $K(m = m)$ gives us $K(m = e)$. Yet this does not appear to be how the Babylonians did it.

On the basis of this and other examples, Frege came to insist that terms such as "morning star" have both a denotation (in this case, a particular astronomical object) and a sense (roughly, how the object is designated). Identity of denotation is expressed by $m = e$, but in a non truth-functional context, such as $K(m = e)$, sense is what matters. This distinction has given rise to a vast literature, but it is enough here to point out that in *mathematical* logic, all contexts are truth-functional, and the problem fades into the background.

3   Russell, on Denoting

Bertrand Russell, in a famous paper [7], gave a theory of definite descriptions—he showed how to treat expressions like "the positive square root of 3" in a formal language. This was to be of special significance later in *Principia Mathematica* since it allowed classes to be introduced via definite descriptions. Like Frege, Russell too had a guiding example. In his case it was how to assign a truth value to "the King of France is bald," given that France has no King.

His solution involved a distinction between grammatical form and logical form. In this case the grammatical form is $B(f)$ (where B is the bald predicate and f is the King of France). But this cannot be the logical structure since f does not designate. According to Russell's theory the expression should be expanded to the following, where F is a predicate specifying what it means to be King of France: $(\exists x)\{F(x) \wedge (\forall y)(F(y) \supset y = x) \wedge B(x)\}$. Something Kings France, only one thing does so, and that thing is bald. Clearly, then, the sentence is false.

Then what about the sentence, "the King of France is not bald." Russell noted an ambiguity here. Should this be taken to assert the King of France has the non-baldness property, or should it be taken to be the negation of the assertion that the King of France has the baldness property? Using the Russell translation, do we negate the predicate $B(x)$, or do we negate the entire sentence? These are not equivalent formally, and the original English sentences do not seem equivalent informally.

To deal with this, Russell at first introduced a *narrow scope/broad scope* distinction, which I won't discuss further. It turned out to be inadequate, and eventually he was led to introduce an explicit scoping mechanism, one that is used systematically in *Principia Mathematica*. Russell's notation is somewhat fierce, and I won't reproduce it here. But the underlying idea is critical, and it's a wonder it came with such difficulty. It amounts to this. When definite descriptions are translated away, an existential quantifier is introduced. That quantifier has a formal scope, given by the usual quantifier rules. Then it must be that definite descriptions also have scopes within sentences. Generally that scope is implicit in natural language constructs, but machinery can be introduced to make it explicit. In short, *terms* of a formal language can have scopes, just as quantified variables do.

Today Russell's treatment of definite descriptions is well-known, at least among those to whom it is well-known. But the explicit introduction of a scoping mechanism is pushed into the background and generally ignored. In fact, it is a central point.

4 Modal Issues

Modal logic is a standard formal tool today in many areas of computer science and artificial intelligence. Propositional issues are well-known, while first-order ones may be less familiar. It is in the modal setting that the problems with scoping become particularly clear, using the familiar machinery of Kripke possible-world semantics.

Suppose we have a first-order modal frame $\langle \mathcal{G}, \mathcal{R}, \mathcal{D} \rangle$, consisting of a collection $\mathcal{G}$ of *possible worlds*, an *accessibility relation* $\mathcal{R}$ between worlds, and a *domain function* $\mathcal{D}$ assigning non-empty domains for quantification to possible worlds. No special assumptions are needed for the points I am about to make. $\mathcal{R}$ can be an equivalence relation, or transitive only, or have no particular properties. Likewise $\mathcal{D}$ can be the constant function, entirely arbitrary, or something inbetween.

Now, consider the behavior of the formula $\Diamond P(c)$, where c is a constant symbol and P is a predicate symbol. We want to allow c to be *non-rigid*, perhaps designating different things at different worlds. For instance, if "the King of France" were to be represented by a constant symbol in a temporal model, it would designate different people at different times. Of course sometimes it would not designate at all—ignore this point for now. Formally, let us say we have an *interpretation* $\mathcal{I}$ that assigns to constant symbols such as c, and to each possible world $\Gamma \in \mathcal{G}$ some object $\mathcal{I}(c, \Gamma)$—the "meaning" of c at Γ. We also assume $\mathcal{I}$ assigns to each relation symbol and each possible world some actual relation. We thus have a *model* $\mathcal{M} = \langle \mathcal{G}, \mathcal{R}, \mathcal{D}, \mathcal{I} \rangle$, based on the original frame. Finally let us symbolize by $\mathcal{M}, \Gamma \Vdash X$ the notion that the closed formula X is true at world Γ of the model $\mathcal{M}$. Now, just what should

$$\mathcal{M}, \Gamma \Vdash \Diamond P(c) \tag{1}$$

mean? I assume you are generally familiar with the behavior of modal operators in possible-world settings.

First Possibility The formula $\Diamond P(c)$ asserts that, whatever c means, it has the "$\Diamond P$" property. A reasonable way of formalizing this in models is to allow the occurrence of free variables, together with a *valuation* function to assign values to them. Thus we broaden the machinery a bit, and write $\mathcal{M}, \Gamma \Vdash_v X$ to mean: X (which may contain free variables) is true at world Γ of model $\mathcal{M}$, with respect to valuation v which assigns values to the free variables of X. With this extra machinery, (1) could be taken to mean $\mathcal{M}, \Gamma \Vdash_v \Diamond P(x)$ where $v(x) = \mathcal{I}(c, \Gamma)$. That is, whatever c designates at Γ is something for which $\Diamond P(x)$ is true (at Γ).

Since $\Diamond P(x)$ is a formula whose main connective is $\Diamond$, according to the usual behavior of modal models, there is some accessible world $\Delta \in \mathcal{G}$, that is $\Gamma \mathcal{R} \Delta$, such that $\mathcal{M}, \Delta \Vdash_v P(x)$ and this is the case provided $v(x)$ is in the relation $\mathcal{I}(P, \Delta)$, the "meaning" of P at Δ. Tracing back, for (1) to be true, we should have $\mathcal{I}(c, \Gamma) \in \mathcal{I}(P, \Delta)$.

Second Possibility The formula $\Diamond P(c)$ has $\Diamond$ as its main connective, so (1) says there is a possible world $\Omega \in \mathcal{G}$ with $\Gamma \mathcal{R} \Omega$ such that $\mathcal{M}, \Omega \Vdash P(c)$ and, most reasonably, this should be so if $\mathcal{I}(c, \Omega)$ is in the relation $\mathcal{I}(P, \Omega)$. Thus (1) means $\mathcal{I}(c, \Omega) \in \mathcal{I}(P, \Omega)$.

We got two ways of reading (1). The world Δ of the First Possibility and the world Ω of the Second Possibility need not be the same, but for simplicity suppose they are, $\Delta = \Omega$. Then we still have the following alternate readings for (1):

1. $\mathcal{I}(c, \Gamma) \in \mathcal{I}(P, \Omega)$
2. $\mathcal{I}(c, \Omega) \in \mathcal{I}(P, \Omega)$

and these are *not* equivalent. They are not equivalent because we allowed c to be non-rigid, so $\mathcal{I}(c, \Gamma)$ and $\mathcal{I}(c, \Omega)$ can be different.

Once the problem is seen, the solution is obvious. We need Russell's scoping mechanism, and just such a device was introduced into modal logic in [8,9]. What it amounts to is separating the notion of formula and predicate. Using notation based on the Lambda-calculus, we *abstract* from a formula $\varphi(x)$ a *predicate* $\langle \lambda x. \varphi(x) \rangle$. A formal treatment of this is given in the next section, but for now we observe that it separates $\Diamond P(c)$ into the following two distinct formulas:

1. $\langle \lambda x. \Diamond P(x) \rangle(c)$
2. $\Diamond \langle \lambda x. P(x) \rangle(c)$

5   Formal Semantics

Based on the sketchy ideas above, we set out a formal modal semantics. And because of space limitations, we go directly to the most general version—in [2] the treatment is much more leisurely. We have already said what a modal *frame* was above, but the present notion of *model* is considerably broadened.

First, in a modal frame $\langle \mathcal{G}, \mathcal{R}, \mathcal{D} \rangle$, recall that we place no restrictions on the domain function $\mathcal{D}$—it assigns to each world some non-empty set, the *domain* of that world. Members of the domain of a world should be thought of as the things "actually" existing at that world. If something exists at a world Δ, at a different world Γ we can think of that thing as a "possible existent." Also, by the *domain of the frame* we mean $\cup_{\Gamma \in \mathcal{G}} \mathcal{D}(\Gamma)$. Thus the members of the domain of the frame are the things that are actually or possibly existent at every world.

A modal *model* is a structure $\langle \mathcal{G}, \mathcal{R}, \mathcal{D}, \mathcal{I} \rangle$ where $\langle \mathcal{G}, \mathcal{R}, \mathcal{D} \rangle$ is a modal frame and $\mathcal{I}$ is an *interpretation* such that:

1. For each relation symbol P and each $\Gamma \in \mathcal{G}$, $\mathcal{I}(P, \Gamma)$ is a relation on the domain of the frame.
2. For some, not necessarily all, constant symbols c and $\Gamma \in \mathcal{G}$, $\mathcal{I}(c, \Gamma)$ is a member of the domain of the frame.

This can be extended to include function symbols, though because of space limitations in this paper, we do not do so.

Think of $\mathcal{I}(P, \Gamma)$ as the "meaning" of the relation symbol P at the world Γ. Note the important point that it may include things that do not exist at Γ. It is true to say, "Pegasus is a mythological beast," and we interpret this to mean that "Pegasus" designates something in a world other than this one (a make-believe world, if you will), the thing designated does not exist in this world, but in this world the property of being mythological correctly applies to it.

Think of $\mathcal{I}(c, \Gamma)$ as what c designates at the world Γ. As with relation symbols, what is designated need not exist. Thus "the first president of the United States" designates George Washington, who in a temporal sense once existed but no longer does. Note the added complication that for constant symbols $\mathcal{I}$ is allowed to be partial. This gives us the start of a mechanism to deal with "the present King of France."

Before we say which formulas are true at which worlds, we should indicate just what things we take as formulas.

1. As atomic formulas we take expressions of the form $R(x_1, \ldots, x_n)$ where R is an n-place relation symbol and $x_1, \ldots, x_n$ are *variables*. (Recall, formulas are not properties, but rather, properties are abstracted from formulas.)
2. Complex formulas are built up from simpler ones using $\wedge$, $\vee$, $\neg$, $\supset$, $\equiv$, $\Box$, $\Diamond$, $\forall$, and $\exists$ in the usual way, with the usual conventions about free variables.
3. If t is a term (which here means a variable or a constant symbol), φ is a formula, and x is a variable, $\langle \lambda x. \varphi \rangle(t)$ is a formula. Its free variables are those of φ, except for occurrences of x, together with the free variables of t.

Think of $\langle \lambda x. \varphi \rangle(t)$ as asserting of the object t designates that it has the property $\langle \lambda x. \varphi \rangle$, the property abstracted from the formula φ.

Let $\mathcal{M} = \langle \mathcal{G}, \mathcal{R}, \mathcal{D}, \mathcal{I} \rangle$ be a model. A *valuation* in $\mathcal{M}$ is a mapping v assigning to each variable x some member $v(x)$ in the domain of the frame underlying the model. Note that valuations are not world dependent.

We say a term t *designates at* Γ if t is a variable, or if t is a constant symbol and $\mathcal{I}(t, \Gamma)$ is defined. If t designates at Γ we use the following notation:

$$(v * \mathcal{I})(t, \Gamma) = \begin{cases} v(x) & \text{if } t \text{ is the variable } x \\ \mathcal{I}(c, \Gamma) & \text{if } t \text{ is the constant symbol } c \end{cases}$$

Finally we must define $\mathcal{M}, \Gamma \Vdash_v \varphi$: formula φ is true at world Γ of model $\mathcal{M}$ with respect to valuation v. Here is the truth definition, much of which is straightforward.

1. For an atomic formula $R(x_1, \ldots, x_n)$, $\mathcal{M}, \Gamma \Vdash_v R(x_1, \ldots, x_n)$ just in case $\langle v(x_1), \ldots, v(x_n) \rangle \in \mathcal{I}(R, \Gamma)$.
2. $\mathcal{M}, \Gamma \Vdash_v X \wedge Y$ if and only if $\mathcal{M}, \Gamma \Vdash_v X$ and $\mathcal{M}, \Gamma \Vdash_v Y$ (and similarly for the other Boolean connectives).
3. $\mathcal{M}, \Gamma \Vdash_v \Box X$ if and only if $\mathcal{M}, \Delta \Vdash_v X$ for every $\Delta \in \mathcal{G}$ such that $\Gamma \mathcal{R} \Delta$ (and similarly for $\Diamond$).
4. $\mathcal{M}, \Gamma \Vdash_v (\forall x)\varphi$ if and only if $\mathcal{M}, \Gamma \Vdash_{v'} \varphi$ for every valuation v' that is like v except that v' is some arbitrary member of $\mathcal{D}(\Gamma)$ (and similarly for $\exists$).
5. If t does not designate at Γ, $\mathcal{M}, \Gamma \not\Vdash_v \langle \lambda x.\varphi \rangle(t)$.
6. If t designates at Γ, $\mathcal{M}, \Gamma \Vdash_v \langle \lambda x.\varphi \rangle(t)$ if and only if $\mathcal{M}, \Gamma \Vdash_{v'} \varphi$ where v' is like v except that $v'(x)$ is what t designates at Γ.

Item 4 makes explicit the idea that quantifiers quantify over what actually exists—over the domain of the particular possible world only. Such quantifiers are called "actualist" by philosophers. They are not the only version available, but a choice among quantifiers is not the issue just now. Item 5 is the formal counterpart of the informal notion that no assertion about what is designated by a term that fails to designate can be correct. Note that the issue is designation by the term, not existence of the object designated. And item 6 expresses the idea that properties are properties of objects, and so we must know what object a term designates before knowing if a property applies.

One last item before turning to examples. We will always assume our models are *normal*: there is a relation symbol $=$, written in infix position, and $\mathcal{I}(=, \Gamma)$ is the equality relation on the domain of the model, for every world Γ.

6   A Few Examples

We discuss several examples to show the richness these simple ideas provides us. We give more examples once tableau proof machinery has been introduced.

Example 1. We begin with Frege's morning star/evening star problem. Suppose we consider an epistemic model in which the various possible worlds are those compatible with the knowledge possessed by the early Babylonians, with the actual situation among them, of course. The constant symbols m and e are intended to designate the morning and the evening stars respectively. (They designate the same object in the actual world, but need not do so in every possible world.) Also, let us read $\Box$ as "the ancients knew that." How can we have $m = e$ without, by substitutivity of equality, also having $\Box(m = e)$?

There is a certain amount of deception in the paragraph above. Neither $m = e$ nor $\Box(m = e)$ is a formula in our formal system. (Recall, constants

cannot appear in atomic formulas, rather they enter via predicate abstraction.)
The incorrect $m = e$ should be replaced with $\langle \lambda x.\langle \lambda y.x = y \rangle(e) \rangle(m)$, which we
abbreviate as $\langle \lambda x, y.x = y \rangle(m, e)$. More significantly, for $\Box(m = e)$ we have a
choice of replacements: $\Box\langle \lambda x, y.x = y \rangle(m, e)$, or $\langle \lambda x, y.\Box(x = y) \rangle(m, e)$, or even
$\langle \lambda x.\Box\langle \lambda y.x = y \rangle(e) \rangle(m)$. These do not behave the same. And, as a matter of
fact, the formula

$$\langle \lambda x, y.x = y \rangle(m, e) \supset \langle \lambda x, y.\Box(x = y) \rangle(m, e) \tag{2}$$

is valid (true at all worlds of all models) while

$$\langle \lambda x, y.x = y \rangle(m, e) \supset \Box\langle \lambda x, y.x = y \rangle(m, e) \tag{3}$$

is not. (We leave the demonstration to you.) A little thought shows that in
formula (2), $\langle \lambda x, y.\Box(x = y) \rangle(m, e)$ asserts that the ancients knew, of the objects
denoted by m and e (in the actual world) that they were identical. This, in fact,
is so, since they certainly knew that an object is self-identical. But in formula
(3), $\Box\langle \lambda x, y.x = y \rangle(m, e)$ asserts the ancients knew that m and e designated
the same object, and at one time they did not know this. The use of predicate
abstraction serves to disambiguate $(m = e) \supset \Box(m = e)$ into a valid version and
an invalid version, corresponding roughly to Frege's use of reference and sense.

Example 2. The introduction of modal operators is not essential to see the effects
of predicate abstraction. Consider the formula

$$\langle \lambda x.\neg\varphi \rangle(c) \equiv \neg\langle \lambda x.\varphi \rangle(c) \tag{4}$$

If we evaluate the truth of this at a possible world at which c designates, the
equivalence is valid. But if we are at a world at which c does not designate,
the left side of (4) is false, since no abstract correctly applies to a term that
fails to designate. But for the same reason, $\langle \lambda x.\varphi \rangle(c)$ is false, so its negation,
the right side of (4), is true. Thus if c fails to designate, (4) is false. This epito-
mizes precisely the distinction Russell made between the King of France having
the non-baldness property, and the King of France failing to have the baldness
property.

Note that if c does designate, (4) is true. And, whether c designates or not,
$\langle \lambda x.\varphi \wedge \psi \rangle(c) \equiv (\langle \lambda x.\varphi \rangle(c) \wedge \langle \lambda x.\psi \rangle(c))$ is true at each possible world. In classical
logic, it is assumed that terms designate, so the consequence of (4) and this is
to make the effects of predicate abstraction invisible. Only when Russell tried to
treat definite descriptions, that may lack designation, or when Frege considered
non-truth functional contexts, did such effects turned up.

Example 3. A term t may or may not designate at a particular possible world Γ.
If it does, $\langle \lambda x.x = x \rangle(t)$ is true there, since whatever t designates is self-identical.
But also, if t does not designate at a world, $\langle \lambda x.x = x \rangle(t)$ is false there, since

predicate abstracts are false when applied to non-designating terms. Therefore, let us define a "designation" abstract:

$$\mathbf{D} \text{ abbreviates } \langle \lambda x.x = x \rangle.$$

This allows us to move a semantic notion into syntax—as we have seen, $\mathbf{D}(t)$ is true at a world if and only if t designates at that world.

Using our designation abstract, let us return to Example 2. We saw that formula (4) is true if c designates. Consequently we have the validity of the following.

$$\mathbf{D}(c) \supset [\langle \lambda x.\neg\varphi \rangle(c) \equiv \neg\langle \lambda x.\varphi \rangle(c)] \tag{5}$$

Remarkably enough, the equivalence holds the other way around as well. That is, the following is valid.

$$\mathbf{D}(c) \equiv [\langle \lambda x.\neg\varphi \rangle(c) \equiv \neg\langle \lambda x.\varphi \rangle(c)] \tag{6}$$

Example 4. In exactly the same way that the semantic notion of designation could be expressed syntactically, we can express existence as well. We introduce the following abbreviation.

$$\mathbf{E} \text{ abbreviates } \langle \lambda x.(\exists y)(y = x) \rangle.$$

It is easy to show that t designates something that exists at world Γ if and only if $\mathbf{E}(t)$ is true at Γ.

We do not have the validity of: $(\forall x)\varphi \supset \langle \lambda x.\varphi \rangle(t)$. But if we assume that t not only designates, but designates something that exists, things become better. The following is valid: $\mathbf{E}(t) \supset [(\forall x)\varphi \supset \langle \lambda x.\varphi \rangle(t)]$.

In classical logic one cannot even talk about things that do not exist, and the $\mathbf{E}(t)$ antecedent above is unnecessary.

Example 5. Suppose p is a constant symbol intended to designate the President of the United States. Assume we have a temporal model in which possible worlds represent instants of time, and $\Box X$ means X is and will remain true, that is, X is true in the present and all future states. Dually, $\Diamond X$ means X is or will become true. For simplicity, let us assume there will always be a President of the United States, so p designates at all times. Now, consider the following formula.

$$\langle \lambda x.\Diamond\neg\langle \lambda y.x = y \rangle(p) \rangle(p) \tag{7}$$

To say this is true at the current world asserts, of the current President of the United States, that at some future time he will not be identical to the individual who is then the President of the United States. That is, (7) asserts: "someday the President of the United States will not be the President of the United States." This is not a valid formula, but it is satisfiable.

7 Herbrand's Theorem

In classical logic, Herbrand's theorem provides a reduction of the first-order provability problem to the propositional level, plus an open-ended search. It is often considered to be the theoretical basis of automated theorem proving for classical logic. Unfortunately, Herbrand's theorem does not extend readily to non-classical logics. Fortunately, the use of predicate abstraction allows us to prove a reasonable version for first-order modal logics. Since the full statement of the resulting theorem is somewhat complex, I only raise a few of the issues, and refer to [1] for a fuller treatment.

Classically, the first step of formula processing in the Herbrand method involves the introduction of Skolem functions. To cite the simplest case, the formula $(\exists x)P(x)$ is replaced with $P(c)$, where c is a new constant symbol. It is not the case that the two formulas are equivalent, but they are equi-satisfiable—if either is, both are.

One would like to Skolemize modally as well, but consider the following formula: $\Box(\exists x)P(x)$. If this is true at a possible world Γ of some modal model, then $(\exists x)P(x)$ is true at each world accessible from Γ. Say Δ and Ω are two such accessible worlds. Then at Δ, some member of the domain associated with Δ satisfies $P(x)$—let the new constant symbol c designate such an object at Δ. The situation is similar with Ω, so let c designate some member of the domain associated with Ω that satisfies $P(x)$ there. In general, c will designate different things at Δ and Ω, and so will be non-rigid. Thus the Skolemization of $\Box(\exists x)P(x)$ seems to be $\Box P(c)$, where c is a new non-rigid constant symbol. But, as we have seen, once non-rigid constant symbols are admitted, conventional syntax becomes ambiguous. Indeed, it would appear that $\Box P(c)$ should also be the Skolemization of $(\exists x)\Box P(x)$, and this seems quite unreasonable, since the two quantified formulas having a common Skolemization behave quite differently.

Of course, the solution involves using predicate abstraction. The proper Skolemization for $\Box(\exists x)P(x)$ is $\Box\langle\lambda x.P(x)\rangle(c)$, while $(\exists x)\Box P(x)$ has a Skolemization of $\langle\lambda x.\Box P(x)\rangle(c)$, which is behaviorally distinct.

Similar results apply to more complex formulas, but function symbols must be involved, and we do not attempt to give a full presentation here. Suffice it to say that the full force of Herbrand's theorem can be brought to bear, even in a modal setting.

8 Dynamic Logic

One of the interesting applications of multi-modal logic is *dynamic logic*, a logic of programs, [6]. In addition to the usual machinery of modal logic, a class of *actions* is introduced, with the class of actions closed under various operations, such as sequencing, repetition, and so on. For each action α there is a corresponding modal operator, generally written $[\alpha]$. The formula $[\alpha]X$ is informally read: after action α is completed, X will be true. (Since non-determinism is allowed, there may be several ways of completing α.) There is a corresponding semantics in which possible worlds are possible states of a computation. Likewise there is a proof theory, at least for the propositional case. A typical principle of dynamic logic is $[\alpha; \beta]X \equiv [\alpha][\beta]X$, where the semicolon corresponds to sequencing of actions.

Dynamic logic provides an elegant treatment of compound actions, but what about atomic ones? Consider the assignment statement $c := c + 1$—what is its dynamic characterization? We are all familiar with the before/after behavior of assignment statements, where the right-hand side uses the current value of c, while the left-hand side reflects the new value it acquires. To explain $c := c + 1$ in English, we would say something like: "after execution, the value of c is one more than its value before execution."

To formalize this, it is enough to recognize that c is non-rigid—it designates different things at different computational states. Then, assuming arithmetic behaves in the expected way, the essential feature of the assignment statement in question is captured by the following, in which we use $\Box$ as shorthand for $[c := c + 1]$.

$$\langle \lambda x. \Box \langle \lambda y. y = x + 1 \rangle (c) \rangle (c) \tag{8}$$

What this expresses is: it is true of the current value of c that, after $c := c + 1$ is executed, the value of c will be that plus 1.

If we assume, about arithmetic, only that incrementing a number gives us a result unequal to the number, then it is easily shown to be a logical consequence of (8) that

$$\langle \lambda x. \Box \neg \langle \lambda y. (y = x) \rangle (c) \rangle (c) \tag{9}$$

This should be compared with (7). Indeed, both simply amount to assertions that p in the one case and c in the other are non-rigid.

Issues of designation and existence are relevant in dynamic logic as well. Saying c designates at a computational state amounts to saying it has been initialized, in the standard programming sense. Saying c exists at a computational state says something about c's availability—we are in the scope of c. Formal notions here are somewhat unclear, but it would be interesting to work them out fully. Finally, saying that c is *rigid* is simply saying that c is *const*, as in C or C++, or *final* as in Java.

Full first-order dynamic logic is not axiomatizable. What we propose is that the addition of terms, equality, and predicate abstraction to propositional dynamic logic, without adding quantifiers, might serve as a satisfactory strengthening of propositional dynamic logic. It is a subject worth investigating.

9   Tableau Proof Methods

Formal proof rules based on *prefixed* tableaus are quite natural for the constructs discussed above. Here I give rules for varying domain S5—versions for other standard modal logics are presented in [2], but that for S5 is simplest to present.

Because of space limitations, rather than giving examples as we go along, I'll reserve them all until the following section.

For S5, by a *prefix* we simply mean a positive integer—think of it as informally designating a possible world. (For other modal logics besides S5 the structure of prefixes is more complex.) A *prefixed formula* is an expression of the form $n\,X$, where n is a prefix and X is a formula—think of it as saying X holds in world n. A *signed prefixed formula* is an expression of the form $T\,n\,X$ or $F\,n\,X$—the first asserts $n\,X$ and the second denies it.

A *tableau proof* of a formula X (without free variables) is a *closed tableau* for $F\,1\,X$. A *tableau* for a signed prefixed formula is a tree, with that signed prefixed formula at the root, and constructed using the various *branch extension rules* to be given below. A branch of a tableau is *closed* if it contains an explicit contradiction: both $T\,k\,Z$ and $F\,k\,Z$, for some k and Z. If every branch is closed, the tableau itself is *closed*.

Intuitively, when we begin a tableau with $F\,1\,X$ we are supposing there is some possible world, designated by 1, at which X fails to hold. A closed tableau represents an impossible situation. So the intuitive understanding of a tableau proof is that X cannot fail to hold at any world—X must be valid. Proper soundness and completeness proofs can be based on this intuition but there is not space here to present them. Now it remains to give the various branch extension rules—the rules for "growing" a tableau. The propositional connective rules are easily described, and we begin with them.

If $T\,n\,X \wedge Y$ occurs on a tableau branch, $X \wedge Y$ intuitively is true at world n, hence both X and Y must also be true there. Consequently the tableau branch can be extended twice, with the first node labeled $T\,n\,X$ and the second $T\,n\,Y$. If $F\,n\,X \wedge Y$ occurs on a branch, $X \wedge Y$ is false at n, so either X or Y must be false at n. This gives rise to two cases, and so the branch "splits." That is, the last node is given two children, the left labeled $F\,n\,X$ and the right $F\,n\,Y$.

Rules for other binary connectives are similar, while those for negation are simpler. In summary form, here they are.

Negation:

$$\frac{T\,n\,\neg X}{F\,n\,X} \qquad \frac{F\,n\,\neg X}{T\,n\,X}$$

Binary:

$$\frac{T\,n\,X \wedge Y}{\begin{array}{c} T\,n\,X \\ T\,n\,Y \end{array}} \qquad \frac{F\,n\,X \wedge Y}{F\,n\,X \mid F\,n\,Y} \qquad \frac{T\,n\,X \vee Y}{T\,n\,X \mid T\,n\,Y} \qquad \frac{F\,n\,X \vee Y}{\begin{array}{c} F\,n\,X \\ F\,n\,Y \end{array}}$$

$$\frac{T\,n\,X \supset Y}{F\,n\,X \mid T\,n\,Y} \qquad \frac{F\,n\,X \supset Y}{\begin{array}{c} T\,n\,X \\ F\,n\,Y \end{array}}$$

Modal rules are quantifier-like in nature. Here the prefixes play a significant role.

Necessity: In these, k is *any* prefix.

$$\frac{T\, n\,\square X}{T\, k\, X} \qquad \frac{F\, n\,\Diamond X}{F\, k\, X}$$

Possibility: In these, k is a prefix *that is new to the branch.*

$$\frac{T\, n\,\Diamond X}{T\, k\, X} \qquad \frac{F\, n\,\square X}{F\, k\, X}$$

For quantifiers we need some additional machinery. For each prefix n we introduce an infinite alphabet of *parameters* associated with n—typically we write p_n, q_n, etc., for parameters associated with n. Think of the parameters associated with n as (designating) the things that exist at world n. From now on we allow parameters to appear in proofs (though not in the formulas being proved). They follow the syntax rules of free variables, though they are never quantified. Now, here are the quantifier rules.

Universal: In these, p_n is *any* parameter associated with n.

$$\frac{T\, n\, (\forall x)\varphi(x)}{T\, n\, \varphi(p_n)} \qquad \frac{F\, n\, (\exists x)\varphi(x)}{F\, n\, \varphi(p_n)}$$

Existential: In these, p_n is a parameter associated with n that is *new* to the branch.

$$\frac{T\, n\, (\exists x)\varphi(x)}{T\, n\, \varphi(p_n)} \qquad \frac{F\, n\, (\forall x)\varphi(x)}{F\, n\, \varphi(p_n)}$$

The rules so far are fairly standard. To deal with non-rigid constant symbols (and this can be extended to include function symbols too), we again need to extend the machinery. If c is a constant symbol and n is a prefix, we allow c_n to occur in a proof (though again, not in the formula being proved). Think of c_n intuitively as the object that c designates at world n. We have allowed partial designation, that is, a constant symbol may not designate at every world. All this is incorporated rather easily into our rules for predicate abstracts, which we give in a moment.

First, however, a little more notation. For each term t we define $t@n$, which we can think of as what t designates at n. (This gets more complicated when function symbols are present.) For a prefix n:

1. For a parameter p_i, let $p_i@n = p_i$.
2. For a subscripted constant symbol c_i, let $c_i@n = c_i$.
3. For an unsubscripted constant symbol c, let $c@n = c_n$.

Now, here are the abstraction rules.

Positive Abstraction:

$$\frac{T\, n\, \langle \lambda x.\varphi(x)\rangle(t)}{T\, n\, \varphi(t@n)}$$

Negative Abstraction: If $t@n$ already occurs on the branch,

$$\frac{F\ n\ \langle\lambda x.\varphi(x)\rangle(t)}{F\ n\ \varphi(t@n)}$$

Finally we have the rules for equality, and these are quite straightforward. Let us say a term is *grounded on a branch* if it is a parameter or a subscripted constant symbol, and it already occurs on the branch.

Reflexivity: If t is grounded on the branch, $T\ n\ t = t$ can be added to the end, for any prefix n. Briefly,

$$\overline{T\ n\ t = t}$$

Substitutivity: If t and u are grounded on the branch, and $T\ k\ t = u$ occurs on the branch, any occurrences of t can be replaced with occurrences of u. Again briefly,

$$\frac{T\ k\ t = u}{\dfrac{T\ n\ \varphi(t)}{T\ n\ \varphi(u)}} \qquad \frac{T\ k\ t = u}{\dfrac{F\ n\ \varphi(t)}{F\ n\ \varphi(u)}}$$

This completes the full set of tableau rules.

10   More Examples

We give several simple examples of tableau proofs. More can be found in [2].

Example 6. In Section 6 we gave formula (6) as an interesting example of a valid formula involving designation. We now give a tableau proof of part of this,

$$\mathbf{D}(c) \supset [\neg\langle\lambda x.P(x)\rangle(c) \supset \langle\lambda x.\neg P(x)\rangle(c)]$$

Line numbers are for explanation purposes only. Since no modal operators are present, only world 1 is involved throughout.

$$
\begin{array}{ll}
F\ 1\ \mathbf{D}(c) \supset [\neg\langle\lambda x.P(x)\rangle(c) \supset \langle\lambda x.\neg P(x)\rangle(c)] & 1. \\
T\ 1\ \mathbf{D}(c) & 2. \\
F\ 1\ \neg\langle\lambda x.P(x)\rangle(c) \supset \langle\lambda x.\neg P(x)\rangle(c) & 3. \\
T\ 1\ \neg\langle\lambda x.P(x)\rangle(c) & 4. \\
F\ 1\ \langle\lambda x.\neg P(x)\rangle(c) & 5. \\
F\ 1\ \langle\lambda x.P(x)\rangle(c) & 6. \\
T\ 1\ \langle\lambda x.x = x\rangle(c) & 2'. \\
T\ 1\ c_1 = c_1 & 7. \\
F\ 1\ \neg P(c_1) & 8. \\
F\ 1\ P(c_1) & 9. \\
T\ 1\ P(c_1) & 10.
\end{array}
$$

In this, 2 and 3 are from 1, and 4 and 5 are from 3 by an implication rule; 6 is from 4 by a negation rule; 2′ is line 2 unabbreviated; 7 is from 2′ by a positive abstraction rule; then 8 is from 5 and 9 is from 6 by negative abstraction; and 10 is from 8 by negation. The single branch is closed because of 9 and 10.

Example 7. We give a tableau proof of a formula discussed in Example 4 of Section 6. Again this does not involve modal operators.

$$F \ 1 \ \mathbf{E}(c) \supset [(\forall x)P(x) \supset \langle \lambda x.P(x) \rangle (c)] \quad 1.$$
$$T \ 1 \ \mathbf{E}(c) \quad 2.$$
$$F \ 1 \ (\forall x)P(x) \supset \langle \lambda x.P(x) \rangle (c) \quad 3.$$
$$T \ 1 \ (\forall x)P(x) \quad 4.$$
$$F \ 1 \ \langle \lambda x.P(x) \rangle (c) \quad 5.$$
$$T \ 1 \ \langle \lambda x.(\exists y)(y = x) \rangle (c) \quad 2'.$$
$$T \ 1 \ (\exists y)(y = c_1) \quad 6.$$
$$T \ 1 \ p_1 = c_1 \quad 7.$$
$$T \ 1 \ P(p_1) \quad 8.$$
$$T \ 1 \ P(c_1) \quad 9.$$
$$F \ 1 \ P(c_1) \quad 10.$$

In this, 2 and 3 are from 1 and 4 and 5 are from 3 by an implication rule; $2'$ is 2 unabbreviated; 6 is from $2'$ by positive abstraction; 7 is from 6 by an existential rule (p_1 is a new parameter at this point); 8 is from 4 by a universal rule (note that a parameter is involved, as the rule requires); 9 is from 7 and 8 by substitutivity of equality; 10 is from 5 by negative abstraction (note that c_1 already occurs on the tableau branch). Closure is by 9 and 10.

Example 8. Our final example is an interesting modal example. It is a proof of

$$\langle \lambda x.\Box \langle \lambda y.x = y \rangle (c) \rangle (c) \supset [\langle \lambda x.\Box P(x) \rangle (c) \supset \Box \langle \lambda x.P(x) \rangle (c)]$$

In [2] we make a case that the antecedent of this expresses rigidity of c. The consequent asserts that a *de re* usage of c implies the corresponding *de dicto* version. There is no space here to discuss just what this is all about. Just take it as providing an illustrative tableau proof.

$$F \ 1 \ \langle \lambda x.\Box \langle \lambda y.x = y \rangle (c) \rangle (c) \supset [\langle \lambda x.\Box P(x) \rangle (c) \supset \Box \langle \lambda x.P(x) \rangle (c)] \quad 1.$$
$$T \ 1 \ \langle \lambda x.\Box \langle \lambda y.x = y \rangle (c) \rangle (c) \quad 2.$$
$$F \ 1 \ \langle \lambda x.\Box P(x) \rangle (c) \supset \Box \langle \lambda x.P(x) \rangle (c) \quad 3.$$
$$T \ 1 \ \langle \lambda x.\Box P(x) \rangle (c) \quad 4.$$
$$F \ 1 \ \Box \langle \lambda x.P(x) \rangle (c) \quad 5.$$
$$T \ 1 \ \Box P(c_1) \quad 6.$$
$$F \ 2 \ \langle \lambda x.P(x) \rangle (c) \quad 7.$$
$$T \ 2 \ P(c_1) \quad 8.$$
$$T \ 1 \ \Box \langle \lambda y.c_1 = y \rangle (c) \quad 9.$$
$$T \ 2 \ \langle \lambda y.c_1 = y \rangle (c) \quad 10.$$
$$T \ 2 \ c_1 = c_2 \quad 11.$$
$$F \ 2 \ P(c_2) \quad 12.$$
$$T \ 2 \ P(c_2) \quad 13.$$

In this, 2 and 3 are from 1 and 4 and 5 are from 3 by an implication rule; 6 is from 4 by positive abstraction; 7 is from 5 by a possibility rule (the prefix 2 is new to the branch at this point); 8 is from 6 by a necessity rule; 9 is from 2 by positive abstraction; 10 is from 9 by a necessity rule; 11 is from 10 by positive abstraction; 12 is from 7 by negative abstraction (c_2 occurs on the branch already); and 13 is from 8 and 11 by substitutivity of equality.

We leave it to you to provide a proof of

$$\langle\lambda x.\Box\langle\lambda y.x = y\rangle(c)\rangle(c) \supset [\Box\langle\lambda x.P(x)\rangle(c) \supset \langle\lambda x.\Box P(x)\rangle(c)]$$

11 Conclusions

The work above should begin to make it clear that first-order modal logic, with its syntax and semantics properly enhanced, is a very expressive formalism. It relates well to natural language constructs and to those of computer science. And it has a simple proof procedure which is automatable (though this has not been done). Keep it in mind.

References

1. M. C. Fitting. A modal Herbrand theorem. *Fundamenta Informaticae*, 28:101–122, 1996.
2. M. C. Fitting and R. Mendelsohn. *First-Order Modal Logic*. Kluwer, 1998. Forthcoming.
3. G. Frege. Uber Sinn und Bedeutung. *Zeitschrift fur Philosophie und philosophische Kritik*, 100:25–50, 1892. "On Sense and Reference" translated in [4].
4. G. Frege. *Translations from the Philosophical Writings of Gottlob Frege*. Basil Blackwell, Oxford, 1952. P. Geach and M. Black editors.
5. J. W. Garson. Quantification in modal logic. In D. Gabbay and F. Guenthner, editors, *Handbook of Philosophical Logic*, volume 2, pages 249–307. D. Reidel, 1984.
6. D. Harel. Dynamic logic. In D. Gabbay and F. Guenthner, editors, *Handbook of Philosophical Logic*, volume 2, pages 497–604. D. Reidel, 1984.
7. B. Russell. On denoting. *Mind*, 14:479–493, 1905. Reprinted in Robert C. Marsh, ed., *Logic and Knowledge: Essays 1901-1950, by Bertrand Russell*, Allen & Unwin, London, 1956.
8. R. Stalnaker and R. Thomason. Abstraction in first-order modal logic. *Theoria*, 34:203–207, 1968.
9. R. Thomason and R. Stalnaker. Modality and reference. *Nous*, 2:359–372, 1968.

Representing and Reasoning with Context

Richmond H. Thomason

Abstract. This paper surveys the recent work in logical AI on context and discusses foundational problems in providing a logic of context. As a general logic of context, I recommend an extension of Richard Montague's Intensional Logic that includes a primitive type for contexts.

1 Introduction

Naturally evolved forms of human communication use linguistic expressions that are rather highly contextualized. For instance, the identity of the speaker, the table and the computer that are mentioned in the following sentence depend on the context of utterance. The time and location of the utterance, as well as the imagined orientation to the table, are left implicit, and also depend on the context.

(1.1) 'I put the computer behind the table yesterday'.

These forms of contextualization are relatively familiar, and people are able to deal with them automatically and, if necessary, prepared to reason about them explicitly. Other forms of contextualization, especially those that affect the meanings of words (see for instance, [Cru95,PB97]) can be more difficult to isolate and think about.

I doubt that the level of contextualization humans find convenient and appropriate is the only possible way of packaging information, but for some reason we seem to be stuck with it. We are made in such a way that we need contextualized language; but this contextualization can be an obstacle to understanding. There are psychological experiments indicating that highly verbose, decontextualized language is unusual; and it is reasonable to suppose that this sort of language is relatively difficult to comprehend. [WH90], for instance, is concerned with task-oriented instructions. Decontextualized verbalizations of plans leave out steps that the hearer can be assumed to know about; here, the context consists of routines for instantiating abstract plans. [1] Other studies show that highly contextualized discourse can be relatively unintelligible to someone who is not familiar with the context.[2]

The challenge of contextualization arises not only in natural communication between humans, but in computerized transactions: in software, databases, knowledge bases, and AI systems. The reasons for contextualization in these cases are similar. In particular, (i) these systems are human products, and a

---

[1] See [You97].
[2] See [CS89].

Jacques Calmet and Jan Plaza (Eds.): AISC'98, LNAI 1476, pp. 29–41, 1998.
© Springer-Verlag Berlin Heidelberg 1998

modular design makes them more intelligible and maintainable for the human designers; and (ii) information and procedures are in general limited to the applications that a designer envisions. But a modular design will contextualize the modules: for instance, the modularization of LATEX into style and document files means that a LATEX formatted document will only work in combination with with right supplementary files. The limits referred to in (ii) mean that software performance can't be predicted, and good performance can't be expected, outside of the limits envisaged by the designer.

In cases of human communication, it can be useful in anticipating and diagnosing misinterpretations to reason explicitly about context dependence; for instance, if we find a message on the answering machine saying

(1.2) 'This is Fred; I'll call back at 2:00'

we may recall that Fred lives in a different time zone and wonder whether he meant 2:00 his time or 2:00 our time. Similarly, it might be useful for software systems to be able to perform this sort of reasoning. It is this idea that led John McCarthy, in a number of influential papers, [3] to pursue the formalization of context. In motivating this program, McCarthy proposes the goal of creating automated reasoning that is able to transcend its context of application. McCarthy's suggestions have inspired a literature in logical AI that is by now fairly extensive.[4]

The sort of transcendence that would be most useful—an intelligent system with the ability to recognize its contextual limitations and transcend them—has to be thought of as a very long-range goal. But there are less ambitious goals that can make use of a formal logic of context, and the formalizations that have been emerged over the last ten years have led to useful applications in knowledge representation and in the integration of knowledge sources. For the former sort of application, see [Guh91], where explicit reference to context is used to modularize the creation of a large-scale knowledge base. For the latter sort of application, see [BF95]. There has been a good deal of interest in applications of these ideas in natural language processing, with at least one conference devoted to the topic,[5] and perhaps we can hope to see these ideas put to use, but I am not aware of a natural language processing system that was made possible by these ideas.

In this paper, I will discuss the logical foundations of the theory of context. The recent work on context in formal AI presents foundational issues that will seem familiar to logicians, since they were addressed in classical work on the semantic paradoxes and on modal logic. But some elements of this work are new.

---

[3] [McC86,McC93,MB95]

[4] At http://www.pitt.edu/~thomason/bibs/context.html, there is a bibliography to this literature.

[5] For the proceedings of this conference, see [BI97].

2   The *Ist* Predicate

The nature and logical treatment of "propositional arguments" of predicates has been thoroughly debated in the logical literature on modal logic and propositional attitudes; see, for instance, [Chu51,Qui53,Qui56]. Similar questions arise in the logic of context.

In its simplest form, the language of a theory of context is an extension of first-order logic containing constructions of the form

(2.1) $ist(c, p)$.

The *ist* predicate is used to explicitly state the dependencies between contexts and assertions. The following example, from [BF95], illustrates its use in stating dependencies between data bases:

(2.2) ist(UA-db, $(\forall d)$[Thursday(d) $\rightarrow$
passenger-record($921, d$, McCarthy) $\wedge$
flight-record(921,SF,7:00,LA,8:21)])

The formula is taken from a hypothetical example involving a translation between entries in John McCarthy's diary and the United Airlines data base. It says that the United Airlines database contains an entry to the effect that McCarthy is booked every Thursday on Flight 921 from San Francisco to Los Angeles.

The logical type of the first argument of (2.1) (the *contextual* argument) is unproblematic; the constant c simply refers to a context. Part of an explicit theory of contexts will have to involve a reification of contexts, and it is perfectly reasonable to treat contexts as individuals and to allow individual constants of first-order logic to refer to them.

We can't treat the second argument (the *propositional argument*) in this way, however, since formulas do not appear as arguments of predicates in first-order logic. In very simple applications, an expression of the form (2.1) could be taken to be an abbreviation of the first-order formula

(2.3) $ist_p(c)$,

where for each formula p that we are interested in tracking we introduce a special first-order predicate ist_p.

This technique, however, is very awkward even in very simple cases. Moreover, it will not allow us to state generalizations that involve quantifiers binding variable positions in p—and these generalizations are needed in developing a theory of the interrelationships between contexts. For instance, [BF95] contains the following "lifting rule" relating statements in McCarthy's diary to United Airlines data base entries:

(2.4) $\forall x \forall d[ist(\text{diary}(x), \text{Thursday}(d)) \leftrightarrow$
$ist(\text{UA-db}, \text{Thursday}(d))]$.

This axiom calibrates the calendars of individual diaries with the United Airlines calendar by ensuring that a date is a Thursday in anyone's diary if and only if it is a Thursday in the United Airlines data base. We could accommodate the quantifier indexed to 'x' in (2.4) in a first-order reduction of ist, but not the quantifer indexed to 'd'. This variable occurs in the sentential argument to ist, and would have nothing to bind if Thursday(d), the formula containing this variable, were absorbed into the predicate.

The applications envisaged for a logic of context, then, call for a logical form in which the propositional argument is referential; but what sort of thing does it refer to? The logical literature on the foundations of modal logic and propositional attitudes offers two main alternatives:

(2.5) The argument is a syntactic object, a formula.

(2.6) The argument is an abstract object, a proposition or something like a proposition.

3   The Quotational Approach and the Semantic Paradoxes

There are two main problems with the *quotational* approach of (2.5): first, if it is formulated naively it leads to paradox, and second, despite the apparent naturalness of this approach, I believe that it doesn't do justice to the nature of the ist relation.

It is natural to use a first-order theory of syntax to formulate the quotational theory.[6] Expressions are treated as first-order individuals, the language has constants referring to the syntactic primitives of its own language and is able to express the concatenation function over expressions. In such a language we get a name n_ϕ for each formula ϕ of $\mathcal{L}$.

In a theory of this kind, it is possible to display simple, plausible axioms for ist that are inconsistent. The instances of the following schemes constitute such a list.

AS1: $ist(c_0, n_\phi)$ for every formula ϕ that is an axiom of FOL.

AS2: $[ist(c_0, n_\phi \wedge ist(c_0, n_{\phi \to \psi})] \to ist(c_0, n_\psi)$, for every formula ϕ and ψ.

AS3: $ist(c_0, n_\phi \to \phi)$, for every formula ϕ.

The third axiom scheme is plausible if we imagine that we are axiomatizing the assertions that hold in the context c_0. It may be objected that it is unnecessary in any natural application of the theory to allow a context to reflect on itself. But more complex forms of the paradox can be constructed using cycles in which contexts reflect on each other, and cases such as this can come up in

---

[6] It is also possible to achieve the effects of syntactic reflection—the ability to reason about the syntactic expressions of ones own language—indirectly. Gödel's arithmetization of syntax is an indirect way of achieving this.

very simple applications—e.g., where we are considering two contexts, each of them capable of modeling the other.

This inconsistency result is due to Richard Montague: see [Mon63]. It uses a general form of the syntactic reflection techniques developed by Gödel and appeals to the formalization of the so-called "knower's paradox," which involves the following sentence:

KP: I don't know this,

where the word 'this' refers to KP.

It is possible to develop a quotational theory of *ist* that avoids these paradoxes by using a solution to the semantic paradoxes such as that of [Kri75]. But there is a price to be paid here in terms of the underlying logic. (For instance, a nonclassical logic is needed for the Boolean connectives.) There is room for debate as to how high this price is, but I do not find it worth paying in this case, since I don't find a quotational theory of *ist* to be very compelling or useful. For instance, the first-person pronoun is one of the most thoroughly investigated context-dependent linguistic forms. At a first approximation, its contextual semantics is very simple; the first-person pronoun refers in the context of an utterance to the agent of that utterance. But the item that is interpreted in this way is not a simple quoted expression. In English, the first person pronoun has two quotational realizations, 'I' and 'me'. This variation in the expression's syntactic form is entirely irrelevant to the meaning.

It seems necessary to infer some abstract level of representation here, at which there is a single element that receives the appropriate interpretation. Of course, you could treat this element as an expression in some sort of abstract language. But, since abstract levels of representation are needed anyway, I prefer to invoke a level that corresponds directly to a context-dependent meaning, and to treat the propositional argument of *ist* as referring to this. As long as these abstract representations are not themselves expressions, the semantic paradoxes are not a problem.

This approach to *ist* is very similar to the standard treatments of propositional arguments in modal logic.

4   The Propositional Approach: *Ist* as a Modality

On the simplest forms of the *propositional approach* of (2.6), *ist* is an indexed modal operator that does not differ to any great extent from the operators used to model reasoning about knowledge; see [FHMV95] for details. The models for these modal logics use possible worlds. The propositional argument of *ist* receives a set of possible worlds as its semantic interpretation.

This idea provides a useful perspective on many of the existing formalisms for and applications of context. And, of course, since modal operators have been thoroughly investigated in the logical literature, this perspective allows us to bring a great deal of logical work to bear on contexts.

Just one way of illustrating this point is the following: it is natural to think of contexts as the epistemic perspectives of various agents. When knowledge is drawn from different databases, for instance, or from different modules of a large knowledge base, it is tempting to speak of what the various modules "know". If we take this metaphor seriously, we can apply modeling techniques from distributed systems (which in turn originate in epsitemic logic, a branch of modal logic) to these cases. As Fagin et al. [FHMV95] shows in considerable detail, these modeling techniques are very fruitful.

Suppose that Alice, Bob, Carol and Dan are talking. Alice's husband is named 'Bob'; Carol and Bob know this, Dan doesn't. If Alice says 'Bob and I are going out to dinner tonight', Carol and Bob are likely to think that Alice and her husband are going out. Dan will probably think that Alice and the man standing next to him are going out. The inferences that different hearers will make from the same utterances depend crucially on the hearer's knowledge; this point is well made in many of Herbert H. Clark's experimental studies; see, for instance, [CS89]. This sort of case provides yet another reason for using epistemic logic to model context.

Modal operators exhibit formal characteristics of the sort that are wanted in a general theory of context. For instance, modal operators can provide a way of adjusting a single agent's reasoning to contextual changes in the accessible information. The classical modalities like $\Box$ are not the best examples of this. But consider a modal operator of the form $[A_1, \ldots, A_n]$, where $[A_1, \ldots, A_n]B$ means that B follows from whatever assumptions apply in the outer context, together with the explicitly mentioned auxiliary hypotheses $A_1, \ldots, A_n$. This is not a great departure from standard modal logic, and is very close to the mechanisms that are used for formalizing context.

These commonalities between modal logic and the theory of context have been exploited in several of the recent theoretical approaches to context. Examples include

[BBM95,NL95,AP97,GN97].

It should perhaps be added that McCarthy himself is opposed to the use of modal logic, in this and other applications; see, for instance, [McC97]. I believe that this is mainly a matter of taste, and I know of no compelling reason to avoid the use of modal logics in applications such as these.

5   Modal Type Theories

It has proved to be very useful as a methodological and conceptual technique to use types in applications to programming languages and to natural language semantics. The types provide a useful way of organizing information, and of providing a syntactic "sanity check" on formulas. In natural language semantics, for instance, types were an important part of Richard Montague's approach,[7] and provided useful and informative constraints on the mapping of natural language

---

[7] See [Mon74].

syntactic structures to logical formulas. For arguments in favor of typing in programming languages, see, for instance, [Sch94].

Types are historically important as a way of avoiding the impredicative paradoxes, but I do not stress this reason for using them because these paradoxes can be avoided by other means. I do not deny that in natural language, as well as programming languages, it can be very useful to relax typing constraints or to dispense with them entirely; for a useful discussion of the issues, with references to the literature, see [Kam95]. But I do believe that a typed logic is the natural starting point, and that we can work comfortably for a long time within the confines of a type theory.

Together, these considerations provide good motivation for using Intensional Logic[8] as a framework for reasoning about context. A further benefit of this approach is that there will be connections to the large literature in natural language semantics that makes use of Intensional Logic.[9]

Intensional Logic is a version of higher-order logic that contains identity and typed lambda abstraction as syntactic primitives. The basic types are the type t of truth values, the type e of individuals, and the type w of possible worlds. Complex types produced by the following rule: if σ and τ are types then so is $\langle \sigma, \tau \rangle$, the type of functions from objects of type σ to objects of type τ.[10] With these few resources, a full set of classical connectives can be defined, as well as the familiar modal operators and quantifiers over all types.

Ordinary modal logic does not treat modalities as objects, and makes available only a very limited variety of modalities. In Intensional Logic, modalities can be regarded as sets of propositions, where propositions are sets of possible worlds. Intensional Logic does not distinguish between a set and its characteristic function; so the type of a modality is $\langle \langle w, t \rangle, t \rangle$. It is reasonable to require modalities to be closed under necessary consequence. So the official definition o a modality in Intensional Logic would be this:

(5.1) $\forall x_{\langle \langle w,t \rangle, t \rangle}[Modality(x) \leftrightarrow$
$\qquad \forall y_{\langle w,t \rangle} \forall z_{\langle w,t \rangle} [[x(y) \wedge \forall u_w[y(u) \rightarrow z(u)]] \rightarrow x(z)]]$

In Intensional Logic, modalities are first-class objects: they are values of variables of type $\langle \langle w, t \rangle, t \rangle$. And the apparatus of Intensional Logic permits very general resources for defining various modalities.

This approach captures some of the important constructs of the AI formalizations of context. In particular, *ist* is a relation between an object of type $\langle \langle w, t \rangle, t \rangle$ (a context) and one of type $\langle w, t \rangle$ (a proposition or set of possible worlds), so it will have type

(5.2) $\langle \langle \langle w, t \rangle, t \rangle, \langle \langle w, t \rangle, t \rangle \rangle$.

---

[8] See [Mon74] and [Gal75].

[9] See [PH96].

[10] Montague formulated the types in a slightly different way, but Gallin showed that this formulation is equivalent to the simpler one I sketch here.

And we can define it as follows.

$$(5.3) \quad ist = \lambda c_{\langle\langle w,t\rangle,t\rangle} \lambda p_{\langle w,t\rangle} [c(p)].$$

This definition makes $ist(c,p)$ equivalent to $c(p)$. Here the second argument to ist does not refer to a formula; it refers to a set of possible worlds.

The fact that higher-order logic is not axiomitizable may induce doubts about its computational usefulness. But the unaxiomatizability of higher-order logic has not prevented it from being used successfully as a framework for understanding a domain. Using a logic of this sort in this way means that implementations may fail to perfectly reflect the logical framework. But often enough, this is true even when the framework is axiomatizable or even decidable. And we can always hope to find tractable special cases.

6   Beyond Modality: The Need for a Theory of Character

A purely modal approach to the theory of context is inadequate in one major respect. The logic of modality is a way of accounting for how variations in the facts can affect the semantic behavior of expressions with a fixed meaning. A sentence like:

(6.1) 'The number of planets might be greater than 9'

can be represented as true because there is a possible world in which there are more than 9 planets. But a sentence like

(6.2) '8 might be greater than 9'

can't, and shouldn't, be represented as true by presenting a possible world in which '8 is greater than 9' means that 8 is less than 9. There is no such possible world, even though it is possible that 'less' might mean what 'greater' in fact means.

In using possible worlds semantics to interpret an expression ξ, we work not with ξ, but with a fixed, disambiguated meaning of ξ. It is relative to such a meaning that we associate a semantic interpretation to ξ. (If ξ is a sentence, the interpretation will be a set of possible worlds, i.e. it will have type $\langle w,t\rangle$. Other sorts of expressions will have interpretations of appropriate type.)

However, many of the desired applications of the theory of context involve cases in which the meaning of an ambiguous expression is resolved in different ways, cases in which the meanings are allowed to vary as well as the facts. An important application of context in the CYC project, for instance, involves cases in which expressions receive different interpretations in different "microtheories"; see [Guh91] for examples. In applications of context to knowledge integration, it is of course important that different data bases have different views of the same facts; but it is equally important that they use expressions in different ways. [MB95] works out the details of a case in which 'price' takes on different meanings in different data bases, and these differences have to be reconciled in order to formalize transactions between these data bases. And some applications of the theory of context are directly concerned with the phenomenon of ambiguity; see [Buv96].

In order to accommodate applications such as these, we need to generalize the framework of modal type theory.

7   Contextial Intensional Logic

[Kap78] provides a good starting point for developing a generalization of Intensional Logic that is able to deal with contexts. Kaplan treats contexts as indices that fix the meanings of context-dependent terms, and concentrates on a language in which there are only three such terms: 'I', 'here', and 'now'. In this case, a context can be identified with a triple consisting of a person, a place, and a time. The truth-value of a context-dependent sentence, e.g. the truth-value of

(7.1) 'I was born in 1732',

will depend on both a context and a possible world, and the interpretation will proceed in two stages. First, a context assigns an intensional value to (7.1); this value will be a set of possible worlds. In case the context assigns 'I' the value George Washington, then (7.1) will be the set of possible worlds in which George Washington was born in 1732. Kaplan introduces some useful terminology for discussing this two-stage interpretation. The set of possible worlds is the *content* of an interpreted sentence. In general, we can expect contents to be functions from possible worlds to appropriate values; in the case of a sentence, the function will take a possible world to a truth value. We can now look at the context-dependent meaning of an expression, or its *character,* as a function from contexts to contents. The character of (7.1), for instance, will be a function taking each context into the set of possible worlds in which the speaker of that context was born in 1732. In these applications, contexts are to be regarded as fact-independent abstractions that serve merely to fix the content of expressions. When an utterance is made, it is of course a fact about the world who the speaker is; but in order to separate the role of context and possible world in fixing a truth value, we need to treat the speaker of the context independently from these facts.

To formalize these ideas in the type-theoretic framework, we add to the three basic types of Intensional Logic a fourth basic type: the type i of *c-indices.* The c-indices affect contextual variations of meaning by determining the characters of expressions.

For instance, suppose that we wish to formalize the contextual variation of the term ACCOUNT across two corporate data bases; in one data base, ACCOUNT refers to active customer accounts, in the other, it refers to active and inactive customer accounts. We create two c-indices i_1 and i_2. Let I_1 be the intension that we want the predicate to receive in i_1 and I_2 be the intension in i_2; these intensions will have type $\langle w, \langle e, t \rangle \rangle$.[11]

The type-theoretic apparatus provides a type of functions from c-indices to intensions of type $\langle w, \langle e, t \rangle \rangle$; this will be the type of characters of one-place predicates, such as ACCOUNT. We represent the behavior of the predicate ACCOUNT by choosing such a function F, where $F(i_1) = I_1$ and $F(i_1) = I_2$. The interpretation of the language then assigns the character F to the predicate ACCOUNT. The following paragraphs generalize this idea.

Suppose that the *content type* of a lexical item ξ is $\langle w, \tau \rangle$. (The content type of ACCOUNT, for instance, will be $\langle e, t \rangle$.) Then the character type of ξ will be $\langle i, \langle w, \tau \rangle \rangle$. An interpretation of the language will assign an appropriate character (i.e., something of type $\langle i, \langle w, \tau \rangle \rangle$) to each lexical item with content type $\langle w, \tau \rangle$. This is my proposal about how to deal with the content-changing aspect of contexts.

To capture the insight of the modal approach to context, that contexts also affect the local axioms, or the information that is accessible, we assume that an interpretation also selects a function *Info* from contexts to modalities. That is, *Info* has type $\langle i, \langle \langle w, t \rangle, t \rangle \rangle$. In the current formalization, with a basic type for contexts, the type of *ist* will simply be $\langle i, \langle \langle i, \langle w, t \rangle \rangle, t \rangle \rangle$; *ist* inputs a context and a propositional character, and outputs a truth-value. We want $ist(c, p)$ to tell us whether the proposition assigned to p in c follows from the information available in c. The general and revised definition of *ist*, then, is as follows.

$$(7.2) \quad ist = \lambda c_i \lambda p_{\langle i, \langle w, t \rangle \rangle} Info(c)(p(c))$$

8   Conclusions

This paper has been primarily concerned with the motivation and formulation of a type-theoretic approach to context that generalizes the uses of intensional type theories in natural language semantics, and that is capable of dealing with all (or almost all) of the applications that have been envisaged for the theory of context in the recent AI literature.

The presentation here is in some ways a sketch. The underlying logic itself is a very straightforward generalization of logics that have been thoroughly investigated (see, for instance, [Gal75]), and the logical work that needs to be done here is a matter of showing how to use the apparatus to develop appropriate formalizations of some reasonably complex examples. I plan to include such formalizations in future versions of this work.

There are several dimensions in which the logical framework that I have presented needs to be generalized in order to obtain adequate coverage:

(8.1) The logic needs to be made partial, to account for expressions which simply lack a value in some contexts.

---

[11] This type corresponds to a function that takes a possible world into a set of individuals.

(8.2) The logic needs dynamic operators of the sort described in McCarthy's papers; e.g., an operator which chooses a context and *enters* it.

(8.3) To account for default lifting rules, we need a nonmonotonic logic of context.

Since we have a general sense of what is involved in making a total logic partial, in making a static logic dynamic, and in making a monotonic logic nonmonotonic, I have adopted the strategy of first formulating an appropriate base logic to which these extensions can be made. Briefly, for (8.1) there are a number of approaches partial logics; see [Mus96] for an extended study of how to modify Intensional Logic using one of these approaches. For (8.2), I favor an approach along the lines of [GS91]; essentially this involves relativizing satisfaction not to just one context, but to a pair of contexts, an input context and an output context. For (8.3), it is relatively straightforward to add a theory of circumscription to Intensional Logic, and to the extension that I have proposed here. (Circumscription is usually formulated in second-order extensional logic, but the generalization to intensional logic of arbitrary order is straightforward.)

None of these logical developments is entirely trivial, and in fact there is material here for many years of work. I hope to report on developments in these directions in future work.

References

[AP97]   Gianni Amati and Fiora Pirri. Contexts as relative definitions: A formalization via fixed points. In Sasa Buvač and Łucia Iwańska, editors, *Working Papers of the AAAI Fall Symposium on Context in Knowledge Representation and Natural Language*, pages 7–14, Menlo Park, California, 1997. American Association for Artificial Intelligence, American Association for Artificial Intelligence.

[BBM95]   Saša Buvač, Vanja Buvač, and Ian Mason. Metamathematics of contexts. *Fundamenta Mathematicae*, 23(3), 1995. Available from http://www-formal.stanford.edu/buvac.

[BF95]   Saša Buvač and Richard Fikes. A declarative formalization of knowledge translation. In *Proceedings of the ACM CIKM: the Fourth International Conference in Information and Knowledge Management*, 1995. Available from http://www-formal.stanford.edu/buvac.

[BI97]   Sasa Buvač and Łucia Iwańska, editors. *Working Papers of the AAAI Fall Symposium on Context in Knowledge Representation and Natural Language*. American Association for Artificial Intelligence, Menlo Park, California, 1997.

[Buv96]   Saša Buvač. Resolving lexical ambiguity using a formal theory of context. In Kees van Deemter and Stanley Peters, editors, *Semantic Ambiguity and Underspecification*, pages 100–124. Cambridge University Press, Cambridge, England, 1996.

[Chu51]   Alonzo Church. The need for abstract entities in semantic analysis. *Proceedings of the American Academy of Arts and Sciences*, 80:100–112, 1951.

[Cru95]   D.A. Cruse. Polysemy and related phenomena from a cognitive linguistic viewpoint. In Patrick Saint-Dizier and Evelyne Viegas, editors, *Computational Lexical Semantics*, pages 33–49. Cambridge University Press, Cambridge, England, 1995.

[CS89]   Herbert H. Clark and Michael Schober. Understanding by addressees and overhearers. *Cognitive Psychology*, 24:259–294, 1989.

[FHMV95]   Ronald Fagin, Joseph Y. Halpern, Yoram Moses, and Moshe Y. Vardi. *Reasoning About Knowledge*. The MIT Press, Cambridge, Massachusetts, 1995.

[Gal75]   Daniel Gallin. *Intensional and Higher-Order Logic*. North-Holland Publishing Company, Amsterdam, 1975.

[GN97]   Dov Gabbay and Rolf T. Nossum. Structured contexts with fibred semantics. In Saša Buvač and Łucia Iwańska, editors, *Working Papers of the AAAI Fall Symposium on Context in Knowledge Representation and Natural Language*, pages 48–57, Menlo Park, California, 1997. American Association for Artificial Intelligence, American Association for Artificial Intelligence.

[GS91]   Jeroen Groenendijk and Martin Stokhof. Dynamic predicate logic. *Linguistics and Philosophy*, 14:39–100, 1991.

[Guh91]   Ramanathan V. Guha. Contexts: a formalization and some applications. Technical Report STAN-CS-91-1399, Stanford Computer Science Department, Stanford, California, 1991.

[Kam95]   Fairouz Kamareddine. Are types needed for natural language? In László Pólos and Michael Masuch, editors, *Applied Logic: How, What, and Why? Logical Approaches to Natural Language*, pages 79–120. Kluwer Academic Publishers, Dordrecht, 1995.

[Kap78]   David Kaplan. On the logic of demonstratives. *Journal of Philosophical Logic*, 8:81–98, 1978.

[Kri75]   Saul Kripke. Outline of a theory of truth. *Journal of Philosophy*, 72:690–715, 1975.

[MB95]   John McCarthy and Saša Buvač. Formalizing context (expanded notes). Available from http://www-formal.stanford.edu/buvac., 1995.

[McC86]   John McCarthy. Notes on formalizing contexts. In Tom Kehler and Stan Rosenschein, editors, *Proceedings of the Fifth National Conference on Artificial Intelligence*, pages 555–560, Los Altos, California, 1986. American Association for Artificial Intelligence, Morgan Kaufmann.

[McC93]   John McCarthy. Notes on formalizing context. In *Proceedings of the Thirteenth International Joint Conference on Artificial Intelligence*, pages 81–98, Los Altos, California, 1993. Morgan Kaufmann.

[McC97]   John McCarthy. Modality si! Modal logic, no! *Studia Logica*, 59(1):29–32, 1997.

[Mon63]   Richard Montague. Syntactical treatments of modality, with corollaries on reflection principles and finite axiomatizability. *Acta Philosophica Fennica*, 16:153–167, 1963.

[Mon74]   Richard Montague. *Formal Philosophy: Selected Papers of Richard Montague*. Yale University Press, New Haven, CT, 1974.

[Mus96]   Reinhard Muskens. *Meaning and Partiality*. Cambridge University Press, Cambridge, England, 1996.

[NL95]   P. Pandurang Nayak and Alan Levy. A semantic theory of abstractions. In Chris Mellish, editor, *Proceedings of the Fourteenth International Joint Conference on Artificial Intelligence*, pages 196–203, San Francisco, 1995. Morgan Kaufmann.

[PB97]   James Pustejovsky and Brian Boguraev, editors. *Lexical Semantics: The Problem of Polysemy.* Oxford University Press, Oxford, 1997.

[PH96]   Barbara H. Partee and Herman L.W. Hendriks. Montague grammar. In Johan van Benthem and Alice ter Meulen, editors, *Handbook of Logic and Language*, pages 5–91. Elsevier Science Publishers, Amsterdam, 1996.

[Qui53]   Willard V. Quine. Three grades of modal involvement. In *Proceedings of the XIth International Congress of Philosophy, Volume 14*, pages 65–81, 1953.

[Qui56]   Willard V. Quine. Quantifiers and propositional attitudes. *The Journal of Philosophy*, 53:177–187, 1956.

[Sch94]   David A. Schmidt. *The Structure of Typed Programming Languages.* The MIT Press, Cambridge, Massachusetts, 1994.

[WH90]   D. Wright and P. Hull. How people give verbal instructions. *Journal of Applied Cognitive Psychology*, 4:153–174, 1990.

[You97]   R. Michael Young. *Generating Concise Descriptions of Complex Activities.* Ph.d. dissertation, Intelligent Systems Program, University of Pittsburgh, Pittsburgh, Pennsylvania, 1997.

From Integrated Reasoning Specialists to "Plug-and-Play" Reasoning Components [*]

Alessandro Armando and Silvio Ranise

DIST – Università di Genova
Via all'Opera Pia 13 – 16145 Genova – Italy
armando,silvio@mrg.dist.unige.it

Abstract. There is an increasing evidence that a new generation of reasoning systems will be obtained via the integration of different reasoning paradigms. In the verification arena, several proposals have been advanced on the integration of theorem proving with model checking. At the same time, the advantages of integrating symbolic computation with deductive capabilities has been recognized and several proposals to this end have been put forward. We propose a methodology for turning reasoning specialists integrated in state-of-the-art reasoning systems into reusable and implementation independent reasoning components to be used in a "plug-and-play" fashion. To test our ideas we have used the Boyer and Moore's linear arithmetic procedure as a case study. We report experimental results which confirm the viability of the approach.

Keywords: integration of decision procedures, integration of deduction and symbolic computation, automated theorem proving.

1 Introduction

There is an increasing evidence that a new generation of reasoning systems will be obtained via the integration of different reasoning paradigms. In the verification arena, several proposals have been advanced on the integration of theorem proving with model checking. At the same time, the advantages of integrating symbolic computation with deductive capabilities have been recognized and several proposals to this end have been put forward.

As pointed out in [3], the approaches to integrating reasoning systems can be classified in homogeneous integration and heterogeneous integration. *Homogeneous integration* amounts to embedding new reasoning paradigms into existing reasoning systems. Examples of such an approach are the integration of deductive

---

[*] We wish to thank Fausto Giunchiglia for very helpful discussions. We are also grateful to Alan Bundy and Alessandro Coglio for comments on an early draft of this paper. The authors are supported in part by *Conferenza dei Rettori delle Università Italiane (CRUI)* in collaboration with *Deutscher Akademischer Austaunschdienst (DAAD)* under the *Vigoni Programme*.

Jacques Calmet and Jan Plaza (Eds.): AISC'98, LNAI 1476, pp. 42–54, 1998.

capabilities into computer algebra systems as discussed, e.g., in [6,7], or the integration of model checking capabilities into deductive systems as advanced, e.g., in [12,11]. *Heterogeneous integration* aims at building hybrid reasoning systems by combining the reasoning services provided by existing tools. For example, [9] illustrates the combination of interactive proof-development systems with computer algebra systems; [3] describes a generic interface to computer algebra systems for the *Isabelle* prover.

Both approaches to integration have advantages and difficulties. Homogeneous integration requires the effort of implementing a reasoning technique from scratch, but it allows for a higher degree of flexibility as the new reasoning technique can be tailored to the needs of the host system. However—as experienced by Boyer and Moore [5]—turning a reasoning technique into an integrated reasoning specialist can be a challenge. Moreover both the efficiency and the range of functionalities provided by state-of-the-art implementations are very difficult to achieve and maintain. Etherogeneous integration—aiming at using services provided by existing tools—is apparently easier to achieve. However, even if state-of-the-art reasoning systems are built out of a set of carefully engineered reasoning specialists (e.g. simplifiers, constraint solvers, decision procedures, model checkers), they are in most cases conceived and built as stand-alone systems to be used by human users. As a consequence direct access to the services provided by reasoning specialists is rarely made available. This is particularly unfortunate as existing reasoning systems represent a real cornucopia of powerful reasoning specialists.

Our interest is in turning reasoning specialists integrated in state-of-the-art reasoning systems into reusable and implementation independent reasoning components thereby lessening the difficulties in attaining heterogeneous integration. To this end we propose a two-step methodology which amounts to *(i)* modeling (part of) an existing reasoning system as a set of reasoning specialists glued together by means of an integration schema, and *(ii)* lifting the selected reasoning specialists and the integration schema identified in the first step into a set of reasoning components and an interaction schema, respectively. Similarly to [8] by *reasoning components* we mean open architectures capable of exchanging a carefully selected set of logical services [13]. By *interaction schema* we mean a communication protocol governing the interplay among the reasoning components. To test our ideas we have used the Boyer and Moore's linear procedure as a case study. We have chosen this case study because of its significance (the Boyer and Moore's approach to integrating decision procedures is notoriously complex, and a better understanding of how it works is by itself of considerable interest) and because it constitutes one of the most challenging case studies we could think of.

Structure of the paper. Sections 2 and 3 illustrate the two steps of our methodology. The methodology is discussed by showing its application to our case study. Section 4 presents and discusses the experimental results. Some final remarks are drawn in Section 5.

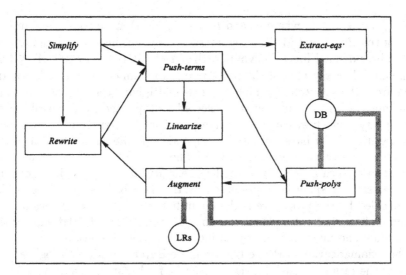

Fig. 1. The reasoning specialists and the integration scheme

2   The Reasoning Specialists and the Integration Schema

The first step of our methodology amounts to opening up a given reasoning system, to identifying the relevant reasoning specialists, the data-flow and the shared data structures between them, and finally to identifying an integration schema. In our case study this required the careful analysis of both the 40 page-long report [5] and the actual implementation code.

As a result, we came up with the rational reconstruction of the Boyer and Moore's integration schema abstractly depicted in Figure 1. Square boxes and black arrows represent the most significant reasoning specialists and their functional dependences respectively. Round boxes denote shared data structured, namely the data base of polynomials (DB) and the data base of the linear rules (LRs). Gray thick edges represent the access to the shared data structures.

The reasoning specialists. *Simplify* takes a clause as input and returns a set of supposedly simpler clauses as output. (A clause is an implicitly disjoint set of literals.) *Simplify* works by successively rewriting each literal of the input clause while assuming the complements of the remaining literals.

Rewrite exhaustively applies a set of conditional rewrite rules of the form:

$$\bigwedge_{i=1}^{n} h_i \to (lhs = rhs)$$

to the input term. A conditional rewrite rule causes *Rewrite* to replace all the instances of *lhs* by the corresponding instance of *rhs* provided each of the instantiated h_i can be proved by a recursive call to *Rewrite*.

The linear arithmetic procedure incrementally maintains an implicitly conjoint set of 'normalized polynomial inequalities' (from here on 'polynomials') in the *data base of polynomials* (DB). A polynomial has the following form:

$a_0 + \sum_{i=1}^{n} a_i * t_i \le 0$, where a_i and t_i $(i = 1, \ldots, n)$ are integer constants and terms of the Boyer and Moore's logic respectively. For instance, if MAX and MIN are two user-defined function symbols, then $1 * \text{MIN}(A) + -1 * \text{MAX}(A) \le 0$ is a polynomial. The key idea of the procedure is to eliminate one multiplicand at a time from the set of polynomials by cross-multiplying coefficients and adding polynomials so as to cancel out the selected multiplicand in all possible ways. This activity – which we call a *macro Fourier step* – is carried out by *Push-polys*. If the resulting data base contains an *impossible polynomial* (that is a polynomial whose constant is greater than 0 and whose coefficients are all greater than or equal to 0, such as $2 \le 0$), then the linear arithmetics procedure concludes that the set of inequalities represented by the data base is unsatisfiable. The logic decided by the procedure is the quantifier-free linear arithmetics over the rationals where $\le$, $+$, $-$, and $*$ denote the 'less-than or equal' relation, addition, subtraction, and multiplication by an integer constant, respectively.

The axiomatization of Peano theory in the Boyer and Moore's logic and the choice of the rational based decision procedure for linear inequalities made necessary the introduction of a linearization routine (*Linearize*) to map the validity problem over the naturals into a satisfiability problem over the rationals. (For our purposes it suffices to know that the language of NQTHM logic contains the equality symbol ($=$), the Peano 'less-than' and 'less-than or equal' relations ($\prec$ and $\preceq$, resp.), the numerals 0, 1, 2, ..., and the Peano addition ($\oplus$) and subtraction ($\ominus$) functions.) The linearization of an inequality is a set of polynomials each one equipped with a set of 'linearization hypotheses' stored in the *hyps* field.[1] Given a set of literals as input, *Push-terms* first invokes *Linearize* to translate (the negation of) the input literals into a set of polynomials which are then added to DB by *Push-polys*. If adding the polynomials to DB yields an impossible polynomial, then *Push-terms* concludes that the literal given as input is true.

The heuristic component stores previously proven lemmas into a data base of *linear rules* (LRs). A linear rule can have one of the following two forms:

$$\bigwedge_{i=1}^{n} h_i \to (lhs \prec rhs) \qquad \bigwedge_{i=1}^{n} h_i \to (lhs \preceq rhs)$$

The *Augment* routine looks for heuristically chosen instances of the linear rules which might contribute to the derivation of an impossible polynomial in DB. Notice that *Augment* may invoke *Rewrite* while attempting to relieve the hypotheses of linear rules, i.e. $h_1, \ldots, h_n$. To see why new polynomials are needed, consider the data base $\text{DB}_0 = \{1 * L + -1 * \text{MIN}(A) \le 0, 1 + -1 * K \le 0, -1 * L + 1 * K + 1 * \text{MAX}(A) \le 0\}$,[2] where MIN (MAX) is the user-defined function returning the minimum (maximum, resp.) element of a list of numbers. After two macro Fourier steps, we get $\text{DB}_1 = \text{DB}_0 \cup \{1 * K + -1 * \text{MIN}(A) + 1 * \text{MAX}(A) \le 0, 1 + -1 * \text{MIN}(A) + 1 * \text{MAX}(A) \le 0\}$. At this stage, no further macro Fourier

---

[1] For the lack of space we omit the explanation of the meaning and role of the linearization hypotheses. See [5] for the details.

[2] DB_0 is the polynomial data base obtained by clausifying and linearizing the formula, $(L \preceq \text{MIN}(A) \land 0 \prec K) \to (L \prec \text{MAX}(A) \oplus K)$.

step can be performed according to our heuristic criterion. This is because the unsatisfiability of DB_0 is not a consequence of linear arithmetic reasoning only. However, if $MIN(X) \preceq MAX(X)$ is an available lemma, then it can be instantiated to $MIN(A) \preceq MAX(A)$. By adding the corresponding linearization, i.e. $1 * MIN(A) + -1 * MAX(A) \leq 0$, to DB_1, a macro Fourier step yields the impossible polynomial $1 \leq 0$.

Finally, *Extract-eqs* detects and returns equalities entailed by DB. For example, if the data base contains the polynomials $1 * X + -1 * Y \leq 0$ and $-1 * X + 1 * Y \leq 0$, then it can be easily seen that $X = Y$ is entailed by the data base.

The integration schema. A more detailed account of the Boyer and Moore's integration schema (w.r.t. Figure 1) is given in Figure 2.

As the work needed to build the polynomial data bases for adjacent literals in a clause is very similar (linearize the negation of the inequalities contained in the clause but the one being rewritten, perform all possible cross-multiply and adds, and augment), *Simplify* sets up a single data base for the clause. If such a data base contains an impossible polynomial the input clause is trivially true and *Simplify* returns this fact. If an impossible polynomial has not been derived, then *Extract-eqs* is invoked and the equalities extracted from DB are added to the clause. *Rewrite* is then asked to rewrite each literal in turn. But when rewriting *lit* we must pay attention not to use the polynomials (and their derivatives) encoding the falsity of *lit*; otherwise we could rewrite *lit* to false. To overcome the difficulty, the data structure representing a polynomial is enriched with a field (called *lits* field) containing the literals from which it derives; the cross-multiply and add routine is then instructed to ignore those polynomials having any literal previously rewritten to false in the *lits* field as well as the literal being rewritten. To this end, *Simplify* keeps track of the literals rewritten to false.

When *Rewrite* is asked to establish a linear inequality, *lit*, it applies a first set of rewrites to *lit* (collectively called *Rewrite1* in Figure 2). If this rewriting leaves *lit* untouched then the linear arithmetic procedure is invoked by pushing the negation of *lit* onto the data base. If the resulting data base DB' contains an impossible polynomial, then *Rewrite* is entitled to rewrite *lit* to true; otherwise a second set of rewrites is applied to *lit* (called *Rewrite2* in Figure 2). Each time a polynomial is pushed into DB and a contradiction is not found, then the linear arithmetic procedure asks *Augment* for additional linear facts. As *Augment* may invoke *Rewrite* when relieving the hypotheses of a linear rule, the rewriting and the augmentation procedures are mutually recursive. It is worth noticing that each literal in the clause is rewritten in the context of the initially set up data base DB, i.e. the temporary data bases created while rewriting the previous literal are discarded before the rewriting of the next literal begins.

```
Function Simplify(cl)
Begin
 DB ← Push-terms(cl, DB);
 if DB contains an impossible polynomial then return true;
 cl ← cl ∪ Extract-eqs(DB);
 cl' ← cl;
 while cl' ≠ {} do
   Begin
     lit ← first(cl'); cl' ← rest(cl');
     lit' ← Rewrite(lit, DB);
     Replace lit with lit' in cl;
   End
 return cl;
End


Function Rewrite(lit, DB)
Begin
 lit' ← Rewrite1(lit);              /* First phase of rewriting */
 if lit' ≠ lit then return lit';
 DB' ← Push-terms({lit}, DB);  /* Invoking the LA specialist */
 if DB' contains an impossible polynomial then return true;
 lit' ← Rewrite2(lit);              /* Second phase of rewriting */
 return lit;
End


Function Push-terms(terms, DB)
Begin
 polys ← {};
 foreach term in terms do polys ← polys ∪ Linearize(¬term);
 return Push-polys(polys, DB);
End


Function Push-polys(polys, DB)
Begin
 foreach poly in polys do
  Begin
    DB ← Cross-multiply-and-add(poly, DB);
    if DB contains an impossible polynomial then return DB;
  End
 DB ← Push-polys(Augment(DB, LRs), DB);
 return DB;
End


Function Augment(DB, LRs)
Begin
 polys ← {};
 foreach new multiplicand m in DB do
   foreach linear rule 'hyps → lr' in LRs do
     if m occurs in lr and Rewrite(hyps, DB) = true
       then polys ← polys ∪ Linearize(lr);
 return polys;
End
```

Fig. 2. The integration schema

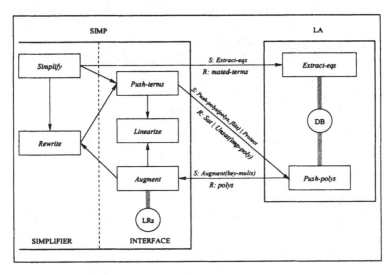

Fig. 3. The reasoning components and the interaction schema

3   The Reasoning Components and Interaction Schema

We now describe the application of the second step of our methodology. In our case study this amounted to scrutinizing the data-flows between the reasoning specialists introduced in Section 2. Data-flows which are too intensive (both in the rate and in the amount of information exchanged) must be avoided as they would compromise the efficiency of the combined system. Furthermore, as we do not allow the sharing of data structures between distinguished reasoning components, the information needed by components which have no longer direct access to a previously shared data structure must be explicitly passed. Also, if the behavior of a reasoning component depend on its internal state, then suitable primitives for managing the state must be identified.

By applying the above analysis to our case study we ended up with the interaction scheme of Figure 3. LA and SIMP are the labels for the linear arithmetic reasoning component and the simplifier respectively. The edges crossing the boundary between the two components have been labeled with service requests (*S:msg*) and service replies (*R:msg*).

The reasoning components. When *Simplify* wants the equalities entailed by DB, it asks the service request *S:Extract-eqs* to the LA and gets back a reply of the form *R:mated-terms*. *Push-terms* can ask either for *Push-polys(polys,flits)* (where *polys* is a list of polynomials and *flits* is the list of the literals previously rewritten to false), or for *Protect* which amounts to asking LA to save the current content of DB for later use. (LA is also equipped with an internal functionality, *Restore*, which allows for the backtracking to a previously saved copy of DB.) A reply *R:Sat* means that no contradiction has been found in DB, whereas a reply *R:Unsat(imp-poly)* means that the impossible polynomial *imp-poly* has been derived. Augmentation is invoked by means of *S:Augment(key-mults)* where

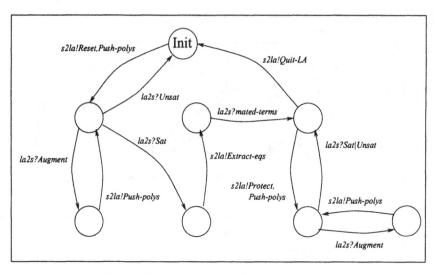

Fig. 4. State transition diagram for SIMP

key-mults are (heuristically chosen) multiplicands occurring in DB. *key-mults* encode all the information of DB needed by *Augment* which therefore no longer needs to directly access DB. *R:polys* are the polynomials sent back by *Augment*.

The interaction schema. Figure 3 provides only an abstract account of the interaction schema between LA and SIMP as it simply specifies the correspondence between the logical services required and those provided by the reasoning components. This level of abstraction is unsatisfactory for many purposes as it leaves unspecified necessary constraints of the sequencing of events. (For instance, a *Restore* must follow a *Protect*, but this in not specified by Figure 3). Therefore we complete the description of the interaction scheme by describing a suitable communication protocol. This amounts to specifying the vocabulary and the format of the messages exchanged, as well as the procedure rules of the protocol [10].

The procedure rules are formalized by the state transition diagrams in Figure 4 and Figure 5, both the vocabulary and the format of the messages can be easily inferred from the labels used in the diagrams and the following description. The labels associated to the edges denote the actions performed when the corresponding transitions take place. *s2la* and *la2s* are channels over which the messages are exchanged. If *ch* is a channel, and *exp* is an expression, then *ch!exp* denotes the action of sending *exp* to *ch*, that is, it appends *exp* to the tail of *ch*; *ch?exp* denotes the action of retrieving a message from the head of *ch* provided that it matches with *exp*. If multiple expressions are transferred per message, they are specified in a comma-separated list: $ch!exp_1, \ldots, exp_n$ and $ch?exp_1, \ldots, exp_n$. An edge labeled with $ch!exp_1 | exp_2$ stands for two alternative edges labeled with $ch!exp_1$ and $ch!exp_2$. Statements are separated by semicolons: if s_1 and s_2 are statements, then $s_1; s_2$ is a statement whose execution amounts to the execution of s_1 followed by the execution of s_2.

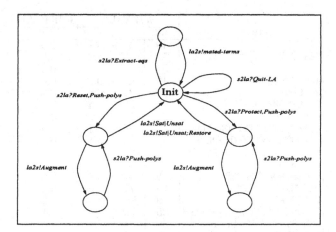

Fig. 5. State transition diagram for LA

We now informally describe how the protocol specified by the state transition diagrams of Figures 4 and 5 reproduce the functionality of *Simplify* given in Figure 2. The protocol starts with SIMP issuing a message of the form *Reset* followed by a *Push-polys(polys)* to LA.[3] Upon receipt of such messages LA resets DB and then pushes *polys* onto DB. If no contradiction is detected and some new multiplicands have been introduced into DB, then LA asks SIMP for new polynomials by means of a message of the form *Augment(key-mults)*. Notice that an arbitrarily long iteration of *Push-polys* and *Augment* may occur. (This is also the case in *Simplify* due to the mutually recursive calls of the functions *Push-terms*, via *Push-polys*, and *Augment*.) As soon as an impossible polynomial *imp-poly* is derived in DB, a message of the form *Unsat(imp-poly)* is sent by LA back to SIMP. If the result of pushing new polynomials does not introduces new multiplicands in DB, then the (supposed) satisfiability of DB is notified to SIMP. At this point, SIMP asks for and gets backs a (possibly empty) set of equalities entailed by DB. The rest of the execution corresponds to the service calls and replies produced during rewriting and everything proceeds similarly to the set up phase. The only difference is that SIMP explicitly asks LA to save the original content of DB (*Protect*) which is later automatically restored by LA (*Restore*). Finally, SIMP issues a message of the form *Quit-LA* in order to start simplifying a new clause. Indeed, upon recepit of such a message LA sets DB to the empty data base.

4   Experimental Results

We have extracted the linear arithmetic procedure from the 1992 version of NQTHM, turned it into an independent reasoning component (LA) and made it interact with the rest of the prover (NQTHM) via the protocol presented in

---

[3] For simplicity, in Figures 4 and 5 we have omitted the arguments of the service requests.

Name	NQTHM+LA			NQTHM(LA)
	calls	info	time	time
EXPT	19	45	0.7	0.0
ZEXPTZ	19	45	0.8	0.0
LEX	53	53	2.8	0.3
ALMOST-EQUAL1	19	47	0.9	0.0
PLUS-0	19	34	0.6	0.0
PLUS-NON-NUMBERP	35	38	1.4	0.1
PLUS-ADD1	21	48	1.2	0.1
COMMUTATIVITY2-OF-PLUS	3	34	0.1	0.0
COMMUTATIVITY-OF-PLUS	3	28	0.1	0.0
ASSOCIATIVITY-OF-PLUS	3	34	0.2	0.0
TIMES-0	25	36	0.7	0.0
TIMES-NON-NUMBERP	25	31	0.7	0.1
DISTRIBUTIVITY	25	71	2.8	0.1
TIMES-ADD1	89	119	16.8	0.2
COMMUTATIVITY2-OF-TIMES	25	64	2.7	0.1
COMMUTATIVITY-OF-TIMES	49	84	7.1	0.3
ASSOCIATIVITY-OF-TIMES	9	53	0.8	0.1
EQUAL-TIMES-0	71	55	6.0	0.3
EQUAL-LESSP	9	15	0.1	0.0

Legenda. *Timings are averaged over five runs on a workstation SUN SPARC10 with 32 MB of RAM. Both* NQTHM *and* LA *were compiled using* GNU Common Lisp 1.0.

Table 1. Experimental results

Section 3. The two reasoning components have been implemented as two distinct processes interacting via a socket interface. In the sequel we use NQTHM(LA) and NQTHM+LA to refer to the implementations of the original and of the combined system respectively.

Clearly NQTHM+LA outperforms NQTHM(LA) in terms of flexibility and reusability. In fact, it is straightforward to replace LA with a new version of the reasoning component (perhaps implemented in a different implementation language), say LA', and obtain an upgraded reasoning system, say NQTHM+LA'. It is also relatively easy to plug LA into a different reasoning system. Of course there is a price to pay for the flexibility offered by the combined system. Indeed a performance degradation is to be expected as a consequence of the fact that function calls have been replaced with calls to an external process.

We have carried out an experimental comparison between NQTHM(LA) and NQTHM+LA to test the practical feasibility of our approach. To this end we have selected a set of definitions and lemmas from the Boyer and Moore's corpus available with the 1992 distribution package of NQTHM. In particular, we focused on the proof of facts contained in the file fortran.events as they represent the kind of proofs which motivated the integration of the linear arithmetic procedure in NQTHM [5].

Table 1 lists the results of our experiments. The first three columns provide data relative to NQTHM+LA: the number of messages exchanged by the two processes (*calls*), the ratio between the total amount of information (in bytes) and the number of messages exchanged (*info*), and the timings (expressed in seconds: 0.0 indicates that the computation time is less than 0.05 seconds). The rightmost column contains the timings of NQTHM(LA).

Our experiments show that (with the only exception of TIMES-ADD1) the current version of NQTHM+LA is ten times slower than NQTHM. In our view this performance degradation is already a fair price to pay for the flexibility and reusability of the reasoning component. However a careful analysis of the experimental data reveals that there is room for significant improvements.

The data show that the performance of NQTHM+LA is influenced by two factors: the number of messages exchanged (*calls*) and the amount of information exchanged per message (*info*). However the latter seems to have a more significant impact. Evidence to this fact is obtained by comparing the values of the *calls* and *info* fields of TIMES-ADD1 and EQUAL-TIMES-0. While the number of messages exchanged is close, the amount of information per message is significantly larger for TIMES-ADD1 and therefore is the (main) culprit for the bad behavior of NQTHM+LA. This consideration is confirmed by comparing DISTRIBUTIVITY with TIMES-NON-NUMBERP, and COMMUTATIVITY-OF-TIMES with LEX.

An analysis of the content of the messages exchanged by NQTHM and LA during the proof of TIMES-ADD1 reveals that the reason of their size is due to the content of the *hyps* and *lits* fields of the polynomials exchanged (see Section 2). Because of the use LA makes of such fields, it is possible for NQTHM to map distinct data structures (contained in such fields) into distinct identifiers and send the latter instead of the lengthy printed representation of the data structures (as the current implementation does). This allows for a much more compact representation of the *hyps* and *lits* fields. We expect this to dramatically improve the performance of *info*-intensive proofs.

In the light of the previous analysis, it is reasonable to expect that a new version of the protocol incorporating the optimization hinted above will provide us with a considerably faster combined reasoning system. Furthermore, we envisage that standard optimization techniques borrowed from the field of distributed computing (e.g. pipelining) can yield a combined reasoning system outperforming its original implementation when run on a multi-processor architecture.

5   Conclusions and Future Work

In this paper we have proposed a methodology for lifting tightly integrated reasoning specialists into "plug-and-play" reasoning components. This allows for the reuse of the sophisticated reasoning functionalities currently embedded in existing state-of-the-art reasoning systems. We believe this is a first but significant step toward the ultimate objective of building reasoning systems (or enhancing existing ones) by combining reasoning components in a "plug-and-play" fashion as envisaged in [8].

The viability of the approach is shown by applying the methodology to a challenge case study: the lifting of the Boyer and Moore's linear arithmetic decision procedure into a stand-alone reasoning component. Experiments conducted on a prototype implementation indicate that the resulting loss of performance is a fair price to pay for the gained flexibility and reusability. Optimizations capable of making the prototype system to compete with (or even to outperform) the original implementation are discussed.

The work described in this paper is part of a wider project, called *Open Mechanized Reasoning Systems* (OMRS) [2,8,14], aiming at the definition of a specification framework for describing *logical services* [13]. [4] extends the OMRS framework to deal with the more general notion of *mathematical services*.

As a final remark it is worth pointing out that our work shares many of the design goals of the OpenMath project [1]. However we regard our contributions as complementary to OpenMath's. In particular, while OpenMath mainly focuses on the mathematical information exchanged by reasoning systems and its standardization, following [8] we aim at a broader characterization taking into account also of the control and of the interaction components of the reasoning systems.

References

1. J. Abbott, A. Díaz, and R. S. Sutor. A report on OpenMath. A protocol for the exchange of mathematical information. *SIGSAM Bulletin (ACM Special Interest Group on Symbolic and Algebraic Manipulation)*, 30(1):21–24, March 1996.
2. A. Armando, P. Bertoli, A. Coglio, F. Giunchiglia, J. Meseguer, S. Ranise, and C. Talcott. Open Mechanized Reasoning Systems: a Preliminary Report. In *Workshop on Integration of Deduction Systems (CADE-15)*, 1998.
3. C. Ballarin, K. Homann, and J. Calmet. Theorems and Algorithms: An Interface between Isabelle and Maple. In *International Symposium on Symbolic and Algebraic Computation*. ACM Press, 1995.
4. P.G. Bertoli, J. Calmet, F. Giunchiglia, and K. Homann. Specification and Combination of Theorem Provers and Computer Algebra Systems. In *4th International Conference Artificial Intelligence And Symbolic Computation*, Plattsburgh, NY, USA, 1998.
5. R. S. Boyer and J S. Moore. Integrating Decision Procedures into Heuristic Theorem Provers: A Case Study of Linear Arithmetic. *Mach. Intel.*, (11):83–124, 1988.
6. B. Buchberger, T. Jebelean, F. Kriftner, M. Marin, E. Tomuta, and D. Vasaru. A Survey of the *Theorema* Project. In *International Symposium on Symbolic and Algebraic Computation*, Hawaii, USA, 1997.
7. E. Clarke and X. Zhao. Analytica – a Theorem Prover for Mathematica. Tech. Rep. CS-92-117, Carnegie Mellon University, 1992.
8. F. Giunchiglia, P. Pecchiari, and C. Talcott. Reasoning Theories: Towards an Architecture for Open Mechanized Reasoning Systems. Tech. Rep. 9409-15, IRST, 1994.
9. J. Harrison and L. Théry. A Sceptic's Approach to Combining HOL and Maple. To appear in the J. of Automated Reasoning, 1997.
10. G. J. Holzmann. *Design and Validation of Computer Protocols*. Prentice Hall, 1990.
11. O. Müller and T. Nipkow. Combining Model Checking and Deduction for I/O-automata. In *Tools and Algorithms for the Construction and Analysis of Systems*, 1995.
12. S. Owre, J. Rushby, and N. Shankar. Integration in PVS: Tables, Types, and Model Checking. In *Tools and Algorithms for the Construction and Analysis of Systems*, Enschede, The Netherlands, 1997.

13. I. Sutherland and R. Platek. A Plea for Logical Infrastructure. In *TTCP XTP-1 Workshop on Effective Use of Automated Reasoning Technology in System Development*, 1992.
14. The OMRS Taskforce. The Open Mechanized Reasoning Systems Project WWW Page. http://www.mrg.dist.unige.it/omrs/.

Reasoning About Coding Theory: The Benefits We Get from Computer Algebra

Clemens Ballarin and Lawrence C. Paulson

Computer Laboratory, University of Cambridge, Cambridge CB2 3QG, UK
{Clemens.Ballarin, Larry.Paulson}@cl.cam.ac.uk
http://www.cl.cam.ac.uk/users/{cmb33, lcp}

Abstract. The use of computer algebra is usually considered beneficial for mechanised reasoning in mathematical domains. We present a case study, in the application domain of coding theory, that supports this claim: the mechanised proofs depend on non-trivial algorithms from computer algebra and increase the reasoning power of the theorem prover.

The unsoundness of computer algebra systems is a major problem in interfacing them to theorem provers. Our approach to obtaining a sound overall system is not blanket distrust but based on the distinction between algorithms we call sound and *ad hoc* respectively. This distinction is blurred in most computer algebra systems. Our experimental interface therefore uses a computer algebra library. It is based on theorem templates, which provide formal specifications for the algorithms.

Keywords. Computer algebra, mechanised reasoning, combining systems, soundness of computer algebra systems, specialisation problem, coding theory.

AISC topics. Integration of logical reasoning and computer algebra, automated theorem provers.

1 Motivation

Is the use of computer algebra technology beneficial for mechanised reasoning in and about mathematical domains? Usually it is assumed that it is. Many works in this area have, however, either only little reasoning content, or the contribution of symbolic computation is only the simplification of expressions. Exceptions are Analytica [Clarke and Zhao, 1993] and work by [Harrison, 1996]. Both these approaches do not scale up. The former trusts the computer algebra system too much, the latter, too little. Computer algebra systems are not logically sound reasoning systems, but collections of algorithms.

Apart from the verification of numerical hardware and software, linking mechanised reasoning and computer algebra gives insight into the design of logically more expressive frameworks for computer algebra, has applications in educational software and is a step towards the development of mathematical assistants. Among the applications, geometry theorem proving is a prospective candidate. For a survey on this, see [Geddes *et al.*, 1992, section 10.6].

This work presents a case study that requires hard techniques from both sides. The proofs we mechanise require algorithms from computer algebra in

Jacques Calmet and Jan Plaza (Eds.): AISC'98, LNAI 1476, pp. 55–66, 1998.
© Springer-Verlag Berlin Heidelberg 1998

order to be solved effectively. They also rely on the formalisation of natural numbers, sets and lists, which are available in Isabelle, and make heavy use of advanced proof procedures.

The outline of this article is as follows. In section 2 we briefly describe the context of interactive theorem proving and the prover Isabelle. We then present an analysis of the soundness problems in computer algebra and based on this describe the design of an interface. The rest of the paper is devoted to our case study. Section 3 introduces the mathematical background along the lines of its mechanisation in Isabelle. Section 4 is a brief introduction to coding theory and section 5 presents the mechanised proofs. Section 6 reviews important details of the implementation and in section 7 we draw conclusions.

2   Interface Between Isabelle and Sumit

The interface we present is between the prover Isabelle and the computer algebra library Sumit. See [Paulson, 1994] and [Bronstein, 1996] respectively.

2.1   Isabelle

Isabelle is a natural deduction-style theorem prover. Proofs are carried out interactively by the user by applying tactics to the proof state and so replacing subgoals by simpler ones until all the subgoals are proved. Isabelle provides tactics that perform single inference steps and also highly automated proof procedures, like the simplifier and a tactic that implements a tableau prover.

Isabelle, like other LCF-style theorem provers, allows the user to program arbitrary tactics, which can implement specialised proof procedures. The design of Isabelle ensures that unsoundness cannot be introduced to the system through these procedures. This is achieved by using an abstract datatype thm for theorems. Theorems can only be generated by operations provided by the datatype. These operations implement the primitive inference rules of the logic.

Isabelle also provides an oracle mechanism to interface trusted external provers. An oracle can create a theorem, *i.e.* an object of type thm, without proving it through the inference rules. This, of course, weakens the rigour of the LCF-approach, but theorems proved later on can record on which external theorems they depend.

We use Isabelle's object logic HOL, which implements Church's theory of simple types, also known as higher order logic. This is a typed version of the λ-calculus. The logic has the usual connectives $(\land, \lor, \longrightarrow, \dots)$ and quantifiers $(\forall, \exists)$. Currying is used for function application. We write $f\,a\,b$ instead of $f(a, b)$. Equality $=$ on the type bool is used to express if-and-only-if. For definitions we use $\equiv$, and $\Longrightarrow$ expresses entailment in a deduction rule. Some definitions require Hilbert's ϵ-operator, which is actually a quantifier: $\epsilon x.P\,x$ denotes the unique value for which the predicate P holds. The notation for formulae in this paper is close to their representation in Isabelle. We have omitted all type information from formulae to improve their legibility. If type information is necessary, we give it informally in the context.

2.2   Soundness in Computer Algebra

Computer algebra systems have been designed as tools that perform complicated algebraic computations. Their soundness or, as some authors might prefer to say, unsoundness has become a focus, see [Harrison, 1996, Homann, 1997] for examples. A systematic presentation of more examples is [Stoutemyer, 1991]. We have identified the following reasons for unsoundness in the design of computer algebra systems:

- They present a misleadingly uniform interface to collections of algorithms. An object, which is used with a particular meaning in one algorithm, can be used with a different meaning in another algorithm. Particularly problematic are symbols, which are used as formal indeterminates in polynomials and as variables in expressions. Interfacing to a computer algebra system through its user interface is therefore problematic.
- They have only limited capabilities for handling side conditions or case splits, if they exist at all. An example is $\int x^n \, dx$. Computer algebra systems return $\frac{x^{n+1}}{n+1}$. Substituting $n = -1$ yields an undefined term, while the solution of the integral is $\ln x$. This problem is known as specialisation problem, but hardly ever referred to in the literature, see [Corless and Jeffrey, 1997].
- Many of the algorithms that are implemented in computer algebra systems rest on mathematical theory and their correctness is well established: proofs for their correctness have been published. Examples for these are factorisation algorithms for polynomials, Gaussian elimination and Risch's method for integration in finite terms. The design of other algorithms is less rigorous. Simplification rules like $(x^2)^{\frac{1}{2}} = x$ are cause for some of the reported soundness problems. [Corless and Jeffrey, 1996] argue that a satisfactory treatment of these requires extending the underlying mathematical model. In this case Riemann surfaces are appropriate. We call the former sort of algorithms sound and the latter *ad hoc*. See [Calmet and Campbell, 1997, section 2] for a historic perspective on this distinction.

Of course, computer algebra systems also contain implementation errors. Depending on how rigorous one wants to be, one can reject any result of a computer algebra system without formal verification in the prover. Considering the amount of work required to re-implement these algorithms in a theorem prover, and the poor efficiency one could expect, we decide to live with possible bugs but look for ways of avoiding the systematic errors.

2.3   Design of the Interface

The interface obviously needs to translate objects between Isabelle's and the computer algebra system's representation. The translation cannot be performed uniformly, but needs to take into account which algorithm the objects are passed to or returned from. As we can only use a selection of algorithms of the system safely, we need to interface to these directly rather than to the system as a whole.

Unfortunately, it turns out to be difficult to tell sound algorithms from *ad hoc* ones in large, multipurpose computer algebra systems. Without lengthy code inspections one cannot be sure that a piece of otherwise sound code depends on a module that is *ad hoc*. We have therefore chosen the rather small computer algebra library Sumit, which is written in the strongly typed language Aldor, originally designed for the computer algebra system Axiom. References to the literature for the algorithms this library implements are available. From these, formal specifications can be extracted.

The implementation of a prototype interface between Isabelle and Sumit is straightforward. We provide stubs that translate between Isabelle's λ-terms and Sumit's algebraic objects. More than one stub is provided for Sumit types that are used for several mathematical domains. This is, for example, the case for Sumit's type **Integer**, which is used to represent both natural numbers and integers. Arguments and results of the computation are then composed to a λ-term representing a theorem. This is done using what we call a theorem template: at this experimental stage, simply a piece of code. The generated theorem is an instance of the algorithm's formal specification. The algebraic algorithms, stubs and theorem templates are wrapped to a server dealing with Isabelle's requests. The server we obtain this way is only a skeleton: stubs and theorem templates are added incrementally for algorithms that are to be used.

3 Polynomial Algebra

The algebraic approach to cyclic codes is based upon the theory of polynomial rings. We sketch this theory briefly and also show to what extent it has been formalised within Isabelle/HOL. The type system of this logic supports simple types extended by axiomatic type classes, which we use to represent abstract algebraic structures. Subtyping has to be made explicit using suitable embedding functions.

3.1 The Hierarchy of Ring Structures

One obtains various kinds of rings by imposing conditions on the ring's multiplicative monoid. *Integral domains*, or domains for short, do not contain any zero divisors other than zero: formally, $a \neq 0$ and $b \neq 0$ implies $a \cdot b \neq 0$.

An element a is said to *divide* b, if there is an element d such that $a \cdot d = b$. We write $a \mid b$. Two elements are *associated* $a \sim b$, if both $a \mid b$ and $b \mid a$. An element that divides 1 is called a *unit*. Associated elements differ by a unit factor only. An element is called *irreducible* if it is nonzero, not a unit and all its proper factors are units. Formally, $\mathrm{irred}\, a \equiv a \neq 0 \wedge a \nmid 1 \wedge (\forall d.\, d \mid a \longrightarrow d \mid 1 \vee a \mid d)$. An element is called *prime* if it is nonzero, not a unit and, whenever it divides a product, it already divides one of the factors. This is, formally, $\mathrm{prime}\, p \equiv p \neq 0 \wedge p \nmid 1 \wedge (\forall a\, b.\, p \mid a \cdot b \longrightarrow p \mid a \vee p \mid b)$. The factorisation of an element x into irreducible elements is defined by the following predicate:

$$\mathrm{Factorisation}\, x\, F\, u \equiv (x = \mathrm{foldr} \cdot F\, u) \wedge (\forall a \in F.\, \mathrm{irred}\, a) \wedge u \mid 1 \qquad (1)$$

F is the list of irreducible factors and u is a unit element. The list operator foldr combines all the elements of a list, here by means of the multiplication operation ".". The product of the elements of F and of u is x.

An integral domain R is called *factorial* if the factorisation of the elements into irreducible factors is unique up to the order of the factors and associated elements. This is equivalent to R satisfying a divisor chain condition and every irreducible element of R being prime. The divisor chain condition is not needed in our proofs. We therefore formalise factorial domains only using the second condition, which is also called the *primeness condition*. Fields are commutative rings where every non-zero element has a multiplicative inverse.

3.2   Polynomials

Polynomials are a generic construction over rings. For every ring R there is a ring of polynomials $R[X]$. The symbol X is called the indeterminate of the polynomial ring. Further to the ring operations there is the embedding const : $\left\{ \begin{array}{c} R \to R[X] \\ a \mapsto aX^0 \end{array} \right\}$. We derive the representation theorem $\deg p \le n \implies \sum_{i=0}^{n} p_i X^i = p$, where the p_i denote the coefficients of p.

Polynomials must not be confused with polynomial functions.[1] Their relation is described in terms of the evaluation homomorphism. Given another ring S and a homomorphism $\phi : R \to S$ we define EVAL $\phi\, a\, p \equiv \sum_{i=0}^{\deg p} \phi p_i \cdot a^n$. EVAL $\phi\, a : R[X] \to S$ is a homomorphism as well and evaluates a polynomial in S substituting $a \in S$ for the indeterminate and mapping the coefficients of p to S by ϕ.

3.3   Fields and Minimal Polynomial

The field $\mathbb{F}_2 = \{0, 1\}$ is fundamental in an algebraic treatment of binary codes. Codewords are represented as polynomials in $\mathbb{F}_2[X]$. Note that associated elements are equal in these domains.

Let h be an irreducible polynomial of degree n. The residue ring obtained from $\mathbb{F}_2[X]$ by "computing modulo h" is a field with 2^n elements. For our purpose we do not carry out this quotient construction of a field extension explicitly, as we only need it to define the notion of minimal polynomial. Let G be an extension field of F and $a \in G$. The nonzero polynomial $m \in F[X]$ of smallest degree, such that m evaluated at a is zero, is the minimal polynomial. Our definition of the minimal polynomial is as follows:

$$\text{minimal}\, g\, S \equiv g \in S \wedge g \ne 0 \wedge (\forall v \in S.\, v \ne 0 \longrightarrow \deg g \le \deg v) \quad (2)$$

$$\text{min\_poly}\, h\, a \equiv \epsilon g.\, \text{minimal}\, g\, \{p.\, \text{EVAL const}\, a\, p\, \text{rem}\, h = 0\} \quad (3)$$

Note that here $a \in \mathbb{F}_2[X]$ and hence the embedding const is needed to lift the coefficients of p to $\mathbb{F}_2[X]$. The computation is carried out modulo h by means of the remainder function rem associated with polynomial division.

---

[1] Polynomial functions are a subtype of $R \to R$ and not isomorphic to $R[X]$ when R is finite: for $\mathbb{F}_2$ we have $|\mathbb{F}_2[X]| = \infty$, but $|\mathbb{F}_2 \to \mathbb{F}_2| = 4$.

4   Coding Theory

This discipline studies the transmission of information over communication channels. In practice, information gets distorted because of noise. Coding theory therefore seeks to design codes that allow for high information rates and the correction of errors introduced in the channel. At the same time, fast encoding and decoding algorithms are required to permit high transmission speeds.

The following presentation of coding theory follows [Hoffman et al., 1991]. The codes we are interested in for the purpose of this case study belong to a class of binary codes with words of fixed length, so called block codes. n-error-detecting codes have the capability to detect n errors in the transmission of a word; n-error-correcting codes can even correct n errors. The distance between two codewords is the number of differing bit-positions between them. The distance of a code is the minimum distance between any two words of that code.

Definition 1 *A code is* linear *if the exclusive or of two codewords is also a codeword. It is* cyclic *if for every codeword $a_0 \cdots a_n$ its cyclic shift $a_n a_0 \cdots a_{n-1}$ is also a codeword.*

Codes that are linear and cyclic can be studied using algebraic methods. Linear codes are $\mathbb{F}_2$-vector spaces. A code with 2^k codewords has dimension k and there is a basis of codewords that span the code. It is convenient to identify codewords with polynomials:

$$a_0 \cdots a_{n-1} \quad \longleftrightarrow \quad a_0 + a_1 X + \ldots + a_{n-1} X^{n-1}$$

The cyclic shift of a codeword a is then $X \cdot a \operatorname{rem}(X^n - 1)$, where rem is the remainder function associated with polynomial division.

There is a nonzero codeword of least degree in every linear cyclic code. This is called the generator polynomial. It is unique and its cyclic shifts form a basis for the code. It is important, because a linear cyclic code is fully determined by its length and its generator polynomial. The generator polynomial has the following algebraic characterisation:

Theorem 2 (Generator polynomial) *There exists a cyclic linear code of length n such that the polynomial g is the generator polynomial of that code if and only if g divides $X^n - 1$.*

4.1   Hamming Codes

Hamming codes are linear codes of distance 3 and are 1-error-correcting. They are *perfect* codes: they attain a theoretical bound limiting the number of codewords of a code of given length and distance. For every $r \geq 2$ there are cyclic Hamming codes of length $2^r - 1$.

An irreducible polynomial of degree n that does not divide $X^m - 1$ for $m \in \{n+1, \ldots, 2^n - 2\}$ is called *primitive*.[2] This allows us to state the following structural theorem on cyclic Hamming codes:

---

[2] Note that the term primitive polynomial is used with a different meaning in other areas of algebra.

Theorem 3 (Hamming code) *There exists a cyclic Hamming code of length* $2^r - 1$ *with generator polynomial* g, *if and only if* g *is primitive and* $\deg g = r$.

4.2   BCH Codes

Bose-Chaudhuri-Hocquenigham (BCH) codes can be constructed according to a required error-correcting capability. We only consider 2-error-correcting BCH codes. These are of length $2^r - 1$ for $r \geq 4$ and have distance 5.

An element a of a field F is *primitive* if $a^i = 1$ is equivalent to $i = |F| - 1$ or $i = 0$. Let G be an extension field of $\mathbb{F}_2$ with 2^r elements and $b \in G$ a primitive element. The generator polynomial of the BCH code of length $2^r - 1$ is $m_b \cdot m_{b^3}$, where m_a denotes the minimal polynomial of a in the field extension. If we describe the field extension in terms of a primitive polynomial h, then X corresponds to a primitive element. Note that, because h is irreducible, it is minimal polynomial of X. Therefore we can define BCH codes a follows:

Definition 4 *Let* $h \in \mathbb{F}_2[X]$ *be a primitive polynomial of degree* r. *The code of length* $2^r - 1$ *generated by* $h \cdot \min\_poly\, h\, X^3$ *is called a* BCH *code.*

5   Formalising Coding Theory

We formalise properties of codes with the following predicates. Codewords are polynomials over $\mathbb{F}_2$ and codes are sets of them. The statement code $n\, C$ means C is a code of length n. The definitions of linear and cyclic are straightforward while generator $n\, g\, C$ states that g is generator polynomial of the code C of length n.

$$
\begin{aligned}
\text{code}\, n\, C &\equiv \forall x \in C.\ \deg x < n \\
\text{linear}\, C &\equiv \forall x \in C.\, \forall y \in C.\, x + y \in C \\
\text{cyclic}\, n\, C &\equiv \forall x \in C.\, X \cdot x\, \mathrm{rem}(X^n - 1) \in C \\
\text{generator}\, n\, g\, C &\equiv \text{code}\, n\, C \wedge \text{linear}\, C \wedge \text{cyclic}\, n\, C \wedge \text{minimal}\, g\, C
\end{aligned}
$$

5.1   The Hamming Code Proofs

We now describe our first application of the interface between Isabelle and Sumit. We use it to prove which Hamming codes of a certain length exist. Restricting the proof to a certain length allows us to make use of computational results obtained by the computer algebra system. The predicate Hamming describes which codes are Hamming codes of a certain length. Theorems 2 and 3 are required and formalised as follows:

$$0 < n \longrightarrow (\exists C.\ \text{generator}\, n\, g\, C) = g \mid X^n - 1 \tag{4}$$

$$(\exists C.\ \text{generator}(2^r - 1)\, g\, C \wedge \text{Hamming}\, r\, C) = (\deg g = r \wedge \text{primitive}\, g) \tag{5}$$

These equations are asserted as axioms and are the starting point of the proof that follows. Note that (5) axiomatises the predicate Hamming. The generators of Hamming codes are the primitive polynomials of degree $2^r - 1$. The primitive polynomials of degree 4 are $X^4 + X^3 + 1$ and $X^4 + X + 1$. Thus for codes of length 15 we prove

$$(\exists C. \text{ generator } 15\, g\, C \wedge \text{Hamming } r\, C) = (g \in \{X^4 + X^3 + 1, X^4 + X + 1\}).$$

We now give a sketch of this proof, which is formally carried out in Isabelle. The proof idea for the direction from left to right is that we obtain all irreducible factors of a polynomial by computing its factorisation. The generator g is irreducible by (5) and a divisor of $X^{15} - 1$ by (4). The factorisation of $X^{15} - 1$ is computed using Berlekamp's algorithm:

$$\text{Factorisation}(X^{15} - 1)\, [X^4 + X^3 + 1, X + 1, X^2 + X + 1,$$
$$X^4 + X^3 + X^2 + X + 1, X^4 + X + 1]\, 1$$

Since associates are equal in $\mathbb{F}_2[X]$ every irreducible divisor of $X^{15} - 1$ is in this list. This follows from the lemma

$$\text{irred } c \wedge \text{Factorisation } x\, F\, u \wedge c \mid x \implies \exists d.\, c \sim d \wedge d \in F, \qquad (6)$$

whose proof requires an induction over the list F. It follows in particular that the generator polynomials are in the list above. But some polynomials in the list cannot be generators: $X + 1$ and $X^2 + X + 1$ do not have degree 4 and $X^4 + X^3 + X^2 + X + 1$ divides $X^5 - 1$ and is therefore not primitive. The only possible generators are thus $X^4 + X^3 + 1$ and $X^4 + X + 1$.

It remains to show that these are indeed generator polynomials of Hamming codes. This is the direction from right to left. According to (5) we need to show that $X^4 + X^3 + 1$ and $X^4 + X + 1$ are primitive and have degree 4. The proof is the same for both polynomials. Let p be one of these. The irreducibility of p is proved by computing the factorisation, which is Factorisation p [p] 1, and follows from the definition of Factorisation, equation (1).

The divisibility condition of primitiveness is shown by verifying $p \nmid X^m - 1$ for $m = 5, \ldots, 14$. $\qquad \square$

5.2   The BCH Code Proofs

The predicate BCH is, in line with definition 4, defined as follows:

$$\text{BCH } r\, C \;\equiv\; (\exists h.\; \text{primitive } h \wedge \deg h = r \wedge$$
$$\text{generator}(2^r - 1)\,(h \cdot \text{min\_poly } h\, X^3)\, C) \qquad (7)$$

We prove that a certain polynomial is generator of a BCH code of length 15:

$$\text{generator } 15\, (X^8 + X^7 + X^6 + X^4 + 1)\, C \implies \text{BCH } 4\, C$$

Here is the outline of the proof: $X^8 + X^7 + X^6 + X^4 + 1$ is the product of the primitive polynomial $X^4 + X + 1$ and the minimal polynomial $X^4 + X^3 + X^2 + X +$

1. According to the definition (7) we need to show that the former polynomial is primitive. This has been described in the second part of the Hamming proof. Secondly, we need to show that the latter is a minimal polynomial:

$$\text{min\_poly}(X^4 + X + 1) \, X^3 = X^4 + X^3 + X^2 + X + 1$$

In order to prove this statement, we need to show that $X^4 + X^3 + X^2 + X + 1$ is a solution of

$$\text{EVAL const } X^3 \, p \, \text{rem}(X^4 + X + 1) = 0 \tag{8}$$

of minimal degree, and that it is the only minimal solution.

- Minimal solution: Simplification establishes that $X^4 + X^3 + X^2 + X + 1$ is a solution of the equation. That there are no solutions of smaller degree can be shown as follows:
  Assume $\deg p \le 3$, so $p = p_0 + p_1 X + p_2 X^2 + p_3 X^3$ for $p_0, \dots, p_3 \in \mathbb{F}_2$. We substitute this representation of p in (8) and obtain, after simplification,

$$p_0 + p_1 X^3 + p_2(X^2 + X^3) + p_3(X + X^3) = 0.$$

  Comparing coefficients leads to a linear equation system, which we can solve using the Gaussian algorithm. The only solution is $p_0 = \dots = p_3 = 0$, so $p = 0$. This is a contradiction to the definition of minimal.
- Uniqueness: We need to show that $X^4 + X^3 + X^2 + X + 1$ is the only polynomial of smallest degree satisfying (7). We study the solutions of (8) of degree of ≤ 4 by setting $p = p_0 + \dots + p_4 X^4$ and obtain another equation system

$$p_0 + p_1 X^3 + p_2(X^2 + X^3) + p_3(X + X^3) + p_4(1 + X + X^2 + X^3) = 0.$$

  Its set of solutions, again computed by the Gaussian algorithm, is $\{0, X^4 + X^3 + X^2 + X + 1\}$. The definition of minimality excludes $p = 0$. Therefore there are indeed no other solutions of minimal degree. $\square$

6   Review of the Development

We have mechanised the mathematics outlined in section 3 and the proofs described in section 5 in our combination of Isabelle and Sumit. The mathematical background presented in section 3 has been formalised by asserting definitions for the entities and deriving the required theorems mechanically. This is advisable to maintain consistency. We have not done the same for coding theory. Here we have only asserted the results, namely theorems 2 and 3 and then mechanised the proofs described in section 5. This part is therefore considerably shorter than the development of the mathematical background.

The following table gives an overview on the effort. The figures are, however, misleading in such that developing proof scripts is much harder than ordinary programming.

Isabelle		Sumit	
Interface	23.7	Interface	43.3
Formalisation of algebra	61.8	Stubs and	
Coding theory proofs	14.6	theorem templates	20.4

Size of the development (code sizes in 1000 bytes)

The interface of Sumit is considerably larger, because datatypes for λ-terms and the server functionality are provided as well. The entry "Coding theory proofs" includes the implementation of proof procedures for irreducibility and primitiveness of polynomials, which automatically examine the proof state and retrieve the required theorems from Sumit.

6.1   Contributions of the Prover

We prove theorems about polynomial algebra, which do not have computational content, in Isabelle. We also establish the relation between coding theory and the specifications of the algebraic algorithms. In our informal presentation these translations may appear simple, but some of them are in fact rather difficult.

For the Hamming code proofs take lemma (6), for example, which is proved by list induction. The induction step, after unfolding definitions, is a quantifier expression, which is solved almost automatically by Isabelle's tableau prover. However, it requires search to a depth of six, which means that six "difficult" rules have to be applied, and produces a proof with 221 inferences. A depth of six is unusually deep in interactive proof. The complete proof of (6) is 372 inferences long but only requires 8 invocations of tactics, which resemble the manual proof steps.

In the proofs about BCH codes, reasoning about minimality needs the full power of first order logic. Note that the definition of minimality (2) contains a quantifier and phrases like "x is the only element, such that P" are really statements that involve quantifiers.

6.2   Contributions of Computer Algebra

Sumit computes normal forms for expressions that do not contain variables; here in the domains $\mathbb{N}, \mathbb{F}_2, \mathbb{F}_2[X]$. This includes the decision of equality, inequalities and divisibility over these expressions. Their theorem templates are of the form $a \odot b = B$, where $\odot$ is the corresponding connective and B becomes either *True* or *False*.

Polynomials are decomposed into square-free factors and then factorised over $\mathbb{F}_2[X]$ using Berlekamp's algorithm. We pass a polynomial p to this procedure and obtain a list of irreducible factors $[x_1, \ldots, x_k]$ and a unit element u. These are then assembled to the theorem

$$\text{Factorisation} \, x \, [x_1, \ldots, x_k] \, u.$$

Linear equation systems over $\mathbb{F}_2$ are solved by Gaussian elimination. The matrix $(a_0 | \cdots | a_n)$ is passed to the algorithm, where a_i is the ith column vector.

The algorithm returns a list of vectors $[v_1, \ldots, v_k]$ that span the solution space. The theorem template generates the theorem

$$(\sum_{i=0}^{n} x_i a_i = 0) = (\exists t_1 \cdots t_k. \, x = t_1 v_1 + \ldots + t_k v_k)$$

$$\text{or} \quad (\sum_{i=0}^{n} x_i a_i = 0) = (x = 0), \quad \text{if } k = 0.$$

The t_i are variables in $\mathbb{F}_2$ and the x_i are elements of the vector x. Note that we use polynomials to denote vectors in Isabelle, as indicated in the proof.

Mechanising the proofs in a system that integrates the computer algebra component without trusting it would require to additionally prove the theorems generated by these templates formally. This holds in particular for [Harrison, 1996, chapter 6] and [Kerber et al., 1996], who try to reconstruct the proofs using the result of the computation and possibly further information, which resembles a certificate for the computation.

In the case of our proofs, the irreducibility of the factors, which constitute a factorisation, is hard to establish and also the direction from left to right in the theorems generated by Gaussian elimination.[3] This direction states that the solution is complete, and it is the direction needed in the proofs.

7 Conclusion

Our approach is pragmatic: we trust the computer algebra component in our system rather than reconstruct proofs for the results of computations within the prover's logic. The approach relies on implementations of algorithms that are trustworthy. This can be achieved by restricting the use of computer algebra to algorithms, for which proofs of their correctness have been published. This is sufficient to avoid systematic soundness problems of computer algebra systems. Errors in the implementation of these algorithms still jeopardise the integrity of the prover, but bugs of this sort should not be more frequent in computer algebra systems than in other software (including provers themselves).

Computational results are turned into theorems using theorem templates that can produce arbitrary theorems. This is more flexible than the approach suggested by one of us [Ballarin et al., 1995], which only allowed conditional rewrite rules, because the logical meaning of the result can be exploited more easily.

Our case study shows that theorems that are rather difficult to verify occur naturally in proofs. It presents a challenge to the approach that does not trust the computer algebra component. But it also makes a contribution: it clarifies which theorems need to be certified.

---

[3] Over some domains theorems of this kind can be proved by decision procedures for linear arithmetic. Here, because $|\mathbb{F}_2| = 2$, this could be done by checking all the 2^{n+1} cases.

Our approach avoids Analytica's soundness problems. This means, of course, that we cannot make use of algorithms that are *ad hoc*. In an interactive environment it does not matter too much that these are not complete. They need, however, to be made sound. Expressive formalisms that are able to deal with side conditions and case splits are used in mechanised reasoning. Expertise gained here could prove useful in the redesign of these algorithms as well.

Acknowledgements. This work has been funded in part by the Studienstiftung des deutschen Volkes and by EPSRC grant GR/K57381 "Mechanizing Temporal Reasoning".

References

[Ballarin *et al.*, 1995] Clemens Ballarin, Karsten Homann, and Jacques Calmet. Theorems and algorithms: An interface between Isabelle and Maple. In A. H. M. Levelt, editor, *ISSAC '95: International symposium on symbolic and algebraic computation — July 1995, Montréal, Canada*, pages 150–157. ACM Press, 1995.

[Bronstein, 1996] Manuel Bronstein. Sumit — a strongly-typed embeddable computer algebra library. In Calmet and Limongelli [1996], pages 22–33.

[Calmet and Campbell, 1997] J. Calmet and J. A. Campbell. A perspective on symbolic mathematical computing and artificial intelligence. *Annals of Mathematics and Artificial Intelligence*, 19(3–4):261–277, 1997.

[Calmet and Limongelli, 1996] Jacques Calmet and Carla Limongelli, editors. *Design and Implementation of Symbolic Computation Systems: International Symposium, DISCO '96, Karlsruhe, Germany, September 18–20, 1996: proceedings*, number 1128 in Lecture Notes in Computer Science. Springer-Verlag, 1996.

[Clarke and Zhao, 1993] Edmund Clarke and Xudong Zhao. Analytica: A theorem prover for Mathematica. *The Mathematica Journal*, 3(1):56–71, 1993.

[Corless and Jeffrey, 1996] Robert M. Corless and David J. Jeffrey. The unwinding number. *ACM SIGSAM Bulletin*, 30(2):28–35, 1996.

[Corless and Jeffrey, 1997] R. M. Corless and D. J. Jeffrey. The Turing factorization of a rectangular matrix. *ACM SIGSAM Bulletin*, 31(3):20–28, 1997.

[Geddes *et al.*, 1992] Keith O. Geddes, Stephen R. Czapor, and George Labahan. *Algorithms for Computer Algebra*. Kluwer Academic Publishers, 1992.

[Harrison, 1996] John Robert Harrison. Theorem proving with the real numbers. Technical Report 408, University of Cambridge, Computer Laboratory, November 1996.

[Hoffman *et al.*, 1991] D. G. Hoffman, D. A. Leonard, C. C. Lindner, K. T. Phelps, C. A. Rodger, and J. R. Wall. *Coding Theory: The Essentials*. Number 150 in Monographs and textbooks in pure and applied mathematics. Marcel Dekker, Inc., New York, 1991.

[Homann, 1997] Karsten Homann. *Symbolisches Lösen mathematischer Probleme durch Kooperation algorithmischer und logischer Systeme*. Number 152 in Dissertationen zur Künstlichen Intelligenz. infix, St. Augustin, 1997.

[Kerber *et al.*, 1996] Manfred Kerber, Michael Kohlhase, and Volker Sorge. Integrating computer algebra with proof planning. In Calmet and Limongelli [1996], pages 204–215.

[Paulson, 1994] Lawrence C. Paulson. *Isabelle: a generic theorem prover*. Number 828 in Lecture Notes in Computer Science. Springer-Verlag, 1994.

[Stoutemyer, 1991] David R. Stoutemyer. Crimes and misdemeanors in the computer algebra trade. *Notices of the American Mathematical Society*, 38(7):778–785, 1991.

Automatic Generation of Epsilon-Delta Proofs of Continuity

Michael Beeson

Department of Mathematics and Computer Science
San Jose State University
San Jose, California 95192, USA
beeson@mathcs.sjsu.edu

Abstract. As part of a project on automatic generation of proofs involving both logic and computation, we have automated the production of some proofs involving epsilon-delta arguments. These proofs involve two or three quantifiers on the logical side, and on the computational side, they involve algebra, trigonometry, and some calculus. At the border of logic and computation, they involve several types of arguments involving inequalities, including transitivity chaining and several types of bounding arguments, in which bounds are sought that do not depend on certain variables. Control mechanisms have been developed for intermixing logical deduction steps with computational steps and with inequality reasoning. Problems discussed here as examples involve the continuity and uniform continuity of various specific functions.[1]

1 Context of this Research

Mathematics consists of logic and computation, interwoven in tapestries of proofs. "Logic" is represented by the manipulation of phrases (or symbols) such as *for all x, there exists an x, implies*, etc. "Computation" refers to chains of formulas progressing towards an "answer", such as one makes when evaluating an integral or solving an equation. Typically computational steps move "forwards" (from the known facts further facts are derived) and logical steps move "backwards" (from the goal towards the hypothesis, as in *it would suffice to prove*. The mixture of logic and computation gives mathematics a rich structure that has not yet been captured, either in the formal systems of logic, or in computer programs. The research reported on here is part of a larger research program to do just that: capture and computerize mathematics.

At present, there exist computer programs that can do mathematical computations, such as *Mathematica, Maple,* and *Macsyma.* These programs, however, do not keep track of the logical conditions needed to make computations legal, and can easily be made to produce incorrect results.[2]

---

[1] This research partially supported by NSF Grant Number CCR-9528913.

[2] Just to give one example: Start with the equation $a = 0$. Divide both sides by a. In all the three systems mentioned, you can get $1 = 0$ since the system thinks $a/a = 1$ and $0/a = 0$. Many other examples have been given in the literature [1],[15].

Jacques Calmet and Jan Plaza (Eds.): AISC'98, LNAI 1476, pp. 67–83, 1998.
© Springer-Verlag Berlin Heidelberg 1998

On the other hand, there are theorem-proving programs such as *Otter* [13] (and others too numerous to mention) which perform logical reasoning. These programs are quite limited in their computational abilities, although some of them can perform rewrites using a specified set of equations. The input consists of a file containing axioms, and a goal, usually expressed in clausal form. The program contains no mathematical knowledge except that supplied in the axioms; it only "knows" the laws of logic. Proof-search, such as these programs perform, is not what is meant here by "computation"; although of course in some sense the execution of any algorithm is computation, what we call computation here is more like what an ordinary mathematician means by the word, a sequence of more or less purposeful-appearing steps, with little or no trial-and-error involved.

This paper will present a framework for integrating logic and computation, and report on experiments with the implementation of that framework. The implementation is contained in two computer programs, *Mathpert* and *Weierstrass*. The former has been reported on elsewhere in detail [1,2,3]: it contains implementations of over two thousand mathematical operations, together with logical apparatus to keep track of assumptions that may be required or generated by those operations. *Mathpert* (as in "Math Expert") uses these operations to provide a computerized environment for learning algebra, trigonometry, and calculus. It is the second program, *Weierstrass*, which is used in the research reported here. *Weierstrass* began life as a C-language implementation of a "backwards Gentzen" theorem prover, whose Prolog progenitor was described in [4]. To this backbone has been added a set of control structures, or if you like, implementations of special inference rules, to allow the proper meshing of logical and computational steps. These control structures operate at the top level of *Weierstrass*, but the computational steps themselves can use, in principle, anything that has been implemented in *Mathpert*, which is all of high-school algebra, trigonometry, and one-variable calculus including limits, differentiation, and integration, as well as a good many techniques for rewriting inequalities, and a few advanced algorithms, such as the Coste-Roy algorithm [10], based on Sturm's theorem, for determining whether polynomials have roots in given intervals. The implementations of these operations in *Mathpert* are logically correct, so that they can be used in *Weierstrass* without the risk of inconsistency that would accompany the similar use of *Mathematica*, *Maple*, or *Macsyma*.[3]

The plan of the paper is to describe the control structures used in *Weierstrass*, and then to illustrate their use by giving several examples of proofs produced by *Weierstrass*.

No program has ever before produced an epsilon-delta proof of the continuity of any specific non-linear function. For example, Bledsoe and his student Hines have used the prover STRIVE [6] to prove that the sum of continuous functions is continuous, and that linear functions are continuous, but it lacks the

---

[3] *Weierstrass* is not an interactive program for producing a proof step-by-step. The user supplies axioms and a goal, and *Weierstrass* finds a proof if it can. However, the techniques discussed in this paper could easily be used to build an interactive program.

computational ability to carry out the proofs given here. *Analytica* [9] is linked to the computational facilities of *Mathematica*, but essentially deals only with quantifier-free proofs. The Boyer-Moore prover [7] has proved some impressive theorems of number theory, including the law of quadratic reciprocity, but like *Analytica*, works best with free-variable proofs, and cannot find epsilon-delta proofs. Other directly relevant work includes [11], [16], [8]. A lengthier discussion of these and other projects is precluded by the length limit on papers in this volume.

2   Nature of the Proofs Produced by *Weierstrass*

To avoid confusion, some discussion of nature and purpose of computer-generated proofs is necessary. *Weierstrass* produces (internally) a proof-object, which can be displayed or saved in more than one form. The intention is, to produce a proof that can be read and checked for correctness by a human mathematician; the standard to be met is "peer review", just as for journal publication. By contrast: the purpose of *Weierstrass* is *not* to produce formal proofs in a specified formal system.

Nevertheless, the program does produce a formal proof object. This object can be regarded as a proof in a formal system, but some of the steps in the proof involve more computation than is normal in formal systems. In traditional logical systems, checking that an inference step is correct according to the system is a simple syntactic comparison to the rule used at that inference. In *Weierstrass*, an inference step might involve the use of a mathematical algorithm, even for example a complicated algorithm based on Sturm sequences, so that the correctness of the step might not be obvious by inspection. If the algorithms used have been correctly implemented, then the proofs are formally correct.[4] But certainly we do not have proofs (formal or informal) of the correctness of the specific programs implementing more than two thousand mathematical operations available in *Weierstrass*, so we must rely on human verifications that the actual output of *Weierstrass* is an acceptable proof.

There is a different and more interesting reason why the proofs produced by *Weierstrass* should be judged by the "peer review" standard. Namely, the algorithms it uses represent theorems of all different "levels". For instance, one might protest that the "right way" to prove continuity theorems such as are considered here, is to prove general theorems about the continuity of compositions of functions, etc., and quote them. Indeed, that is the way *Mathpert* proceeds internally, e.g. when it has to verify the continuity of an integrand in evaluating a definite integral. But when trying to prove the continuity of x^3, we don't want a one-line proof based on the continuity of polynomials. To state the point another

---

[4] This is really no different than in a purely logical theorem prover: one does not demand that one should prove the correctness of the theorem-prover before accepting its output as a proof. Otherwise, one would be involved either in an infinite regress or a reflexive situation where a prover would prove its own correctness. And would we believe it then?

way: People interested in foundations of mathematics try to order the concepts and theorems of mathematics so that each one depends only on earlier results, with everything resting on a few self-evident axioms. The body of mathematics available to *Weierstrass* has *not* been so ordered; rather, there is simply a "web" of known facts and algorithms, any of which can be used as required. We see one example of this in the third example presented in this paper: *Weierstrass* doesn't mind using the mean value theorem to prove the continuity of $f(x) = \sqrt{x}$, even though the result seems "simpler" than the tool.

Some of the descendants of this program may become the "mathematician's assistant" of the future, a tool to which a practicing mathematician may turn when a stubborn inequality needs to be proved. The standard of "peer review" is the appropriate one for this type of program. Other descendants of this program may be used in a project to construct a database of formalized mathematics, similar to the Mizar project of today. In that case the questions of formal correctness proofs for the computational steps, and of ordering the results and deriving them from axioms, will eventually arise, but this will be a difficult enterprise.

3   The Logical Backbone of *Weierstrass*

The core of the logical apparatus in *Weierstrass* is a "backwards Gentzen prover". I shall now explain what is meant by that phrase. A *Gentzen sequent*, or just *sequent*, is an expression of the form $A_1, \ldots, A_n \Rightarrow B$, where the A_i and B are logical formulae (in some language).[5] The right side of the sequent symbol $\Rightarrow$ is called the *succedent* and the left side is the *antecedent*. The semantic interpretation of a sequent is that the conjunction of the A_i implies B. We allow **true** and **false** as atomic propositions. The *sequent calculus* is a set of inference rules for deducing one sequent from another. One standard reference for these rules is Kleene's book [12]. In that reference, an empty list can appear in the succedent; we use **false** for this purpose, so that a formula always stands in the succedent.

When the sequent calculus is implemented in *Weierstrass*, most of the antecedent is kept in a list of *assumptions*, which could in principle be quite long. If axioms of induction are used, for example, new instances of the axioms can be generated as required; axioms belong in the antecedent, since the sequent calculus is for producing purely logical proofs. The only parts of the antecedent that will be "passed" as function parameters are assumptions that are made temporarily during the argument. For example, to prove an implication $A \to B$, we "assume" A and then try to derive B. This is the implementation of the Gentzen rule

$$\frac{\Gamma, A \Rightarrow B}{\Gamma \Rightarrow A \to B}$$

---

[5] These are sometimes called *intuitionistic sequents*, by contrast with *classical sequents* which allow $B_1, \ldots, B_n$ on the right. *Weierstrass* uses intuitionistic logic, but that is for efficiency and convenience only, and is not essential. Indeed some of the computational steps may not be intuitionistically valid.

In addition to the ordinary variables of logic, we also make use of "metavariables", whose values are expressions (terms or sometimes formulae), rather than numbers or other mathematical objects which are the values of ordinary (object) variables. *Weierstrass* introduces metavariables when it uses the Gentzen rules

$$\frac{\Gamma \Rightarrow A[t/x]}{\Gamma \Rightarrow \exists x A} \qquad \frac{A[t/x], \Gamma \Rightarrow B}{\forall x A, \Gamma \Rightarrow B}$$

For example, to prove $\exists \delta \forall x, y(|x-y| < \delta - |f(x)-f(y)| < \epsilon)$, *Weierstrass* will change δ to a metavariable, and try to prove $\forall x, y(|x-y| < \delta - |f(x)-f(y)| < \epsilon)$. Eventually, δ will be given a value (if the proof is successful), and when that value is put in for the metavariable δ, the result will be a (part of a) proof tree which is legal according to the Gentzen rules. In the meantime, that is before the final unifications take place, what is being constructed is something slightly more general than a Gentzen proof tree; it is a Gentzen proof tree in which metavariables are allowed, and the metavariables may have values, and the values may be expressions involving other metavariables. (That is, the metavariables can be "partially instantiated".) A formal definition of an *extended derivation*, and some related theorems, can be found in [4].

In general, *Weierstrass* starts by loading an axiom file, which contains (zero or more) axioms and a goal. The goal is placed in the succudent, and the axioms in the assumption list, which implements the antecedent. Weierstrass then attempts to construct an extended derivation of this sequent. Unification is used in order to instantiate metavariables introduced as described above. Unification is also used to control the selection of the next rule to be applied. Some of the logical rules are broken into subcases, and the order in which they are tried is important. These matters are discussed in [4]. For the present work, it suffices to note that the logical apparatus, functioning on its own, is a decent theorem-prover. While it makes no attempt to compete with Otter, the logic required for ordinary mathematics is comparatively simple, and in no example has the logical apparatus required revision beyond what was described in [4] seven years ago.

The implementation of a metavariable X includes a data structure designed to keep track of a list of variables that are "forbidden" to X; this means that X cannot have a value that contains variable forbidden to X. We make this part of the definition (and implementation) of unification; see [5] for theoretical reasons. This method provides an efficient way to keep track of the conditions on variables that accompany the quantifier rules that introduce metavariables. Instead of giving a metavariable a forbidden value, and carrying out another long subproof, only to discard the result because the conditions on the quantifier rule are violated, the unification will fail instead. I credit Natarajan Shankar for first telling me something similar to this. The idea has been fruitful beyond this improvement in efficiency, since we can use it in connection with computational steps where we want to bound a certain quantity in terms of some bound that does not depend on certain variables; this will be discussed below.

4 Computational Methods in a Quantifier-Free Prover

The theorem prover, or logical apparatus, in *Mathpert* is responsible for maintaining the correctness of computations; it must block incorrect steps and ensure that the assumptions do support the steps that are taken. The program *Weierstrass* began by combining the methods of [4] for handling first-order logic, with the methods of *Mathpert* for handling computations in a quantifier-free setting. More precisely, the logical apparatus in *Mathpert* deals with sequents composed of formulae which involve no implication or negation, but only disjunctions and conjunctions of equalities and inequalities. When we say "quantifier-free" below, this is what we mean. We describe those methods, as implemented in *Weierstrass*, in this section.

4.1 Logical and Mathematical Simplification

Computational methods can be applied either to a mathematical expression (term), or to a logical expression (proposition). That is, we can treat rewriting propositions according to the laws of Boolean algebra on the same footing as rewriting algebraic expressions according to the laws of eighth-grade algebra. We apply the term "simplification" to describe the application of both algebraic and logical operations.

4.2 Operations More General than Rewrite Rules

The term *operation* is here used to mean an algorithm that transforms an expression of a certain form into another expression, which is equal (algebraically or logically) to the input, possibly under certain "side conditions". For example, $\sqrt{x^2} = x$ represents an operation that transforms an expression of the form $\sqrt{A^2}$, where A is any expression, into A; but it has the side condition $0 \leq A$. Operations may be rewrite rules, but they may well be more general than rewrite rules. For example, the operation named *collect powers* can be used to rewrite $x^2 x^3$ as x^5, which is not a rewrite rule since arithmetic on the exponents is involved. The same operation can be used to collect powers separated by other terms, as in $x^2 y x^3$, and to collect any number of powers, as in $x^2 y x^3 x^4$. But, like rewrite rules, operations can be applied to any subterm.

4.3 Operations and Side Conditions

Weierstrass keeps the assumption list (the antecedent) in simplified form, so it is not necessary to look for operations to apply to the antecedent.[6]

---

[6] When an operation is applied to the succedent, any occurrences of the same formula that was changed in the succedent are also changed in the antecedent, after which some simplifications in the antecedent may be performed to keep the antecedent in simplified form.

When the succedent is quantifier-free, *Weierstrass* will try to simplify it, trying a large number of logical and algebraic operations. These operations, for example, may simplify Boolean combinations of inequalities, or simplify certain inequalities to **true** or **false**. Purely algebraic simplification will also be performed, but not, for example, factoring or common denominators; except that greatest common divisors may be cancelled out of fractions.

This last example brings up the interesting question of the relationship between the antecedent (assumptions) and the simplifications performed in the succudent. Consider, for example, the proposition

$$\frac{x^4 - 1}{x^2 - 1} > 0$$

If we cancel $x^2 - 1$ from numerator and denominator, we arrive at the proposition $x^2 + 1 > 0$, which will simplify to **true**. But, at some point we must assume $x^2 - 1 \neq 0$; otherwise the result is incorrect. The points to consider here are two: (1) the original expression is not defined for all values of x, and (2) the domain changes as a result of the application of an operation, which *on the common domain* preserves equivalence. The problem of "partial terms" (terms which can be undefined) is thus closely related to the problem of "side conditions" of symbolic operations.

4.4   Partial Terms, Domains, and Side Conditions

There are two natural ways to make the assumption $x^2 - 1 \neq 0$ in the above example: either at the outset, or when the cancellation is performed. Plan A would be to analyze the domain of the goal when the problem is set up, and put the domain conditions into the antecedent. According to plan A, the condition $x^2 - 1$ would have been assumed at the outset, and hence could have been inferred when required as the side condition for cancelling the common factor of numerator and denominator. Plan B would be to allow potentially undefined terms in the partially constructed proof, and only assume $x^2 - 1 \neq 0$ when it is required as a side condition for an operation. Plan B was mentioned in earlier publications such as [4], but in the practical implementation of *Mathpert*, Plan A was found to be more efficient; after all, we certainly need to assume that the goal is defined to prove anything sensible at all. Therefore, Plan A has been adopted in *Weierstrass* as well.

However, Plan B is not thereby consigned to the dustbin of history. There remain situations in which a symbolic operation may have a side condition that is not necessarily implied by the domain.[7] In such situations, Plan B will still be used. An example would be the application of the operation $\sqrt{x^2} = x$, whose

---

[7] We use the word *domain* to mean a proposition giving the conditions under which a term is defined; thus the domain of $\sqrt{x}$ is $0 \leq x$. If propositions are thought of as Boolean-valued functions, and sets are also thought of as Boolean-valued functions, as Church suggested, then this coincides with the usual usage that the domain of $\sqrt{x}$ is $\{x | 0 \leq x\}$.

side condition $0 \le x$ is not implied by the domain.[8] In the proof tree formalism of logic, Plan B would entail copying the new assumption to the antecedents all the way down the tree from the place where the operation is applied to the conclusion. In the implementation, however, most of the antecedent is kept in the assumption list, rather than duplicated at every line of the proof tree, and so adding the new assumption once suffices.

4.5 Infer, Refute, Assume

When an operation has a side condition, there are two choices: either the operation can try to *infer* the side condition, and fail if the inference fails, or it can try to *check* the side condition, which means that it will try to infer it, and if that fails, it will try to refute it, and if that fails it will simply assume it. Thus, if we try to simplify an expression of the form $\sqrt{A^2}$ to A, the side condition $0 \le A$ will be checked. If, for example, A is 3, the condition $0 \le 3$ will be successfully inferred, so the simplification takes place without an assumption. If, on the other hand, A is -3, then the condition will be refuted, and the simplification will not take place. If, however, A is an expression such that $0 \le A$ can neither be inferred nor refuted, then it will be assumed.

Consider the example mentioned in the introduction, of dividing both sides of $a = 0$ by a. The side condition for dividing both sides of an equation by a is that $a \ne 0$. Can we infer this? No. Can we refute it? Not officially, since $a = 0$ is in the succedent, rather than the antecedent. Then, we will assume it, and obtain the logically correct but useless proof

$$\frac{a \ne 0 \Rightarrow 1 = 0}{a \ne 0 \Rightarrow a = 0}$$

To prevent this sort of thing, *refute* is also allowed to use the antecedent as a temporary assumption. That way, if the side condition is inconsistent with the goal, we will avoid making a contradictory assumption. When this is done, the attempt to divide $a = 0$ by a will result in an error message to the effect that you can't divide by zero. This can be seen working in *Mathpert*.

Note that the choice whether *infer* or *check* is used is specified in the operation itself. That is, there will be two different operations represented loosely by the equation $\sqrt{x^2} = x$. One of them will *infer* the side condition (or fail) and the other will *check* the side condition. In practice, it seems to work best to avoid using *check* in the elementary simplification that are automatically applied in *Weierstrass*; after all, if we fail to prove the desired theorem because we failed to list all the assumptions, we can run *Weierstrass* again after adding the omitted assumption in the axiom file. But, the method is used to good advantage in *Mathpert*, and may prove of value in future applications of systems like *Weierstrass*.

To make this scheme work, *infer* and *refute* must be guaranteed to terminate, and hence must be incomplete; that is, sometimes a true side condition will not be

---

[8] In practice, few such operations are applied automatically in *Weierstrass*, but an interactive prover based on these principles would certainly use Plan B extensively.

inferred, or a false one not refuted. We may wind up making a false assumption. For example, if $p(x)$ is an expression which is really identically zero, but can't be simplified to zero by the means of simplification used by *infer*, then we might be led to make the assumption $p(x) \neq 0$, e.g. to divide both sides of an equation by $p(x)$. This could lead to logically correct but senseless results. This is, however, unavoidable, as the problem of determining whether mathematical expressions $p(x)$ are identically zero is recursively unsolvable [14].

A related situation arises in solving equations. For example, consider the equation $x^2 - x = 0$. If we divide both sides by x, we make the assumption $x \neq 0$ and find the solution $x = 1$. This is logically correct, but we didn't achieve the goal of finding all solutions of the original equation. This may not be a logical error, but it is a mathematical error, and hence has been blocked in *Mathpert*, but in an interactive system based on *Weierstrass*, it would not necessarily be blocked.

4.6   Using the Assumptions in a Computation

Suppose we try to simplify $0 \leq x$, while $0 < x$ is in the assumption list. Then it is efficient to allow $0 \leq x$ to simplify to **true**. For example, if $0 \leq x$ occurs inside a disjunction in the succedent, the whole succedent may simplify to **true**, completing (that branch of) a proof. Similarly, a side condition involving $x \neq 0$ should be reduced to $0 < x$ if $0 \leq x$ is in the assumption list.

4.7   Computation Within the Scope of a Bound Variable

Even though *Weierstrass* applies simplification only to quantifier-free formulae, sometimes it is still necessary to compute inside the scope of a bound variable, since variables can be bound by definite integrals or indexed sums. For example, we want to conclude that $\sum_{k=1} \rightarrow 5x^k$ is everywhere defined, even though the condition for x^k to be defined (for an integer k) requires $k > 0 \vee x \neq 0$. But $k > 0$ holds because the lower limit of the sum is positive. In the case of definite integrals and indexed sums, this is handled by making temporary assumptions out of the limits of the sum or integral, while the focus of computation is in the scope of the sum or integral. Limit terms are handled similarly, but the assumptions to made involve infinitesimals and the use of non-standard analysis; this much more complicated algorithm is discussed in detail in [3]. Computation within the scope of bound variables will not be discussed further in this paper.

4.8   What Formal System Has Been Implemented?

It is an interesting question to formulate precisely a language and rules of inference that could be said to be implemented by *Weierstrass*. One such language has been specified in [3]; it essentially allows variables for integers and real numbers, equality and inequality, and symbols for all the elementary functions used in calculus. A complete and precise grammar for such a language can be found in [3]. This language also allows the formation of integrals, derivatives, indexed sums, and limit terms; definite integrals, indexed sums, and limit terms can bind variables. *Weierstrass* also allows the formation of λ-terms which are not specified in [3].

We turn now to rules of inference. A single additional rule schema describes the simplest way to add computation to a quantifier-free prover:

$$\frac{\Gamma \Rightarrow B \qquad B, \Gamma \Rightarrow A\sigma}{\Gamma \Rightarrow A}$$

where $A\sigma$ denotes the result of applying some mathematical or logical operation to A, or to a subterm of A, replacing the subterm by the result of the operation. In the rule, B is the side condition of the operation, if any; if the operation has no side condition, the premise $\Gamma \Rightarrow B$ does not occur. In principle an operation could also have more than one side condition, in which case there might be more than two premises.

The control strategy for applying this rule is this: whenever A contains no quantifiers or implications or negations, try this rule, with a certain selection of operations in a certain pre-specified order. But, the second premise $\Gamma \Rightarrow B$ representing the side condition is not passed recursively to the main theorem-prover, but must be derived by very limited means. This is to prevent long delays or even infinite regresses attempting to verify the side conditions of mathematical operations; in other words, a practical rather than a theoretical consideration.

This rule of inference does not, however, adequately describe the technique of using the antecedent in simplification as described above. One way to do so, although it is admittedly not very elegant, is to generalize the rule to this:

$$\frac{C, \Gamma \Rightarrow (C \to A)\sigma \qquad C, \Gamma \Rightarrow B}{C, \Gamma \Rightarrow A}$$

Here C is one assumption, and σ is an operation that can work on an implication (usually of inequalities). For example, σ might simplify $a < c \to a \leq c$ to **true**. In both *Mathpert* and *Weierstrass*, we never use more than one assumption at once in the simplification process.

The above rules still don't adequately describe *Weierstrass* or even *Mathpert*, because they do not account for keeping the assumption list in simplified form. To describe this we need to add the rule

$$\frac{\Gamma\sigma \Rightarrow A}{\Gamma \Rightarrow A}$$

where $\Gamma\sigma$ represents the result of simplifying the assumption list Γ. Since simplification generally can use formulas in the assumption list, *Weierstrass* has to be careful when simplifying assumptions, or each assumption would simplify to **true**! Each assumption is temporarily removed from the assumption list, then simplified (possibly using the other assumptions), and the result replaces the original assumption. This process is continued until nothing changes. The result of these simplifications is $\Gamma\sigma$.

5   Combining Computation with First-Order Logic

In previous sections, we have considered the backwards-Gentzen framework for a theorem-prover, and the means of adding computation (simplification) to the quantifier-free fragment of such a prover. We now take up the additional features which were added to *Weierstrass* to allow it to handle epsilon-delta proofs.

The first point is that we must, under certain circumstances, allow *Weierstrass* to factor, or even use trig factor identities. This is a question of control, and not of something new in principle: since factoring preserves mathematical equality, it can be treated exactly like the other computation rules discussed above. It is just a question of factoring when it is useful, and not factoring when it is not useful. To achieve this, we simply put it at the bottom of the list of things to try; that is, below all the things that have been discussed above. It will thus not be tried unless without it, the proof would fail. That will dispose of the problem of factoring when it is not useful.

The other new features can be represented as additional inference rules, which are, like the Gentzen rules, to be applied "backwards" with the aid of unification. We shall describe several of these rules. Like all the rules in sequent calculus, the premises and conclusion of these rules are sequents; but in all cases, the antecedent is unchanged from premise to conclusion, so when writing the rules below, we shall omit $\Gamma \Rightarrow$ in both premises and conclusion.

5.1   Finding Upper and Lower Bounds

Every mathematician knows that many a proof boils down to finding a suitable bound for some expression that does not depend on certain variables. We have implemented a pair of algorithms called *UpperBound* and *LowerBound*. *UpperBound* takes as input a term t to be bounded, and a list of variables on which the bound may not depend. Otherwise put, it tries to find a legal value for a metavariable M such that $|t| \leq M$ could be derived, with the specified list of variables forbidden to M. For example, *UpperBound* knows that $|\sin x| \leq 1$. A better example: if *UpperBound* is asked to bound x by a bound not depending on x, and the current assumptions include $a < x$ and $x < b$, then it will return the bound $|x| \leq \max(|a|, |b|)$. *UpperBound* is probably as good as a very good calculus student at what it does. *LowerBound* is similar, but it tries to find M such that $M \leq |t|$. The two algorithms are defined by mutual recursion.

UpperBound is added directly to *Weierstrass* as a rule of inference with no premises. That is, when we have a goal of the form $\alpha < M$, where M is a metavariable and α is some expression, we can directly terminate that proof branch, instantiating M to the expression produced by *UpperBound*, supplying as the second argument to *UpperBound* the list of variables forbidden to M.

5.2   Factor Bounding

The second new inference rule to be added is called *FactorBounding*. It says that if you want to prove $\beta\gamma$ is small, one way to do it is to prove that γ is small and give a bound for β. The following rule is state for simplicity using only two factors, but the rule is implemented for a product of any number of factors:

$$\frac{\Gamma, |\alpha| < \delta \Rightarrow \gamma \leq M \qquad \Gamma, |\alpha| < \delta \Rightarrow |\beta| < \epsilon/(M+1)}{\Gamma, |\alpha| < \delta \Rightarrow |\beta\gamma| < \epsilon}$$

When this rule is implemented, we take M to be a fresh metavariable, and forbid to M all the variables that are forbidden to δ. In the present implementation of *Weierstrass*, the rule is used only when δ is a metavariable. The implementation also provides an algorithm for deciding which of several factors to bound: it first identifies the quantity in the antecedent that must be less then δ, and then looks for a factor which has a nonzero finite limit as that quantity tends to zero. Limit calculations are performed by symbolic code from *Mathpert*. These limit calculations do not enter the actual proof; they are only used to select the factor to try to bound.

At this point, you might want to turn to Example 1 in the next section, to see how *UpperBound* and *FactorBounding* are used to prove the continuity of $f(x) = x^3$.

5.3   Inequality Chaining

A notorious difficulty in inequality proving is the necessity of using transitivity chains, and the difficulty of finding the right chain of inequalities in an exponentially large search space. However, many useful chains are of length two, based on some standard "known" inequality. For example, if we want to prove $|\sin x| < \epsilon$, it will suffice to prove $|x| < \epsilon$ in view of the known inequality $|\sin x| < |x|$. Weierstrass implements this idea in an algorithm *UsefulBounds*. Described as in inference rule, this just looks like transitivity:

$$\frac{\alpha \leq \beta \qquad \beta < \epsilon}{\alpha < \epsilon}$$

When implemented, $\alpha \leq \beta$ is one of a list of specific known inequalities that have been supplied to *Weierstrass*. For example, a special case of the rule would be

$$\frac{|\sin x| \leq |x| \qquad |x| < \epsilon}{|\sin x| < \epsilon}$$

This rule of inference is needed by Weierstrass to prove the uniform continuity of $\sin x$. See the discussion of this example in the next section.

UpperBound is also capable of controlling some transitivity chaining through the inequalities present in the antecedent. For example, if it is trying to solve $x < M$, where x and y are forbidden to M, and the antecedent contains $x < y$ and $y < b$, the bound $x < b$ will be found, and M will get the value b.

5.4   Mean Value Theorem

Weierstrass can use the mean value theorem to prove an inequality. This is an interesting rule of inference, because it reduces a quantifier-free goal to a subgoal involving quantifiers. The purely logical rules of *Weierstrass* use the cut-free rules of sequent calculus, which always reduce goals to logically simpler subgoals. Here is the rule of inference *MVT*:

$$\frac{\forall z(x \leq z \leq y \rightarrow f'(x) \leq M) \qquad |x - y| \leq \epsilon/M}{|f(x) - f(y)| \leq \epsilon}$$

There is another rule under the same name, in which the conclusion and the second premise have strict inequality. When implemented, M is a freshly-created metavariable, and x and y are added to the list of variables forbidden to M. Note that this would not be the case if the rule were stated with an existential quantifier over M in the premise (combining the two premises into a conjunction). It is by controlling the list of variable forbidden to M that Weierstrass is induced to look for a bound independent of x and y. Now, in general such a bound cannot exist unless the range of x and y is restricted by further inequalities, so some inequality chaining will generally be needed to find the bound M. As an example of such a proof, we consider in the next section, a proof of the uniform continuity of $\sqrt{x}$ on closed intervals $[a, b]$ with $a > 0$.

6   Examples of Proofs That Weierstrass Can Find

In this section we describe the key points of certain illustrative example proofs. The strict length limit does not permit the inclusion of the actual output of Weierstrass.

6.1   Uniform Continuity of $f(x) = x^3$ on Closed Intervals

. This example illustrates the use of *UpperBound* and *FactorBounding*. When *Weierstrass* is asked to prove the uniform continuity of $f(x) = x^3$ on closed intervals $[a, b]$, it soon arrives at the problem of finding a value for the metavariable δ such that, assuming $|x - y| < \delta$, we could derive $|x^3 - y^3| < \epsilon$. Factoring, this reduces to $|x - y||x^2 + xy + y^2| < \epsilon$. At this point, the above rule will be used (in reverse, with $\alpha = |x - y|$), creating the two new goals $|x^2 + xy + y^2| \leq M$ and $|x - y| < \epsilon/(M + 1)$. The first one will be solved by using *UpperBound*, instantiating the metavariable M to $3 \max(|a|, |b|)$ and the second will be solved by the axiom rule $\Gamma, A \Rightarrow A$, where A is the assumption $|x - y| < \delta$, instantiating the metavariable δ to $\epsilon/(M + 1)$.[9]

6.2   Uniform Continuity of $\sin x$ and $\cos x$

These two theorems are proved by *Weierstrass* in a way similar to the above example. However, there are two new twists to the argument. First, *Weierstrass* needs to use the trig factoring operations, not just polynomial factoring, in order

---

[9] *Weierstrass* will be able to handle the case of $f(x) = x^n$ similarly, where n is an integer variable, as soon as *UpperBound* is extended to handle indexed sums, since an indexed sum arises when $x^n - y^n$ is factored.

to write $\sin x - \sin y$ as $2\sin(1/2(x - y))\cos((1/2)(x + y))$. Then in order to instantiate δ, it must use *UsefulBounds* to apply the known inequality $|\sin u| \leq |u|$, since *FactorBounding* will produce the subgoal $\sin(1/2(x-y)) < \epsilon/(M+1)$, which does not unify directly with $|x - y| < \delta$. Even after $|\sin u| \leq |u|$ is used, the 2 in the denominator still requires another step, which however *Weierstrass* takes without difficulty, since an inequality can be simplified by multiplying both sides by 2. This is an example of computation applied to a proposition rather than a mathematical term.

6.3   Continuity of $f(x) = \sqrt{x}$

More precisely, the example is the uniform continuity of $\sqrt{x}$ on closed intervals $[a, b]$ with $0 < a$. To handle the continuity of $\sqrt{x}$ by factoring, we would have to get *Weierstrass* to write

$$|\sqrt{x} - \sqrt{y}| = \frac{|x - y|}{\sqrt{x} + \sqrt{y}}$$

It would certainly be possible to do this, but it would be *ad hoc*, as the kind of computation rule that would do this would cause trouble elsewhere, so it would have to be added as a logical inference rule for this special sort of inequality. Rather than add an *ad hoc* rule, we chose to use this example as an illustration of the use of the Mean Value Theorem. *Weierstrass* will compute the derivative of $\sqrt{x}$ and bound it. Specifically, the inference rule *MVT* described above will introduce a new metavariable M and create the subgoals, $|x - y| < \epsilon/M$ and $\forall z(x \leq z \leq y \rightarrow |(1/2)z^{-2}| \leq M$. Note that the derivative is evaluated. The variables x, y, and z are forbidden to M. When *UpperBound* tries to bound z^{-2}, it calls *LowerBound* to bound z, and successfully finds the transitivity chain $a \leq x \leq z$, arriving at the bound $a \leq z$.

References

1. Beeson, M.: Logic and computation in *Mathpert*: an expert system for learning mathematics, in: Kaltofen, E., and Watt, S. M. (eds.), *Computers and Mathematics*, pp. 202–214, Springer-Verlag (1989).
2. Beeson, M.: Design Principles of Mathpert: Software to support education in algebra and calculus, in: Kajler, N. (ed.) *Human Interfaces to Symbolic Computation*, Springer-Verlag, Berlin/ Heidelberg/ New York (1996).
3. Beeson, M.: Using nonstandard analysis to ensure the correctness of symbolic computations, *International Journal of Foundations of Computer Science* **6**(3) (1995) 299-338.
4. Beeson, M.: Some applications of Gentzen's proof theory in automated deduction, in: Shroeder-Heister, P., *Extensions of Logic Programming*, Springer Lecture Notes in Computer Science **475**, pp. 101–156, Springer-Verlag (1991).
5. Beeson, M.: Unification in lambda-calculus, to appear in *Automated Deduction: CADE-15 - Proc. of the 15th International Conference on Automated Deduction*, Springer-Verlag, Berlin/Heidelberg (1998).

6. Bledsoe, W. W.: Some automatic proofs in analysis, pp. 89–118 in: W. Bledsoe and D. Loveland (eds.) *Automoated Theorem Proving: After 25 Years*, volume 29 in the *Contemporary Mathematics* series, AMS, Providence, R. I. (1984).

7. Boyer, R., and Moore, J.: *A Computational Logic*, Academic Press (1979).

8. Buchberger, B.: History and basic features of the critical-pair completeion procedure, *J. Symbolic Computation* **3**:3–88 (1987).

9. Clarke, E., and Zhao, X.: Analytica: A Theorem Prover in Mathematica, in: Kapur, D. (ed.), *Automated Deduction: CADE-11 - Proc. of the 11th International Conference on Automated Deduction*, pp. 761–765, Springer-Verlag, Berlin/Heidelberg (1992).

10. Coste, M., and Roy, M. F.: Thom's lemma, the coding of real algebraic numbers, and the computation of the topology of semi-algebraic sets, in: Arnon, D. S., and Buchberger, B., *Algorithms in Real Algebraic Geometry*, Academic Press, London (1988).

11. Harrison, J., and Thery, L.: Extending the HOL theorem prover with a computer algebra system to reason about the reals, in *Higher Order Logic Theorem Proving and its Applications: 6th International Workshop, HUG '93*, pp. 174–184, Lecture Notes in Computer Science **780**, Springer-Verlag (1993).

12. Kleene, S. C., *Introduction to Metamathematics*, van Nostrand, Princeton, N. J. (1952).

13. McCune, W.: Otter 2.0, in: Stickel, M. E. (ed.), *10th International Conference on Automated Deduction* pp. 663-664, Springer-Verlag, Berlin/Heidelberg (1990).

14. Richardson, D., Some unsolvable problems involving elementary functions of a real variable, J. Symbolic Logic **33** 511–520 (1968).

15. Stoutemeyer, R.: Crimes and misdemeanors in the computer algebra trade, *Notices of the A.M.S* **38**(7) 779–785, September 1991.

16. Wu Wen-Tsum: Basic principles of mechanical theorem-proving in elementary geometries, *J. Automated Reasoning* **2** 221-252, 1986.

A   Appendix: Output of *Weierstrass* on the Examples

Weierstrass produces an internal proof object, which can be viewed in either "trace view" or "proof tree view". These views both use two-dimensional display of formulas on the screen. When you choose File | Save As, you save a text representation of the proof, either as trace or as proof tree. Formulas are written in a parseable form, similar to TEX, but without backslashes, and enclosed in dollar signs. In the future, I intend to use these files with WebTEXto post proofs to the Web. For purposes of these appendices, I have simply included these files verbatim (inserting only some line breaks) to avoid any errors introduced by transcribing them into TEX, and to demonstrate exactly what the program produces. I have used trace view, since the files are more readable than with proof tree view. Even so, these files do not convey the process of proof construction well, since the metavariables are replaced by their final values; for example, we don't see how and when δ is found, but instead it appears to be "pulled out of a hat" near the beginning of the proof. It is interesting that this very phenomenon is often a problem in the presentation of proofs produced by human mathematicians!

A.1   Continuity of $f(x) = x^3$

```
Assuming $epsilon > 0$
Trying $exists(delta,all(x,y,a <= x,x <= b,a <= y,y <= b,
         abs(x-y) < delta->abs(x^3-y^3) < epsilon))$
 Trying $all(x,y,a <= x,x <= b,a <= y,y <= b,
         abs(x-y) < X->abs(x^3-y^3) < epsilon)$
   Assuming $a <= x,x <= b,a <= y,y <= b,abs(x-y) < X$
   Trying $abs(x^3-y^3) < epsilon$
   Factoring, it would suffice to prove:
       $abs(x-y) abs(x^2+x y+y^2) < epsilon$
   We have the following bound:
       $abs(x^2+x y+y^2) <= 3(max(abs(a),abs(b)))^2$
   So it would suffice to prove:
       $abs(x-y) < epsilon/(3(max(abs(a),abs(b)))^2+1)$
   Aha!  we have
       $abs(x-y) < epsilon/(3(max(abs(a),abs(b)))^2+1)$
   success
   Discharging
 success
 success
 Discharging
```

A.2   Continuity of $f(x) = \sin x$

```
 Trying $epsilon > 0->exists(delta,all(x,y,abs(x-y) < delta->
                abs(sin(x)-sin(y)) < epsilon))$
 Assuming $epsilon > 0$
 Trying $exists(delta,all(x,y,abs(x-y) <
         delta->abs(sin(x)-sin(y)) < epsilon))$
  Trying $all(x,y,abs(x-y) < 1/2 epsilon->
         abs(sin(x)-sin(y)) < epsilon)$
   Assuming $abs(x-y) < 1/2 epsilon$
   Trying $abs(sin(x)-sin(y)) < epsilon$
   Using trigonometry, it would suffice to prove:
       $2abs(sin(x-y)/2) abs(cos(x+y)/2) < epsilon$
   Dividing both sides, it would suffice to prove:
       $abs(sin(x-y)/2) abs(cos(x+y)/2) < 1/2 epsilon$
   We have the following bound:
       $abs(cos(x+y)/2) <= 1$
   So it would suffice to prove:
       $abs(sin(x-y)/2) < epsilon/4$
   In view of the known inequality |sin x| < |x| we have:
       $abs(sin(x-y)/2) <= abs((x-y)/2)$
   it would therefore suffice to prove:
       $abs((x-y)/2) < epsilon/4$
```

```
    Simplifying, it would suffice to prove:
        $2abs(x-y) < epsilon$
    Dividing both sides, it would suffice to prove:
        $abs(x-y) < 1/2 epsilon$
    Aha!  we have $abs(x-y) < 1/2 epsilon$
    success
    Discharging
  success
success
Discharging
success
```

A.3   Continuity of $f(x) = \sqrt{x}$

```
  Assuming $a > 0,epsilon > 0$
  Trying $exists(delta,all(x,y,a <= x,x <= b,a <= y,y <= b,
        abs(x-y) < delta->abs(sqrt(x)-sqrt(y)) < epsilon))$
   Trying $all(x,y,a <= x,x <= b,a <= y,y <= b,abs(x-y) <
        epsilon/(1/2 a^(-1/2))->abs(sqrt(x)-sqrt(y)) < epsilon)$
    Assuming $a <= x,x <= b,a <= y,y <= b,
        abs(x-y) < epsilon/(1/2 a^(-1/2))$
    Trying $abs(sqrt(x)-sqrt(y)) < epsilon$
    Simplifying, it would suffice to prove:
        $abs(x^(1/2)-y^(1/2)) < epsilon$
    By the mean value theorem applied to $fz = z^(1/2)$
    it would suffice to prove:
        $all(z,x <= z,z <= y->abs(1/2 z^(-1/2))
        <= 1/2 a^(-1/2)),abs(x-y) < epsilon/(1/2 a^(-1/2))$
     Trying $all(z,x <= z,z <= y->abs(1/2 z^(-1/2))
                                        <= 1/2 a^(-1/2))$
      Trying $x <= z,z <= y->abs(1/2 z^(-1/2)) <= 1/2 a^(-1/2)$
      Assuming $x <= z,z <= y$
      Trying $abs(1/2 z^(-1/2)) <= 1/2 a^(-1/2)$
      We have the bound: $abs(1/2 z^(-1/2)) <= 1/2 a^(-1/2)$
      success
      Discharging
     success
     success
     Trying $abs(x-y) < epsilon/(1/2 a^(-1/2))$
     Aha!  we have $abs(x-y) < epsilon/(1/2 a^(-1/2))$
     success
    success
    Discharging
   success
  success
  Discharging
```

Finite Model Search for Equational Theories (FMSET)

Belaid Benhamou and Laurent Henocque

Laboratoire d'Informatique de Marseille
Centre de Mathématiques et d'Informatique
39, rue Joliot Curie - 13453 Marseille cedex 13, France
phone number : 91.11.36.22
Benhamou@gyptis.univ-mrs.fr
henocque@esil.univ-mrs.fr

Abstract. Finite model and counter model generation is a potential alternative in automated theorem proving. In this paper, we introduce a system called FMSET which generates finite structures representing models of equational theories. FMSET performs a satisfiability test over a set of special first order clauses called "simple clauses". Several experiments over a variety of problems have been pursued. FMSET uses symmetries to prune the search space from isomorphic branches with very competitive performances in the domain.

Topics: Computer Algebra Systems and Automated Theorem Provers.
Keywords: Finite model, equational theories, symmetry.

1 Introduction

Model generation is well known as a difficult problem in mathematical logic, undecidable in the general case. In this paper, we study finite model generation for equational theories.

Equational theories provide a great number of difficult problems. Zhang in [5] defines a set of problems which form the challenge of finite model search systems. Several open problems were solved with different approaches: FALCON [6], FINDER [3], MGTP-G [1], LDPP, SATO [4], and MACE [2]. FALCON is the most recent and most efficient method for equational theories and serves as a basis of comparison.

An equational theory is a set of axioms: first order logic formulas involving equality (ex: $\forall x, \forall y, \forall z : h(f(x,y)) = f(z,x)$). Finding a finite model for such a theory amounts to finding an interpretation of functional symbols over a finite domain D_n which satisfies all the axioms. The existence of a model proves the consistency of the theory. The existence of a counter model may prove refutation of a conjecture. This is why model generation is a possible approach to automated theorem proving.

Jacques Calmet and Jan Plaza (Eds.): AISC'98, LNAI 1476, pp. 84–93, 1998.

In this paper, we present a new finite model generator, the system FMSET (Finite Model Search for Equational Theories). FMSET translates the set of axioms in an equational theory to an equivalent set of "simple" first order clauses. Model generation operates by variable domain enumeration, in the spirit of the Davis and Putnam procedure. This enumeration is performed directly over the simple clauses obtained in the first phase. It has the double advantage of: 1) generating only the propositional clauses involved in the proof, 2) achieving efficient unit clause propagation as in the propositional case.

Symmetry detection is used to eliminate isomorphic branches in the search tree, thus improving FMSET's performance. A heuristic for choosing literals similar to the one used in Zhang[6] helps keeping symmetries as long as possible during the search. The paper is organised as follows:

Section 2 defines equational theories. Section 3 describes the translation of equational theories to simple first order clauses. The enumeration procedure, the mechanism of propagations and the symmetry elimination procedure are introduced in section 4. Experimental results are listed in section 5 and section 6 concludes the work.

2 Equational Theories

An equational theory is a set of axioms such as $t_1 = t_2$, and $t_1 \neq t_2$, where t_1 and t_2 are terms. Each term is recursively built upon functional symbols and universally quantified variables.

A model of an equational theory is a structure consisting in a non empty set of individuals and functions that satisfy the axioms. We only consider finite models, characterized by their number n of individuals. The domain D_n of a model of size n is $D_n = \{0, 1, ..., n-1\}$.

Example 1. Let T_1 be the equational theory based on the following axioms:

- A_1: $\forall x, h(x, x) = x$
- A_2: $\forall x, \forall y, h(h(x, y), x) = y$

T_1 has a model of size 4 :

$$h(0,0) = 0, \; h(0,1) = 2, \; h(0,2) = 3, \; h(0,3) = 1$$
$$h(1,0) = 3, \; h(1,1) = 1, \; h(1,2) = 0, \; h(1,3) = 2$$
$$h(2,0) = 1, \; h(2,1) = 3, \; h(2,2) = 2, \; h(2,3) = 0$$
$$h(3,0) = 2, \; h(3,1) = 0, \; h(3,2) = 1, \; h(3,3) = 3$$

3 Translation of an Equational Theory to a Set of Simple First Order Clauses

In the sequel, except when necessary, the word *equation* designates equations as well as disequations.

Definition 1. – *The number of functional symbol occurrences in an equation defines its degree.*
- *An equation of degree 1 is a simple equation (which is a literal).*
- *A simple first order clause is a disjunction of simple equations.*

The translation of equational theories to simple first order clauses is based on the systematic replacement of axioms by simple equations, by means of auxiliary variables. The principle is the following: when applied to the axiom $A_2 : (h(h(x, y), x) = y)$ from the theory T_1 in example 1, it produces the clause $h(x, y) \neq v_1 \lor h(v_1, x) = y$. The innermost term $h(x, y)$ is replaced by an auxiliary variable v_1 so as to turn the axiom into a disjunction of simple equations. The result can be read "for all x, y, v_1 if $h(x, y) = v_1$ then $h(v_1, x) = y$". Applied to an axiom of initial degree 3: $f(h(x, g(y)), x) = y$, we obtain the clause $g(y) \neq v_1 \lor h(x, v_1) \neq v_2 \lor f(v_2, x) = y$, which can be read "for all x, y, v_1, v_2 if $g(y) = v_1$ and if $h(x, v_1) = v_2$ then $f(v_2, x) = y$".

The translation of an equational theory is performed in two steps :

- Normalisation : consists in changing the original axioms into clauses where literals have the form $t = x$, x being a variable and t a term.
- Simplification : consists in transforming the clauses resulting from the normalisation into simple clauses.

If $var(t)$ denotes the variable which replaces the term t, then the translation algorithm uses the following set of rules:

1. the equation $x = t_1$ becomes $t_1 = x$ (variable x is moved to the right)
2. the equation $x \neq t_1$ becomes $t_1 \neq x$ (variable x is moved to the right)
3. the equation $t_1 = t_2$ becomes $((t_1 \neq var(t_1) \lor t_2 = var(t_1)) \land ((t_2 \neq var(t_2) \lor t_1 = var(t_2))$ (a conjunct of two clauses)
4. the equation $t_1 \neq t_2$ becomes $((t_1 \neq var(t_1) \lor t_2 \neq var(t_1)) \land ((t_2 \neq var(t_2) \lor t_1 \neq var(t_2))$ (a conjunct of two clauses)
5. the equation $f(t_1, t_2) = z$ becomes $(t_1 \neq var(t_1)) \lor (t_2 \neq var(t_2)) \lor (f(var(t_1), var(t_2)) = z)$
6. the equation $f(t_1, t_2) \neq z$ becomes $(t_1 \neq var(t_1)) \lor (t_2 \neq var(t_2)) \lor (f(var(t_1), var(t_2)) \neq z)$

Normalisation is the result of a single application of one of the rules 1 to 4 and simplification results from the repeated application of rules 5 and 6 to the equations until fixed point is reached.

In the rules 3 to 6, $var(t)$ is the variable which replaces the term t. If t itself is a variable then $var(t)$ is nothing but t.

This algorithm terminates because rules 1 to 4 are applied at most once, and rules 5 and 6 strictly reduce the degree of their argument.

Remark 1. This algorithm produces simple Horn clauses, containing at most one positive literal ($t_1 \neq x$ is the negation of $t_1 = x$).

4   The Enumeration Procedure

We have tried two enumeration techniques inspired from the Davis and Putnam procedure.

4.1   Working with Propositional Clauses

Given a domain D_n, it is straightforward to translate a set of simple clauses C into a set of propositional clauses. It simply requires to generate the terminal instances $c[..., e_i/x_i, ...]$ for all $c \in C$, and for every possible substitution $< e_i/x_i >$ where e_i belongs to D_n and x_i is a variable that occurs in c. For a domain of size n, and a clause having k variables, the number of propositional clauses produced is n^k in the worst case.

For instance, given a domain of size 2, the clause $(h(x, y) \neq v_1 \lor h(v_1, x) = y)$ representing the axiom A_2 in the theory T_1 (example 1 before) is expressed by the following set of propositional clauses:

$$1: \neg h(0,0) = 0 \lor h(0,0) = 0, \ 2: \neg h(0,0) = 1 \lor h(1,0) = 0$$
$$3: \neg h(0,1) = 0 \lor h(0,0) = 1, \ 4: \neg h(0,1) = 1 \lor h(1,0) = 1$$
$$5: \neg h(1,0) = 0 \lor h(0,1) = 0, \ 6: \neg h(1,0) = 1 \lor h(1,1) = 0$$
$$7: \neg h(1,1) = 0 \lor h(0,1) = 1, \ 8: \neg h(1,1) = 1 \lor h(1,1) = 1$$

to which the set of clauses describing the mutual exclusion of function values must be added. For instance, take $h(0,0)$, and a domain of size 3 or more :

$$h(0,0) = 0 \lor h(0,0) = 1 \lor h(0,0) = 2 \lor ...$$
$$\neg h(0,0) = 0 \lor \neg h(0,0) = 1$$
$$\neg h(0,0) = 0 \lor \neg h(0,0) = 2$$
$$\neg h(0,0) = 1 \lor \neg h(0,0) = 2$$
$$...$$

This simple example shows that such a set of clauses allows for monoliteral propagation even in the presence of negative facts. For instance, if $h(1,1) \neq 0$ is true, clause 6 propagates $\neg h(1,0) = 1$. Such propagations are not performed by the algorithm described in [6], which only reacts to the introduction of positive facts and thus loses in search efficiency.

The number of clauses grows quickly as the domain sizes grow, then using a classical model search procedure (like Davis and Putnam) for such sets of propositional clauses must be intractable, except when the domain sizes are small. Actually, even simple theories generate a huge number of clauses.

For instance, the theory T_2 described in figure 1 expresses a non commutative group. For a domain of size 6, the translation of the axioms in figure 1 to propositional clauses requires 252 literals and nearly 100000 clauses. Memory consumption, and the time needed to simply generate the clauses renders this approach irrealistic. This is why we prefer to use directly the first order clauses.

$$h(x, 0) = x$$
$$h(0, x) = x$$
$$h(x, g(x)) = 0$$
$$h(g(x), x) = 0$$
$$h(h(x, y), z) = h(x, h(y, z))$$
$$h(1, 2) \neq h(2, 1)$$

Fig. 1. Non commutative group

4.2   Working with the Simple First Order Clauses

The clauses produced by the translation algorithm of section 3 contain at most one positive literal (Horn simple first order clauses). As soon as a positive literal propagates (for instance $f(0) = 0$), implied negative facts are numerous (here: $f(0) \neq 1$, $f(0) \neq 2$, ... $f(0) \neq n - 1$). Many propositional clauses are removed immediately (because they are true), and would have been created needlessly with the first approach. It is thus realistic to envision the dynamic creation of the only useful propositional clauses.

Our enumeration procedure $FMSET$ is described by the recursive function of figure 2.

function FMSET(F : a set of clauses):boolean;
begin
        for each non assigned literal b on the stack pmod
        begin
          Assign(b);
          if the empty clause appeared then return(false);
          if all the clauses in F are satisfied then return(true);
        end
        choose next literal a
        push a on the stack pmod
        if FMSET(F) return(true)
        push $\neg a$ on the stack pmod
        return(FMSET(F))
end

Fig. 2. The enumeration procedure

We use an intermediate representation between first order and propositional clauses, producing exactly the same propagations as the latter, at a lower cost. Propagations occur when a clause is shortened so that its length becomes equal to one. To obtain these propagations without using propositional clauses, it is enough to keep the simple clauses which generate the clauses shortened by the current interpretation.

For instance : let c_1 be the clause $f(z) \neq y \vee h(x) \neq y \vee g(y) = x$. When the literal $f(0) = 1$ becomes true, we generate the substitution $\sigma = \{< 1/y >, < 0/z >\}$ with which we produce the clause $c_2 : h(x) \neq 1 \vee g(1) = x$. Later on, when $g(1) = 3$ becomes false, we build $c_3 : h(3) \neq 1$ of length one. This forces the propagation of $h(3) \neq 1$.

More interesting, the simple clause $c_4 : f(y) \neq z \vee h(x) = y$, when the literal $f(0) = 0$ becomes true, produces the monoliteral clause $c_5 : h(x) = 0$ which forces the literals $h(0) = 0, h(1) = 0, h(2) = 0 ... h(n - 1) = 0$ to be true.

The simple clauses generated as a result of literal propagation are called pseudo clauses in FMSET. They are well described by a substitution of part of the set of variables of the original clause by values in the domain. Formally:

Definition 2. *Let $c \in C$ be a simple clause, V_c the set of variables occuring in c, V a subset of V_c, and σ a substitution $\{< e_i/x_i >| \ e_i \in D, x_i \in V\}$. The application of σ to the variables of c produces the pseudo-clause c_σ.*

Initially, every simple clause c maps to the pseudo-clause $c_\emptyset$ where $\emptyset$ is the empty substitution.

Definition 3. *Two substitutions σ_1 and σ_2 are incompatible iff there exists $< a/x >\in \sigma_1$ and $< b/x >\in \sigma_2$ such that $a \neq b$.*

Procedure Assign($a \in literals$)
begin
      for each pseudo clause c_σ
         for each simple equation $s \in c_\sigma$
            if unifiable(s,a) then
            begin
               let $\acute{\sigma}$ be the substitution due to unification of a and s
               if σ and $\acute{\sigma}$ are compatible then
               begin
                  build the pseudo clause $\acute{c} = (c_\sigma - s)_{\sigma \cup \acute{\sigma}}$
                  if $\acute{c}$ is of length 1 then
                    push on stack all monoliterals due to $\acute{c}$
                  else if not a tautology($\acute{c}$) then
                    add $\acute{c}$
            end
         end
end

Fig. 3. Propagation procedure

The function $FMSET$ described in figure 2 uses the propagation procedure *assign* described in figure 3.

4.3 Removing Symmetries

Many symmetries exist in equational theories. They slow down the algorithms because of unwanted exploration of isomorphic branches within the search space.

As it was done in [6], some symmetries can be removed at low cost because they are due to trivial symmetries in the set of domain individuals. Especially, at the beginning of the search, all individuals in the domain are interchangeable.

Definition 4. *A set of simple first order clauses C is symmetrical with respect to a subset D_{sym} of the domain D_n iff it remains unchanged under any permutation of the individuals in D_{sym}.*

In particular, if D_n is the set of integers $\{0, 1, \ldots, m, m+1, \ldots, n-1\}$ and $\{0, 1, \ldots, m\}$ is the subset of D_n used by the literals in the current partial model *pmod* that were set by a choice point, then the property of symmetries is characterised by the following theorem :

Theorem 1. *The set of simple clauses C_{pmod} generated by the partial model pmod from C is symmetrical with respect to $D_{sym} = \{m+1, \ldots, n-1\}$*

Proof. There are only two possibilities of occurrence of an individual bigger than m in *pmod*:
- because of the assignment of a positive literal of the form $t = k$, with $k \leq m$. In that case, for all $k2 > m$, the literal $t \neq k2$ is also in *pmod*. Hence this propagation keeps the symmetry of *pmod* wrt. $D_{sym} = \{m+1, \ldots, n-1\}$.
- because of the propagation of a simple clause of length one. Such a clause being universally quantified, the propagation naturally maintains the symmetry of *pmod* wrt. $D_{sym} = \{m+1, \ldots, n-1\}$. Thus, in both cases, the system C_{pmod} is itself symmetrical wrt. D_{sym}.

This theorem is exploited in FMSET by a cut in the search and a heuristic for literal selection. We use a number m equal to the maximum integer used in literals chosen for assignment in *pmod*.

The cut operates as follows: when we choose a literal $t = k$ for assignment, we compute $m_{t=k}$ the greatest integer occurring in $t = k$, we set $m = max(m, m_{t=k})$, and we add to *pmod* the set of $t \neq l$, for all $l \in [m+2, n-1]$. Of course, we try to minimize the resulting value of m by a heuristic.

In a similar way as in [6], this heuristic consists in choosing for assignment a literal $t = k$ such that the value $m_{t=k}$ is lower or equal to m. When this is impossible, the heuristic selects a literal which minimises the value of $m_{t=k}$.

5   Experimentation

5.1   Description of the Problems

The problems that we have experimented are described in [6]. AG is an abelian group. NG is a non commutative group (see figure 1). RU is a ring with unit.

GRP is a non commutative group satisfying the additional axiom $(xy)^4 = x^4y^4$. RNA is a ring, plus a counter example of associativity, the existence of models proves the independence of associativity. RNB is a non boolean ring plus the axiom $x^7 = x$.

These six problems are not open problems, but provide a good comparison basis for FMSET and FALCON. Zhang in [6] shows that FALCON outperforms both MACE and FINDER on these problems.

5.2   Results

FMSET is developped in C++, our computation times in seconds are obtained on a Pentium 133. All of MACE, FINDER and FALCON are writen in C, the corresponding CPU times are the ones given in [6] which are obtained on a Sparc 2, $(t_0 + t_1)$ for both MACE and FINDER and (t_2) for FALCON. t_0 is a preprocessing time, t_1 and t_2 are search times. FALCON's preprocessing time is negligible as is FMSET's. To the best of our knowledge, a Sparc 2 is about three times slower than a Pentium 133.

The CPU time listed for problems AG, NG and RU corresponds to the search for all models. The CPU time listed for GRP, RNA and RNB is only the search time for model existence.

The branch count provides a better comparison basis for the two systems. In FALCON (column b1 in table 1), the detection of a bad value through assignment propagation is not counted as a branch. In FMSET (column b2 in table 1) all branches are counted.

Our approach is very different, and can be generalized to arbitrary first order logic theories. It is noteworthy that our execution times compare well to FALCON's times, and outperform the ones of MACE and FINDER, which fail to conclude on several of these problems. Our branch counts are also close to those from FALCON. In the results table 1, the column "clauses" displays the total count of simple clauses and mutual exclusion clauses used to express the corresponding problems. The time spent generating these clauses is negligible, never above 0.1 second. Note that '*' indicates that the program runs out of memory, and '+' indicates that the execution time of the program exceeds one hour. A '?' means that the information is not available.

6   Conclusion

We have implemented a system for searching finite models for equational theories. It is based on the direct use of first order clauses within a standard propositional logic model enumerator. This hybrid approach produces more propagations, without the cost of either the propositional clauses, or the full first order logic. In a sense, the algorithm combines model enumeration and unification based resolution.

In that framework, symmetries can be efficiently detected and improve FMSET's efficiency. Our results compete with the best known system FALCON.

Problems	MACE	FINDER	FALCON		FMSET			
	time	time	time	b1	literals	clauses	time	b2
Abelian groups								
AG.4	0.37	0.07	0.05	4	80	28	0.13	2
AG.5	1.44	0.39	0.22	7	150	38	0.27	9
AG.6	6.50	8.95	0.68	13	252	50	1.03	22
AG.7	28.78	253,20	1.97	20	392	64	2.23	40
AG.8	493.24	+	4.73	33	576	80	7.5	63
AG.9					810	98	12.8	101
Non commutative groups								
NG.4	0.37	0.07	0.05	4	80	28	0.08	6
NG.5	1.36	0.70	0.17	6	150	38	0.15	14
NG.6	5.13	12.95	0.48	7	252	50	0.56	24
NG.7	13.30	530.63	1.30	11	392	64	0.63	29
NG.8	198.28	+	3.37	20	576	80	1.5	41
NG.9	*	+	7.70	31	810	98	3.1	75
Unit rings								
RU.4	3.95	0.46	0.25	7	144	49	0.6	2
RU.5	19.64	6.00	1.27	9	275	68	1.4	8
RU.6	83.54	2078.23	3.50	15	468	91	7.5	29
RU.7	*	+	9.43	22	735	118	18.8	44
Non commutative groups satisfying $(xy)^4 = x^4 y^4$								
GRP.4	40.22	0.7	0.08	?	112	43	0.58	6
GRP.5	309.16	1.82	0.27	?	200	55	1.78	14
GRP.6	*	499.78	0.65	?	324	69	9.76	26
GRP.7		+	1.53	?	490	85	15.9	29
ring not associative								
RNA.5	25.60	+	24.72	?	275	70	5.43	28
RNA.6	*		107.72	?	468	93	52.4	172
RNA.7			562	?	735	120	426	605
RNA.8			350	?	1088	151	11.3	9
non boolean ring								
RNB.4	3.81	3.32	0.2	?	144	51	0.46	3
RNB.5	18.33	0.24	0.68	?	273	70	0.22	4
RNB.6	67.83	2.11	1.67	?	468	93	0.45	4
RNB.7	*	3.58	3.37	?	735	120	3.7	12
RNB.8	*	+	4.22	?	1088	151	10.9	3
RNB.9					1539	186	23.5	34

Table 1. Results obtained by FALCON and FMSET

We plan to generalize this approach to other categories of problems for which the translation to propositional logic is not tractable and extend it's input to any first order logic formula. More symmetries can be detected in a set of simple clauses. The fact that this algorithm propagates more than other techniques makes it a potentially good third party algorithm in the field of constraint programming, in combination with other solvers.

References

1. J. Slaney M. Fujita and F. Bennett. Automatic generation of some results in finite algebra. *In proceedings of the 13th Internationnal Joint Conference on Artificial Intelligence, Chambery, France*, pages 52–57, 1993.
2. W. McCune. A Davis Putnam program and its application to finite fist order model search : quasi-group existence problems. Technical Report ANL/MCS-TM-1994, Argonne National Laboratory, 1994.
3. J. Slaney. Finder: Finite domain enumerator. version 3.0 notes and guide. Technical report, Austrian National University, 1993.
4. H. Zhang and M. Stickel. Implementing the Davis and Putnam algorithm by tries. Technical report, University of IOWA, 1994.
5. J. Zhang. Problems on the generation of finite models. *in proceedings of CADE-12, Nancy, France*, pages 753–757, 1994.
6. J. Zhang. Constructing finite algebras with FALCON. *Journal of automated reasoning, 17*, pages 1–22, 1996.

Specification and Integration of Theorem Provers and Computer Algebra Systems

P.G. Bertoli[1], J. Calmet[2], F. Giunchiglia[3], and K. Homann[4]

[1] bertoli@itc.it - ITC-IRST - Trento (Italy)
[2] calmet@ira.uka.de - University of Karlsruhe (Germany)
[3] fausto@itc.it - ITC-IRST - Trento and DISA - University of Trento (Italy)
[4] karsten.homann@pn.siemens.de - Siemens Corporation - Munich (Germany)

Abstract. Computer algebra systems (CASs) and automated theorem provers (ATPs) exhibit complementary abilities. CASs focus on efficiently solving domain-specific problems. ATPs are designed to allow for the formalization and solution of wide classes of problems within some logical framework. Integrating CASs and ATPs allows for the solution of problems of a higher complexity than those confronted by each class alone. However, most experiments conducted so far followed an ad-hoc approach, resulting in tailored solutions to specific problems. A structured and principled approach is necessary to allow for the sound integration of systems in a modular way. The Open Mechanized Reasoning Systems (OMRS) framework was introduced for the specification and implementation of mechanized reasoning systems, e.g. ATPs. The approach was recasted to the domain of computer algebra systems. In this paper, we introduce a generalization of OMRS, named OMSCS (Open Mechanized Symbolic Computation Systems). We show how OMSCS can be used to soundly express CASs, ATPs, and their integration, by formalizing a combination between the Isabelle prover and the Maple algebra system. We show how the integrated system solves a problem which could not be tackled by each single system alone.

Topics: Integration of Logical Reasoning and Computer Algebra, Computer Algebra Systems and Automated Theorem Provers.

Keywords: Computer Algebra Systems, Theorem Provers, Integration, Formal Frameworks.

1 Introduction

Automated theorem provers (ATPs) are used in the formal verification and validation of systems, protocols, and mathematical statements. These systems are complex software packages, designed in a stand-alone way, and each implementing a certain fixed logic which must be adopted to formalize and solve problems. Computer Algebra Systems (CASs) have become a standard support tool for performing complex computations, or for representing functions. Similarly to

Jacques Calmet and Jan Plaza (Eds.): AISC'98, LNAI 1476, pp. 94–106, 1998.

ATPs, these tools have been designed in a stand-alone way, and implement customary syntaxes; the user may only perceive them as black boxes, and is forced to trust them "blindly".

ATPs and CASs exploit complementary abilities. ATPs implement heuristic search procedures; their underlying logics are designed to express wide varieties of problems. CASs can be perceived as extensive libraries of very complex and efficient procedures, tailored to the solution of specific problems within specific domains. Thus, by coupling the efficiency of CASs and the generality of ATPs, it should be possible to obtain systems which are able to solve problems of a higher complexity than those that have been confronted with by stand-alone CASs and ATPs. Several approaches can be followed to combine these two paradigms, e.g. integrating external algorithms into proof structures as oracles or untrusted steps, or extending CASs by reasoning components. These approaches have led to design and implement several integrations between ATPs and CASs. [4] reports an experiment of integration between the Isabelle prover and the Maple algebra system; [7] describes the implementation of an ATP within the Mathematica environment; [9] defines an extension of the HOL prover to reason about real numbers.

However, all the previous attempts are ad-hoc solutions, tailored to solving specific problems. On the opposite, it is desirable to be able to make ATPs and CASs cooperate in a more principled and generic way. This cooperation can only be achieved by formally defining provers and algebra systems as *symbolic mathematical services*. By symbolic mathematical service we mean a software able to engage in useful and semantically meaningful two-way interactions with the environment. A symbolic mathematical service should be structurally organized as an *open architecture* able to provide services like, e.g., proving that a formula is a theorem, or computing a definite symbolic integral, and to be able, if and when necessary, to rely on similar services provided by other tools. In [8], the Open Mechanized Reasoning System (OMRS) architecture was introduced as a mean to specify and implement reasoning systems (e.g., theorem provers) as logical services. In [5], this approach has been recast to symbolic computer algebra systems. In this paper, we show a generalization of this architecture, OMSCS (Open Mechanized Symbolic Computation Systems), which can be used to specify both ATPs and CASs, and to formally represent their integration. Section 2 summarizes the ideas which are at the base of the OMSCS architecture, and provides the formal descriptions of its components. Section 3 describes the specification of an integration between the Isabelle prover and the Maple algebra system. We highlight the synergic effects of the integration by showing how the resulting system solves a problem none of the starting systems could tackle alone. Section 4 is devoted to conclusions and future work.

2   The OMSCS Framework

In the OMRS framework, reasoning systems are presented as logical services. The specification of a service must be performed at various levels. At the lower level, it is necessary to formally define the objects involved in the service, and the basic operations upon them. E.g., for a theorem prover, one must define the kind of assertions it manipulates, and the basic inference rules that can be applied upon them. On top of this "object level", a control level provides a means to define the implementation of the computational capabilities defined at the object level, and to combine them. The control level must include some sort of "programming language" which is used to describe a strategy in the applications of modules implementing basic operations, therefore to actually define the behaviour of the complex system implementing the service. Finally, the way the service is perceived by the environment, e.g. the naming of services and the protocols implementing them, is defined within a further level, called the interaction level. This leads to the following OMRS architectural structure:

$$Reasoning\ Theory = Sequents + Rules$$
$$Reasoning\ System = Reasoning\ Theory + Control$$
$$Logical\ Service = Reasoning\ System + Interaction$$

Analogously, as shown in [5], CASs can be presented as algorithmic services, based upon a definition of computation system and computation theory:

$$Computation\ Theory = Objects + Algorithms$$
$$Computation\ System = Computation\ Theory + Control$$
$$Algorithmic\ Service = Computation\ System + Interaction$$

In order to allow for a unified description of both classes of systems, we synthesize these definitions into that of Symbolic Mathematical Service. It is based upon definitions of symbolic entities and operations which include the previous definitions of sequents and objects, and of rules and algorithms respectively.

$$Symbolic\ Computation\ Theory = Symbolic\ Entities + Operations$$
$$Symbolic\ Computation\ System = Symbolic\ Computation\ Theory + Control$$
$$Symbolic\ Mathematical\ Service = Symbolic\ Computation\ System + Interaction$$

We call this architecture Open Mechanized Symbolic Computation Systems (OMSCS). The following two subsections deal with the formal description of the object and control layers. The interaction layer is the object of ongoing research. In this document, we will consider one single service, namely, asking the system to apply some computation strategy over some mathematical entity. This amounts to the possibility of asking an ATP for a proof of a conjecture, or a CAS for a computation over some algebraic structure. A naive functional perception of this service will be sufficient to our purposes.

2.1   The Object Level

Actual systems implement a variety of computation paradigms, based on a wide spectrum of classes of entities. The object level of the OMSCS framework is designed to allow for the representation of this variety of objects and behaviours. The notion of domain is extended by defining a *system of symbolic entities* as a mean to represent the entities manipulated by a symbolic computation system, and the basic relationships governing them. A system of entities includes a set of symbolic objects, a system of symbolic instantiations and a system of symbolic constraints. Objects, constraints and instantiations are taken as primitive sorts. Objects and constraints may be schematic, to allow for the representation of schematic computations via the instantiation system. The constraint system allows for the representation of provisional computations. Thus a system of symbolic entities is a triple as follows:

$$Esys = <O, Csys, Isys>$$

O is the set of *symbolic objects*. $Csys$ is a constraint system $<C, \models>$, where C is a set of *constraints*, and $\models \subseteq (\mathbf{P}_\omega(C) \times C)$ is a *consequence relation on constraints*. $Isys$ is an instantiation system $<I, \_[\_]>$, where I is the set of *instantiation maps* (or instantiations), and $\_[\_]$ is the operation for *application of instantiations* to objects and to constraints, that is $\_[\_] : [O \times I \to O]$ and $\_[\_] : [C \times I \to C]$. In order to qualify as a system of symbolic entities, $Esys$, and more specifically $\models$, I, and $\_[\_]$ must meet certain requirements, which can be lifted from [8].

The basic operations which can be performed over a system of symbolic entities are defined in a set-theoretic way, as relations between single entities and tuples of entities, instantiations and constraints. These relations are required to be closed w.r.t. instantiation. Namely, let $Esys = <O, Csys, Isys>$ be a symbolic computation system, where $Csys = <C, \models>$, $Isys = <I, \_[\_]>$. Then the set of operations $\mathbf{Op}[Esys]$ over $Esys$ is defined as follows:

$\mathbf{Op}[Esys] =$
$\{Op \subseteq (\mathbf{P}_\omega(O) \times O \times \mathbf{P}_\omega(C)) \mid \forall <\widetilde{o}, o, \widetilde{c}> \in Op : \forall \iota \in I : <\widetilde{o}, o, \widetilde{c}>[\iota] \in Op\}$

In order to conveniently identify operations, we use maps from identifiers (belonging to a set Id) to operations. Thus we define the set of operations sets as follows:

$$\mathbf{Opset}[Esys, Id] = Id \xrightarrow{f} \mathbf{Op}[Esys]$$

Based on the definition of a system of entities and of a set of operations, the computation capabilities of a system are described by means of a *symbolic computation theory*, a triple containing a system of entities $Esys$, a set of identifiers Id, and a set of operations $\widetilde{Op} \in \mathbf{Opset}[Esys, Id]$:

$$SCth = <Esys, Id, \widetilde{Op}>$$

A symbolic computation theory entails a generalized notion of computation which is represented within *symbolic mathematical structures*. Intuitively, symbolic mathematical structures are defined as nested directed graphs containing two kinds of nodes, representing symbolic objects and links between them. A link between an object o and a set of objects $\widetilde{o}$ represents the fact that $\widetilde{o}$ has been computed from o, or vice versa (depending on the orientation of the arcs); the link node provides a justification, which may consist in a basic operation, or in another symbolic mathematical structure. We point to [8] for a complete formal description of (nested) structures, instantiated to the class of reasoning systems. Standard derivations and proofs, and functional computations are represented by specific subclasses of mathematical structures. This allows for the decoupling the specification of derivability or computability from the control strategies for executing computations, and gives greater flexibility for algorithm design and for definition of high-level control abstractions.

The following two examples demonstrate the features of the OMSCS object level, by formalizing sections of two state-of-the-art systems. In particular, we describe (some of) the reasoning capabilities of the Isabelle prover, and (some of) the computational capabilities of the Maple computer algebra system.

Example 2.1 (Isabelle at the object level). Isabelle is a tactical theorem prover. Its basic deductive capabilities are implemented by means of modules called primitive tactics. Primitive tactics receive and produce assertions, implementing base inference steps. Some primitive tactics may manipulate some additional information, which they use to compute their output. Primitive tactics are combined by means of a tactical language. Isabelle uses a meta-theoretical resolution tactic to refine a subgoal by unifying it with an instance of a ground rule contained within the current theory. Other relevant primitive tactics of Isabelle include an "axiomaticity" tactic, which removes a goal whose value is the logical constant **True**, and an "assumption" tactic, which removes a goal which matches some assumption via a unifier. Several predefined ground theories can be exploited in Isabelle. For instance, the theory of natural numbers includes Peano's axioms and an induction rule whose first-order axiomatization can be presented as follows:

$$(induct)\quad [P(a) \wedge (a \leq n) \wedge \forall x : (x \in N \wedge (x \geq a) \wedge P(x)) \longrightarrow P(x+1)] \longrightarrow P(n)$$

The theory can be enriched, e.g by a classical first-order axiomatization of (some of) the properties of the less-than relation, such as transitivity, reflexivity, and distributivity w.r.t. the sum and product operations.

We consider this theoretical setting of Isabelle to start our OMSCS formalization by describing its object level by means of a symbolic entity system $Esys_I = \langle O_I, Csys_I, Isys_I \rangle$. The symbolic objects we consider are first-order sequents which represent goals under a set of assumptions :

$$O_I = \{TH \vdash_I g\}$$

where TH is a sequence of first-order formulas, and g is a single first-order formula. The I subscript to the sequent symbol identifies this specific sort of object. The constraints we consider are equalities between first-order terms:

$$Csys_I = \langle t_1 = t_2, \models \rangle$$

where t_1 and t_2 are first-order terms, and $\models$ is defined to obey monotonicity, axiomaticity and cut. The simple system of instantiation we consider is based upon maps between schematic objects and first-order terms, designed as pairs

$\langle Sc, t \rangle$ where Sc is a schematic object and t is a first-order term. The instantiation application mechanism for these pairs is defined as a substitution of t for Sc:

$$Isys_I = \langle I_I, \_[\_] \rangle \quad : \quad \begin{cases} I_I = \{\langle Sc, t \rangle\} \\ O[\langle Sc, t \rangle] = O_{|Sc \leftarrow t} \end{cases}$$

The deductive capabilities of the prover within the previous theory setting are represented by OMSCS operation generators. Since the underlying system of entities describes the ground assertions manipulated by Isabelle, the operations will describe the manipulation of such entities. Thus, they will describe the induction rule and the less-than formalization presented above, and the axiomaticity and assumption rule presented by the prover as primitive tactics. No formalization of Isabelle's meta-resolution will appear. What follows is an informal representation of the generators, which resembles the classical presentation of logical rules found in, e.g., [11]. Within such description, $g(f)$ indicates a goal g containing a subformula f. In the context of a rule, $g(f')$ will be an abbreviation for $g(f)_{|f \leftarrow f'}$.

Induct
$$\frac{TH \vdash_I g(a) \quad TH \vdash_I a \leq n \quad TH \vdash_I \forall x \in N : ((x \geq a) \wedge g(x)) \rightarrow g(x+1)}{TH \vdash_I g(n)}$$

Ax
$$\frac{}{TH \vdash_I True} \qquad \textbf{Trans} \quad \frac{TH \vdash_I g(x \leq y) \quad TH \vdash_I g(y \leq z)}{TH \vdash_I g(x \leq z)}$$

Assume
$$\frac{}{\phi_1, \ldots, \phi_n \vdash_I g} \phi_i s = g \qquad \textbf{Sum} \quad \frac{TH \vdash_I g(x' \leq x'') \quad TH \vdash_I g(y' \leq y'')}{TH \vdash_I g(x' + y' \leq x'' + y'')}$$

Reflexivity
$$\frac{TH \vdash_I g(True)}{TH \vdash_I g(x \leq x)} \qquad \textbf{Prod} \quad \frac{TH \vdash_I g(x' \leq x'') \quad TH \vdash_I g(y' \leq y'')}{TH \vdash_I g(x' * y' \leq x'' * y'')}$$

The `Induct`, `Reflexivity`, `Trans`, `Sum` and `Prod` operations map the Isabelle rules presented when describing the system. The `Ax` and `Assume` operations represent Isabelle's axiomaticity and assumption primitive tactics. The schemas like those above should be thought of as presenting the generators for the operations considered, linking them to operation identifiers. For instance, the first schema links the `Induct` identifier to the following generator:

$$\{\langle [TH \vdash_I g(a), TH \vdash_I a \leq n, TH \vdash_I \forall x \in N :$$
$$((x \geq a) \wedge g(x)) \rightarrow g(x+1)], TH \vdash_I g(n), \emptyset \rangle\}$$

The symbolic computation theory $SCth_I = \langle Esys_I, Id_I, \widetilde{Op}_I \rangle$ represents Isabelle at the object level, where $Id_I \supseteq \{$`Induct`, `Ax`, `Assume`, `Reflexivity`, `Trans`, `Sum`, `Prod`$\}$, and $\widetilde{Op}_I$ maps Id_I to the generators according to the presentation above.

Example 2.2 (Maple at the object level). Maple is a powerful, complex CAS, featuring a number of complex algorithms for solving equations, computing integrals and derivatives, performing operations on matrixes, and so on. However, at a high level of abstraction, Maple can be perceived simply as a term rewrite system (where the terms to take into account are those of the Maple mathematical language). Thus, every calculation performed by the system can be represented by means of a conditional rewrite rule of the form $[P_1, \ldots, P_n] \Longrightarrow t \equiv t'$, where P_i are preconditions to the equivalence between terms t and t'. The preconditions include type declarations of the form $x \in \mathcal{D}$, where x is a Maple variable, and $\mathcal{D}$ is a Maple domain identifier, and equalities between terms.

In particular, among the plethora of capabilities featured by the system, we focus on those that allow Maple to normalize natural powers of binomials or polynomial powers of monomials to their "flat" polynomial form, and to evaluate disequalities between natural values. It is possible to represent these capabilities in a black-box form, making use of two functions, $NormalizePoly$ and $EvalBool$, specifying their semantics in terms of input/output behaviour.

We describe the entities manipulated by Maple at the OMSCS object level via a symbolic entity system $Esys_M = <O_M, Csys_M, Isys_M>$. The objects we consider represent equivalences between first-order terms under a set of declarations, which we may view as sequents:

$$O_M = \{TH \vdash_M t \equiv t'\}$$

where t and t' are Maple terms, and TH contains a set of Maple type declarations. The constraints we consider are equalities between Maple terms:

$$Csys_M = <t_1 = t_2, \models>$$

where t_1 and t_2 are Maple terms, and $\models$ is defined to obey the monotonicity, axiomaticity and cut properties. The system of instantiation is based upon maps between schematic objects and Maple terms, represented as pairs $<Sc, t>$ where Sc is a schematic object and t is a Maple term. The instantiation application mechanism is defined as a substitution of t for Sc:

$$Isys_M = <I_M, \_[\_]> \quad : \quad \begin{cases} I_M = \{<Sc, t>\} \\ O[<Sc, t>] = O_{|Sc \leftarrow t} \end{cases}$$

We represent the computation capabilities of Maple using OMSCS operation generators, which we present using the informal notation used in the previous example.

MapleEval $\dfrac{}{TH \vdash_M t \equiv t'} \quad t' = NormalizePoly(t)$

MapleEvalLess $\dfrac{}{TH \vdash_M (x \leq y) \equiv EvalBool(x, y)} \quad \{x : Bool, y : Bool\} \subseteq TH$

The symbolic computation theory $SCth_M = <Esys_M, Id_M, \widetilde{Op}_M>$ represents Maple at the object level, where $Id_M \supseteq \{\texttt{MapleEval}, \texttt{MapleEvalLess}\}$, and $\widetilde{Op}_M$ maps Id_M to the generators according to the presentation above.

The combination of systems at the object level is performed by gluing symbolic computation theories. Gluing two symbolic computation theories $SCth_1 =$

$\langle Esys_1, Id_1, \widetilde{Op_1}\rangle$ and $SCth_2 = \langle Esys_2, Id_2, \widetilde{Op_2}\rangle$ via bridge operations $\widetilde{Op_b}$ results in a new symbolic computation theory $SCth_g = \langle Esys_g, Id_g, \widetilde{Op_g}\rangle$ with the following features:

- the new system of entities $Esys_g$ is the disjoint union of the starting systems $Esys_1$ and $Esys_2$. Intuitively, the disjoint union of systems of entities results from their set union, where matching objects belonging to different systems are unified; the formal definition is presented in [8].
- the new identifiers and operations are simply the union of the identifiers and operations presented by $SCth_1$ and $SCth_2$ with Id_b and $\widetilde{Op_b}$ respectively. Note that the bridge operations, which specify the way computations in the two systems can be combined, must be defined over $\mathbf{Opset}[Esys_g, Id_b]$.

2.2   The Control Level

The control level of a mathematical service must specify the way the system implements the computing capabilities specified at the object level, and the strategies adopted to combine them to achieve complex behaviours.

OMSCS adopts the tactic-based approach to pursue the first aim. The basic computation abilities of a system are represented using *primitive tactics*. A primitive tactic provides a particular implementation of an operation defined within a symbolic computation theory. Intuitively, a primitive tactic is defined to be a correct implementation of an operation *op* if every tuple describing its input/output behaviour corresponds to some tuple contained within the definition of *op*. A formal definition is given in [6]. Primitive tactics implement OMSCS operations directionally. Tactics may fail, representing the partiality of operation applications. It must be possible to control primitive tactics so that they realize specific instances of operations. This is achieved by exploiting two mechanisms, control arguments and control annotations. Control arguments are additional objects manipulated by the tactics (in addition to symbolic entities) in order to generate values for the output entities. Control annotations are meant as a colouring of the symbolic entities manipulated by tactics, and consisting of additional information, which can be removed via an "annotation removal mapping". Control arguments and control annotations capture the two forms of control (explicit, or environment-driven and implicit, or system-driven) presented by systems. All the features above are taken into account by redefining the notions provided at the object level accordingly. Thus a *system of annotated symbolic entities* is a 5-tuple as follows:

$$Esys_a = \langle O_a, O_c, F, Csys_a, Isys_a\rangle$$

O_c is the set of control objects, F is the set of failures, which contains at least a no-failure element $\mathtt{Ok}$ and a generic failure element $\mathtt{Fail}$; O_a, $Csys_a = \langle C_a, \models\rangle$, $Isys_a = \langle I_a, \_[\_]\rangle$ are the annotated counterparts of the object level definitions. Every annotated element of the components of $Esys_a$ is mapped to an element of the corresponding component of a system of entities $Esys$ via an annotation removal mapping μ which must preserve the behaviour of the instantiation maps and of the consequence relation between constraints. That is, if c_1, c_2 are constraints, and e a generic entity, then $\mu(\iota(e)) = \iota(\mu(e))$ and $\alpha \models \beta \longrightarrow \mu(\alpha) \models \mu(\beta)$.

The definition of primitive tactics at the control level must extend the object level operation definition, taking into account the additional presence of control arguments, and directionality. Therefore, every entity manipulated by a tactic will be marked as an input or output argument. We indicate with O_{aIO}, O_{cIO}, C_{aIO}, I_{aIO} the sets of pairs $<o_a, IO>$, $<o_c, IO>$, $<o_a, IO>$, $<i_a, IO>$ respectively, where $IO \in \{Input, Output\}$, $o_c \in O_c$, $o_a \in O_a$, $i_a \in I_a$. The η mapping retrieves the unannotated, orientation-free content of these pairs, e.g. $\eta(<O_c, IO>) = \mu(O_c)$. Thus, primitive tactics will be defined as follows:

$$\mathbf{Tac}[Esys_a] = \{PTac \subseteq (O^*_{aIO} \times O^*_{cIO} \times O_{aIO} \times C_{aIO^*} \times F)|$$
$$\exists Op \in \mu(Esys_a) :$$
$$\forall <\overline{o_{aIO}}, \overline{o_{cIO}}, o_{aIO}, \overline{c_{aIO}}, f> \in PTac :$$
$$\exists p_1 \in Perm(|\overline{o_{aIO}}|), p_2 \in Perm(|\overline{c_{aIO}}|) :$$
$$(f = \mathtt{Ok}) \longrightarrow <\eta(p_1(\overline{o_{aIO}})), \eta(o_{aIO}), \eta(p_2(\overline{c_{aIO}}))> \in Op\}$$

Similarly to operations, primitive tactics are linked to identifiers within sets of tactics:

$$\mathbf{Tacset}[Esys_a, Id] = Id \xrightarrow{f} \mathbf{Tac}[Esys_a]$$

Thus the control level of a system is described by a symbolic control theory, whose definition lifts from that of computation theory at the object level:

$$SCNth = <Esys_a, Id, \widetilde{Tac}>$$

where $\widetilde{Tac} \in \mathbf{Tacset}[Esys_a, Id]$. The definition of tactic above allows for the representation of various computational paradigms, e.g. standard backward and forward reasoning.

The application of primitive tactics onto object nodes of mathematical structures can be represented as a series of applications of primitive actions onto such structure; see [8] for details. The combination of primitive tactics is realized by defining a control language. The definition of such a language can be performed by specifying, for each construct, the relation between the OMSCS definition of the compound tactic w.r.t. those of the originating tactics. The language must be proven to be correctness-preserving: provided that primitive tactics correctly implement computation capabilities, compound tactics must correctly implement compound computations.

Example 2.3 (Isabelle at the control level). Let us consider the object level formalization of the Isabelle prover, provided by Example 2.1. In order to describe the system at the level of control, we first define the system of annotated symbolic entities. In Isabelle, no implicit notion of control exists; thus, entities are not annotated. Explicit control arguments appear instead; the control arguments O_{Ic} are either natural numbers, used to identify uniquely the factorization of disequality terms within the Sum or Prod rules, or first-order formulas, used to designate a matching candidate in the assumptions of an assertion. The annotated system is simply an extension of $Esys_I$: $Esys_{Ia} = \langle O_I, O_{Ic}, F, Isys_I, Csys_I \rangle$, where $F = \{\texttt{Ok}, \texttt{Fail}\}$. In this case, control arguments do not modify the design of instantiations and constraints. Rather than formally presenting the backward primitive tactics corresponding to the operations described in Example 2.1, we consider their former informal presentation and re-interpret it accordingly. In particular, the tactics we consider are intended as bottom-to-top orientation of the corresponding operations, where the output objects are returned in a left-to-right order. Whenever input arguments do not match the symbolic schema given in the presentation, tactics are supposed to return $\texttt{Fail}$. Tactics may receive control arguments. Namely, $\texttt{InductTac}(a, n)$ is controlled w.r.t. the bound and the term of the induction; $\texttt{AssumeTac}(\phi_i)$ is controlled w.r.t. the term to match with the goal; $\texttt{ReflTac}(x)$ and $\texttt{TransTac}(y)$ receive the term involved in reflexivity and transitivity respectively; $\texttt{SumTac}(x', x'')$ and $\texttt{ProdTac}(x', x'')$ require control arguments to uniquely identify a decomposition of the input disequalities. The control theory $SCNth_I$ which defines Isabelle parallels the computation theory $SCth_I$, based upon the annotated entity system and the tactics described above. Notice that, since Isabelle is a tactical prover based on meta-resolution, and the primitive tactics above implement ground rules, in general, a many-to-one correspondence exists between these tactics and the original Isabelle tactics.

In order to combine primitive backward tactics, it is possible to adopt (a conservative extension of) the language used in LCF and in Isabelle. The formal definition of the variant of the LCF tactical language we adopt is given in [6]. The language is extended in order to allow tactics to handle control arguments; thus, tactical expressions receive an atomic symbolic mathematical structure (a single object node) and a list of control arguments. As a result, either a failure or an updated symbolic mathematical structure is returned.

Example 2.4 (Maple at the control level). The representation of Maple's behaviour at the control level is straightforward, since Maple does not make use of control annotations, nor of control arguments. Thus the system of annotated symbolic entities is a trivial extension of $Esys_M$. The presentation of the primitive tactics $\texttt{MapleEvalTac}$ and $\texttt{MapleEvalLessTac}$ is inherited from Example 2.2. The tactics are intended as bottom-to-top orientations of the operations, returning their results according to the left-to-right ordering in the presentation. The control theory $SCNth_M$ which defines Maple parallels the computation theory $SCth_M$, based upon the annotated entity system and the tactics described above.

The combination of systems at the control level is based on gluing the annotated symbolic computation theories, and on combining the definitions of tactical languages. The first aim is achieved by lifting the definition of gluing from the object level. The definition of a formal framework for the uniform presentation of tactical languages and compound tactical expressions is the subject of ongoing work. In this document, we rely on a common definition of tacticals languages between the components of the integration.

3   An Example: Integrating Isabelle and Maple

In this section, we show how the framework can be used to integrate in a clear and sound manner a computer algebra system and a theorem prover. We consider the OMSCS formalizations provided in the previous examples as a starting point to design a specific combination of Isabelle and Maple, and study how the resulting system can be used to solve a problem none of the two single systems could tackle alone. In [4], the two systems are combined in an ad-hoc way, by enriching Isabelle's simplifier with an additional external invocation rule referring to the Maple system. In this way, Isabelle is configured to act as a master to Maple, which is simply considered as an evaluation engine.

Starting from our previous OMSCS formalizations, we are able to formalize such an integration of Isabelle and Maple at the object level by gluing the symbolic computation theories $SCth_I$ and $SCth_M$ describing each system. To represent the way terms contained within an Isabelle assertion are simplified by using Maple as a rewriter, the gluing must include a bridge rule. Its informal presentation follows:

$$\text{Simplify} \quad \frac{TH_M \vdash_M t \equiv t' \quad TH_I \vdash_I g(t')}{TH_I \vdash_I g(t)} \quad TH_M \subset TH_I$$

At the control level, the gluing between the annotated theories involves a corresponding backward primitive bridge tactic $\texttt{SimplifyTac}(t)$, which receives as a control argument the term submitted to simplification. Its presentation coincides with that of the originating operation. The tactical language adopted in the Isabelle example is used to combine this tactic with Isabelle's and Maple's. This allows for expressing the behaviour of the integrated system via a unique compound tactic, perceiving the integrated system as a unique entity.

As an example of the abilities which derive from combining the two systems, we consider a problem first described in [10]. Namely, we intend to prove that, for every natural greater or equal than 5, it holds that $n^5 \leq 5^n$; that is:

$$[n \in N; 5 \leq n] \longrightarrow n^5 \leq 5^n \tag{1}$$

Albeit a simple conjecture, this theorem cannot be solved by Maple alone, and cannot be solved by Isabelle in an efficient way. Maple does not posses any deductive capability to perform such a proof. Isabelle does not possess any basic capability to expand powers of binomials, and evaluate disequalities. Formalizing these abilities is possible, but would lead to a very lengthy proof search.

The proof of (1) develops along the following lines:

1. Isabelle's induction rule is applied upon the main goal; this leads to three subgoals, two of which define the base step, and the third defining the induction step;
2. The subgoals defining the base step are simply solved by applying basic rules;
3. The subgoal defining the induction step contains a disequality between polynomials in non-normal form; they are normalized by expanding it through

Maple calls. In concrete, this is performed by simplification via the evaluation rules.
4. Finally, by a repeated use of the laws which governate disequalities between products and sums, the induction step is proved. In this phase, additional Maple calls are used to verify disequalities between ground values, e.g. $2 \le 5$.

The compound OMSCS tactic that originates the proof in our formalization closely resembles the series of Isabelle's tactics invocations used in [10] to achieve the result. Its execution results in a (flat) symbolic mathematical structure which represents the proof of the conjecture. The following picture provides a simplified presentation of the structure.

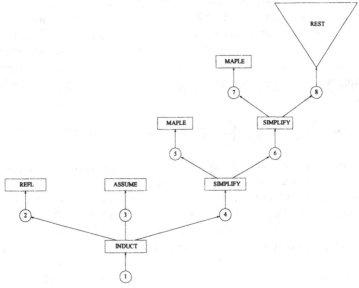

$$
\begin{array}{l}
1 : TH \vdash_I n^5 \le 5^n \\
2 : TH \vdash_I 5^5 \le 5^5 \\
3 : TH \vdash_I n \le 5 \\
4 : TH \vdash_I \forall x : [x \in N \wedge 5 \le x \wedge x^5 \le 5^x] \implies (x+1)^5 \le 5^{(x+1)} \\
5 : x \in N \vdash_M (x+1)^5 \equiv x^5 + 5x^4 + 10x^3 + 10x^2 + 5x + 1 \\
6 : TH \vdash_I \forall x : [x \in N \wedge 5 \le x \wedge x^5 \le 5^x] \implies \\
\qquad\qquad x^5 + 5x^4 + 10x^3 + 10x^2 + 5x + 1 \le 5^{(x+1)} \\
7 : x \in N \vdash_M 5^{(x+1)} \equiv 5 * 5^x \\
8 : TH \vdash_I \forall x : [x \in N \wedge 5 \le x \wedge x^5 \le 5^x] \implies \\
\qquad\qquad x^5 + 5x^4 + 10x^3 + 10x^2 + 5x + 1 \le 5 * 5^x
\end{array}
$$

Circles represent object nodes, whose labels are reported in the table; rectangles represent link nodes, and contain their labels. The complex series of steps corresponding to the final phase of the proof are folded within the triangular REST node. Link nodes labelled with SIMPLIFY identify the points where the systems cooperate to the solution of the problem; namely, where Maple is invoked to expand some polynomial power. Note that the REST folded node hides away several additional Maple calls, meant to perform evaluations of disequalitites.

4   Conclusions

This paper described a framework for the uniform specification of symbolic computation engines and mechanized reasoning systems. The framework can be exploited to specify the integration between computer algebra systems and theorem provers, providing a clear formal description of the way they combine. As a result, the cooperation process between the tools is clearly represented within a symbolic mathematical structure.

Several research directions are open. First of all, the concrete implementation of the OMSCS specification and integration described in the example is in order. Its feasibility is hinted at by [3], which describes a practical example of the use of the framework. Regarding the interaction level of the framework, a standard, extendible set of services must be determined; their formal semantics must be defined, and given a concrete syntax. The work in [2] represents a first step in this direction. The studies undertaken by many research groups, e.g. the OpenMath group [12] and the PosSo/XDR group [1], must be considered as a starting point to the formalization of this architectural component.

References

1. ABBOTT, J. PossoXDR specifications. Internal Posso technical report, 1994.
2. ARMANDO, A., AND RANISE, S. From Integrated Reasoning Specialists to "Plug-and-Play" Reasoning Components. In *Proceedings of AISC'98*, Springer Verlag.
3. BERTOLI, P. *Using OMRS for real: a case study with* ACL2. PhD thesis, Computer Science Dept., University Rome 3, Rome, 1997. Forthcoming.
4. C. BALLARIN, K. HOMANN, J. C. Theorems and Algorithms: An interface between Isabelle and Maple. In *International Symposium on Symbolic and Agebraic Computation, ISSAC '95* (1995), H. M. Levelt, Ed., ACM Press, pp. 150–157.
5. CALMET, J., AND HOMANN, K. Structures for symbolic mathematical reasoning and computation. In *Proceedings of DISCO '96 - Design and Implementation of Symbolic Computation Systems* (1996), J. Calmet and C. Limongelli, Eds.
6. COGLIO, A. Definizione di un formalismo per la specifica delle strategie di inferenza dei sistemi di ragionamento meccanizzato e sua applicazione ad un sistema allo stato dell'arte. Master's thesis, University of Genoa, Italy, 1996.
7. E. CLARKE, X. Z. Analytica - a theorem prover in mathematica. In *Proc. of the 10th Conference on Automated Deduction* (1992), Springer-Verlag, pp. 761–765.
8. GIUNCHIGLIA, F., PECCHIARI, P., AND TALCOTT, C. Reasoning Theories: Towards an Architecture for Open Mechanized Reasoning Systems. Tech. Rep. 9409-15, IRST, Trento, Italy, 1994. Short version in Proc. of the First International Workshop on Frontiers of Combining Systems (FroCoS'96), Munich, Germany, 1996.
9. J. HARRISON, L. T. Extending the HOL prover with a Computer Algebra System to reason about the Reals. In *Higher Order Logic Theorem proving and its Applications*, J. J. Joyce and J. H. Seger, Eds. Springer-Verlag, 1993, pp. 174–184.
10. HOMANN, K.. *Symbolisches Loesen mathematischer Probleme durch Kooperation algorithmischer und logischer Systeme.* PhD thesis, Univ. of Karlsruhe, 1996.
11. PRAWITZ, D. *Natural Deduction: A Proof-theoretical Study.* , 1965.
12. VORKOETTER, S. OpenMath specifications: March 1994, March 1994.

COLETTE, Prototyping CSP Solvers Using a Rule-Based Language

Carlos Castro

LORIA, 615, rue du Jardin Botanique, BP 101, 54602 Villers-lès-Nancy Cedex, France
e-mail: Carlos.Castro@loria.fr

Abstract. We are interested in modelling constraint solving as a deduction process. In previous works we have described how a constraint solver can be viewed as a strategy that specifies the order of application of a set of rules, to transform a set of constraints into a solved form. In this framework, the computation process is associated to a constructive proof of a formula. On one hand, the use of transformation rules allows to prove termination in a very easy way. On the other, the use of strategies allows to describe constraint handling in a very abstract way, prototype new heuristics almost by modifying only the choice of rules, and combine constraint solver in the same theoretical framework. In this paper, we introduce COLETTE, an implementation of these ideas using ELAN, a rule-based language. We exemplify step by step how a flexible strategy language allows to prototype existing algorithms for solving CSPs. The theoretical and practical results show that this approach could be useful for better understand constraint solving.

Keywords: Constraint Satisfaction Problems, Rewriting Logic.

1 Introduction

The problem of finding values in the domain of variables such that a set of constraints is satisfied is called a Constraint Satisfaction Problem (CSP) [12]. Polynomial algorithms have been developed to eliminate values that the variables can take while preserving the set of solutions [11]. The incorporation of these algorithms into an exhaustive search process is the main reason for the success of CSP techniques. The need of distinction between basic transformations and their control has been well-recognised. Now, the challenge is to have an abstract way to control the basic transformations, which is easily modifiable by the user, and to be able to integrate several solvers.

Several languages and libraries are now available to deal with CSPs and they have been successfully applied for solving real-life problems. ILOG Solver [1] allows to deal with constraints through the use of a library of C functions. In the domain of Logic Programming, constraints were introduced giving origin to the Constraint Logic Programming paradigm: during execution of a traditional logic program, constraints are accumulated and solved externally by a solver that can be seen as a *black-box* [9]. Constraint Handling Rules were proposed, and included in the ECLiPSe system, to allow the users to define their own

Jacques Calmet and Jan Plaza (Eds.): AISC'98, LNAI 1476, pp. 107–119, 1998.
© Springer-Verlag Berlin Heidelberg 1998

propagation rules [8]. In this paper, we introduce COLETTE[1], a computational system [10], implemented in ELAN [4], for dealing with CSPs. The main originality of COLETTE with respect to the existing systems is the explicit use of strategies to guide the search process in a flexible way. Indeed, at our best knowledge, COLETTE is the first attempt to implement constraint solvers using a fully rewrite-based approach.

From a theoretical point of view, our original goal was to model constraint solving as an inference process, where a CSP is transformed into another, equivalent one, but more simple. In this way we model constraint solving as a deduction process where the computation process is associated to a constructive proof of a formula, the query, from a set of axioms [2]. In previous works, we have described constraint solving as an inference process [6,7]. We have verified how naturally a constraint solver can be viewed as a strategy that specifies the order of application of a set of rules, to transform a set of constraints into a solved form. From a practical point of view, our main motivation is to provide an environment for prototyping heuristics using the expressive power of strategies. Rewriting Logic and rule-based languages seem to be of greatest interest for implementing these techniques. This is the reason why we are interested in implementing CSP techniques within a language like ELAN . We have implemented local consistency (Arc-Consistency) and exhaustive search algorithms (*Backtracking, Forward-Checking, ...*) for solving constraints over Integers and Finite Domains.

At the beginning, we were just interested in a clean rule-based formalization of constraint solving without taking care about efficiency considerations. Using the interpreter of the first version of ELAN we were able to solve simple sets of constraints, for example, crypto arithmetic puzzles, like $SEND + MORE = MONEY$, using some thousands of rewrite steps in some seconds. However, we have been surprised that using the ELAN compiler, included in the latest version of the language, we have been able to solve random instances of hard combinatorial problems, like Job-Shop, that are traditionally used as benchmarks by the CSP community. These results are mainly due to the efficient compilation of leftmost-innermost normalization and non-determinism carried out by the ELAN compiler [15]. Techniques like many-to-one syntactic matching and reuse of parts of terms allow to apply up to ten millions of rewrite rules per second. Based on a typing of the strategies, smart inferences allow to eliminate choice points improving in this way memory management and, as a consequence, performance in terms of time. Considering the theoretical and practical results, we are convinced that this approach can be of greatest interest to better understand constraint solving.

This paper is organized as follows: in section 2, we briefly present CSPs. In section 3, we present a set of transformation rules that express the basic transformation carried out by existing algorithms used for solving CSPs. Section 4 describes in detail the system COLETTE: we first present the language used for implementing the system, we show the rewrite rules implementing the set of transformation rules, we then describe step by step how we can combine

---

[1] http://www.loria.fr/~castro/COLETTE/index.html.

these rules in order to built several constraint solvers, next we present the use of parametrized strategies to built solvers in a very flexible way, and finally we exemplify how we can integrate several solvers working together to solve a global problem. In section 5, we present some standard benchmarks solved using COLETTE. Finally, in section 6, we conclude the paper.

2 CSP

An *elementary constraint* $c^?$ is an atomic formula built on a denumerable set of variables and a first-order signature [2]. Elementary constraints can be combined with usual first-order connectives. A CSP is any set $C = (c_1^? \wedge \ldots \wedge c_n^?)$. We consider CSPs in which the carrier D of the structure is a finite set. A *solution* of C is a mapping α such that $\alpha(C)$ is true ($\mathbf{T}$). C is *satisfiable* if it has at least one solution. We use the constants $\mathbf{F}$ or **Unsatisfiable** to denote an unsatisfiable CSP. To process C we introduce two kinds of constraints: a relation $x \in^? D_x$, for a variable x and a non-empty set $D_x \subseteq D$, is called a *membership constraint*; a relation $x =^? v$, for a variable x and a value $v \in D$, is called an *equality constraint*.

When solving CSPs we can be interested in finding one solution, all solutions, or the best solution wrt some given criteria. Solving techniques can be described as an interleaving process between local consistency verification and domain splitting. Local consistency verification consists in the elimination of values that the variables can take while preserving the set of solutions. The effect of these eliminations is propagated through the set of constraints until a fixed point is reached: no more values can be eliminated from the domain of the variables. If we cannot derive directly a solution from the set of constraints so obtained, domain splitting is carried out: we choose a variable and split its domain creating in this way two independent subproblems. Then local consistency is verified for each subproblem and this process is repeated until a solution is reached. By iterating this process we obtain a search tree where nodes correspond to subproblems and branches to splitting decisions.

3 Transformation Rules for Solving CSPs

In order to simplify their processing, we only deal with CSPs in Conjunctive Normal Form (CNF.) That is, a CSP is any set $C = (c_1^? \wedge \ldots \wedge c_n^?)$ such that $c_i^?$ is a disjunction of elementary constraints. A *basic form* for a CSP P is any conjunction of formulae of the form:

$$C = \bigwedge_{i \in I} (x_i \in^? D_{x_i}) \wedge \bigwedge_{j \in J} (x_j =^? v_j) \wedge \bigwedge_{m \in M} (c_m^?) \wedge \bigwedge_{k \in K} \bigvee_{l=1,..,n_k} (c_{kl}^?)$$

---

[2] For clarity, constraints are syntactically distinguished from formulae by a question mark exponent on their predicate symbols.

equivalent to P, where $c_m^?$ and $c_{kl}^?$ are elementary constraints, and such that for each variable appearing in the elementary constraints there must be associated a membership constraint or an equality constraint, the sets of variables of the membership constraints and equality constraints are disjoint, and the set associated to each variable in the membership constraints is not empty. Variables which are only involved in equality constraints are called *solved variables* and the others *non-solved variables*. A *solved form* for a CSP P is a conjunction of formulae in basic form equivalent to P and such that all assignments satisfy all constraints.

In the rest of the paper we suppose that at the beginning of the constraint solving process we start with a CSP where a membership constraint has been created for each variable occuring in the set of constraints (the sets D_{x_i} are set up to D) and no equality constraint is present [3].

Given that we accept elementary constraints to be combined with all first-order connectives we need to transform the input set of constraints. This kind of preprocessing can be naturally specified as a normalization step: the following set of rewrite rules is applied eagerly on the input problem:

```
    c1 <-> c2 => (c1 -> c2) & (c2 -> c1)
    c1  -> c2 =>  NOT(c1) V c2
  NOT(NOT(c)) => c
 NOT(c1 & c2) => NOT(c1) V NOT(c2)
 NOT(c1 V c2) => NOT(c1) & NOT(c2)
c V (c1 & c2) => (c V c1) & (c V c2)
```

In [7], we propose the set of transformation rules presented in figure 1 that express the basic operations carried out by the algorithms developed for solving CSP.

[ArcConsistency] $x_i \in^? D_{x_i} \wedge c_m^? \wedge \mathcal{C} \Leftrightarrow x_i \in^? RD(x_i \in^? D_{x_i}, c_m^?) \wedge c_m^? \wedge C$
          if $RD(x_i \in^? D_{x_i}, c_m^?) \neq D_{x_i}$
[Falsity] $x \in^? \emptyset \wedge C$ $\Rightarrow F$
[Instantiation] $x \in^? D_x \wedge C$ $\Rightarrow x =^? v \wedge C$ if $D_x = \{v\}$
[Elimination] $x =^? v \wedge C$ $\Rightarrow x =^? v \wedge C\{x \mapsto v\}$ if $x \in Var(C)$
[SplitDomain] $x \in^? D_x \wedge C$ $\Rightarrow x \in^? D'_x \wedge C$ or $x \in^? D''_x \wedge C$
          if $D'_x \cup D''_x = D_x$ and $D'_x \cap D''_x = \emptyset$ and $D'_x \neq \emptyset$ and
                                                $D''_x \neq \emptyset$
[ChoicePoint] $\bigvee_{l=1,..,n_k}(c_{kl}^?) \wedge C \Rightarrow \bigvee_{l=1,..,l'}(c_{kl}^?) \wedge C$ or $\bigvee_{l=l'+1,..,n_k}(c_{kl}^?) \wedge C$

Fig. 1. Transformation Rules for solving CSP

Formally, $RD(x_i \in^? D_{x_i}, c_m^?)$ stands for the set $D'_{x_i} = \{v_i \in D_{x_i} \mid (\exists v_1 \in D_{x_1}, \ldots, v_{i-1} \in D_{x_{i-1}}, v_{i+1} \in D_{x_{i+1}}, \ldots, v_n \in D_{x_n}) : c_m^?(v_1, \ldots, v_i, \ldots, v_n)\}$ and $C\{x \mapsto v\}$ denotes the conjunction of constraints obtained from C by replacing all occurrenc es of the variable x by the value v. In [7], we prove termination, correction and completeness of this set of rules. For a satisfiable problem,

---

[3] By equality constraints we mean the constraints of the form $x_j =^? v_j$ presented in this section. This does not mean that elementary constraints involving an equality predicate cannot occur in the set of constraints.

the application of these rules gives each solution in a particular solved form: we only have an equality constraint associated to each variable, and all membership constraints and the sets of elementary and disjunctive constraints have been eliminated.

In the next section, we present the implementation of these rules using a rule-based programming language and we show how we can combine them in order to built a constraint solver.

4   COLETTE

In this section, we describe the system COLETTE. We first present the language used for implementing the system, we show the rewrite rules implementing the set of transformation rules presented in section 3, and we then describe step by step how we can combine these rules in order to built several constraint solvers. Next, we motivate how useful it can be to handle parametrized strategies in order to built solvers in a very flexible way, and finally we exemplify how we can integrate several solvers working together to solve a global problem.

4.1   A Rule-Based Programming Language

ELAN is an environment for prototyping computational systems [10]. A computational system is given by a signature providing the syntax, a set of conditional rewrite rules describing the deduction mechanism, and a strategy to guide the application of rewrite rules. Formally, this is the combination of a rewrite theory in rewriting logic [13], together with a notion of strategy to efficiently compute with given rewrite rules. Computation is exactly the application of rewrite rules on a term. The strategies describe the intended set of computations, or equivalently in rewriting logic, a subset of proofs terms. We now explain briefly ELAN syntax and semantics. We describe informally only the features of the language that we need to define our system. A full description of ELAN can be found in [4].

ELAN rules are of the form: $[\ell] : l \Rightarrow r$ **if** *cond* **where** $y := (s)t$, where ℓ is the rule label, l and r the respective left- and right-hand sides, *cond* the condition and y is replaced in l and r by the terms computed through the application of strategy s onto the term t. The constant rule **id** represents identity. For applying rules, ELAN provides several elementary strategy operators. The operator ';' expresses the concatenation of strategies. For instance, the strategy iterator $\mathbf{dk}(\ell_1, \ldots, \ell_n)$, standing for *don't know choose*, returns *all* possible results of the application of ℓ_i, and $\mathbf{first}(\ell_1, \ldots, \ell_n)$ takes the *first*, in textual order, successful branch. The strategy iterator $\mathbf{repeat}^*(s)$, iterates a strategy s until it fails and returns the terms resulting of the last unfailing call of the strategy.

4.2   Data Structure

The object representing a CSP and rewritten by the computation process is defined as a 5-tuple CSP[lmc,lec,EC,DC,store], where lmc is a list containing the membership constraints, lec is a list containing the equality constraints, EC is a list containing the set of elementary constraints to solve, DC is a list containing the set of disjunctive constraints to solve and `store` is a list containing the elementary constraints already verified.

4.3   Rewrite Rules

The rule **ArcConsistency** verifies local consistency based on the algorithm AC-3 [11]: the constraints in EC are verified and stored in `store`. Once a variable's domain is modified we extract all constraints in `store` where the variable is involved in and add them to EC. The operator `ReviseDxWRTc` carries out the verification of the set of values associated to each variable involved in the constraint c. For integer domains we have implemented the predicates $\leq$, $<$, $\geq$, $>$, $=$, and $\neq$. We also implemented predicates like *alldiff*, known as global constraints and widely used for improving constraint propagation and memory use. In fact, to extend our language we just have to define the propagation method for the new predicates, and, as we will see further, their evaluation as a ground formula.

```
[ArcConsistency]            // This is the name of the rule. When applying
CSP[lmc,lec,c.EC,DC,store]  // this rule, a term of the form CSP[,,,,] is
=>                          // replaced by the right-hand side of the rule.
CSP[append(lmc1,lmc2),lec,append(EC1,EC),DC,newstore]
where       [lmc1,lmc2,lv] := ()ReviseDxWRTc(lmc,lec,c)
where       [EC1,remainingstore] := ()GetConstraintsOnVar(lv,store,nil,nil)
choose  try     if ArityOfConstraint(c) == 1
                where newstore := ()remainingstore
        try     if ArityOfConstraint(c) >= 2
                where newstore := ()c.remainingstore
end
```

After application of rule `ArcConsistency` we could get two particular cases: a variable's domain is empty or is reduced to a singleton. The rule `Falsity` detects an empty domain and in that case rewrites the problem to *Unsatisfiable*. The rule `Instantiation` detects a singleton variable's domain, and in that case the membership constraint is deleted and a new equality constraint is added instantiating the variable to its only possible value.

```
[Falsity]                            [Instantiation]
CSP[x in? empty.lmc,lec,EC,DC,store] CSP[x in? D.lmc,lec,EC,DC,store]
=>                                   =>
Unsatisfiable                        CSP[lmc,x = v.lec,EC,DC,store]
                                     if GetCard(D) == 1
                                     where v:= ()GetFirstValueOfDomain(D)
```

Once a variable has been instantiated by creating an equality constraint we can propagate its effect: the rule `Elimination` replaces an instantiated variable by its value through the sets of constraints EC and DC. The rule `Elimination` reduces the arity of constraints where the variable is involved in: unary constraints will become ground formulae whose truth value has to be verified, binary constraints will become unary constraints which will be more easily tested, and so on. When evaluating ground formulae we use two logical rules: $C \wedge \mathbf{T} \Rightarrow C$ and $C \wedge \mathbf{F} \Rightarrow \mathbf{F}$. If one of the ground formulae is evaluated to $\mathbf{F}$ the rule Elimination will return $\mathbf{F}$, that is the CSP has not solution. Otherwise, when ground formulae are evaluated to $\mathbf{T}$ they are eliminated and the simplified set C is returned.

```
[Elimination]
CSP[lmc,x = v.lec,EC,DC,store]
=>
P
if      occurs x in EC or occurs x in DC
where   [l1,l2] := ()GetRemainingConstraintsInEC(x = v,EC,nil,nil)
choose  try if l1 == false.nil
              where P := ()Unsatisfiable
        try if l1 != false.nil
              where [l3,l4,l5]:=()GetRemainingConstraintsInDC(x = v,
                               DC,nil,nil,nil)
              choose try if l3 == false.nil
                         where P := ()Unsatisfiable
                     try if l3 != false.nil
                         where lu := ()append(l1,l3)
                         where lb := ()append(l2,l4)
                         where P:=()CSP[lmc,ShiftConstraint(lec,x = v),
                                    append(lu,lb),l5,store]
              end
end
```

```
[InstantiateFirstValueOfDomain]          [EliminateFirstValueOfDomain]
CSP[x in? D.lmc,lec,EC,DC,store]         CSP[x in? D.lmc,lec,EC,DC,store]
=>                                       =>
CSP[lmc,x = v.lec,EC,DC,store]           CSP[x in? RD.lmc,lec,EC,DC,store]
if   D != empty                          if   GetCard(D) > 1
where v := ()GetFirstValueOfDomain(D)    RD:= ()DeleteFirstValueOfDomain(D)
```

To deal with the non-determinism of the rule **SplitDomain** we decompose it in the rules InstantiateFirstValueOfDomain and EliminateFirstValueOf Domain. The rule InstantiateFirstValueOfDomain creates an equality constraint whose value is the first one in the variable's domain and delete the membership constraint. The rule EliminateFirstValueOfDomain reduces the variable's domain by deleting the first value. Of course, this is a particular instance of domain splitting. We have also implemented rules for enumerating values *from last to first* and for splitting where each new domain has at least two values.

To deal with the non-determinism of the rule **ChoicePoint** we use an associative-commutative operator V and the rule PostDisjunct. This rule extracts a disjunct corresponding to an elementary constraint and add it to the list of elementary constraints EC. When applying this rule with the *don't know choose* operator we get all elementary constraints involved in the disjunctive constraint.

```
[PostDisjunct]
CSP[lmc,lec,EC,c1 V c2.DC,store]
=>
CSP[lmc,lec,c1.EC,DC,store]
if ConstraintIsElementary(c1)
```

4.4   Strategies

The simplest heuristics for solving CSPs is the force brute algorithm *Generate and Test*. The strategy GenerateTest implements this exhaustive search. The heuristics *Backtracking* aims to detect failures earlier: it carries out constraint checking immediately after each variable instantiation. By modifying a little the order of application of the rules in GenerateTest we obtain the new strategy Backtracking.

```
[] GenerateTest =>                       // This is the name of the
repeat*(dk (InstantiateFirstValueOfDomain,// strategy. Each application
         EliminateFirstValueOfDomain));// of dk returns two new
repeat*(Elimination)                     // problems, each one is an
                                         // input for the next iteration.
[] Backtracking =>
repeat* (dk (InstantiateFirstValueOfDomain;  first one (Elimination , id),
         EliminateFirstValueOfDomain))
```

The strategies GenerateTest and Backtracking make *a priori* choices too early. A well known general principle when dealing with combinatorial search problems is to carry out deterministic computations as much as possible before branching. This is the key idea behind local consistency verification. With this idea in mind we design the strategy LocalConsistencyForEC that verifies local consistency for a set of elementary constraints. After reaching a fixed point applying the strategy LocalConsistencyForEC we have to carry out an enumeration in order to traverse the search space. When after each variable instantiation we verify again local consistency for all elementary constraints we are implementing the heuristics *Full Lookahead*. With a simple modification to the strategy Backtracking, a new strategy FullLookAheadForEC can be easily designed.

The strategy FullLookAheadForEC only deals with elementary constraints. We can modify the strategy LocalConsistencyForEC to take into account disjunctive constraints. We can use the widely used *Choice Point* approach: we choose a disjunctive constraint, we extracts a disjunct, and we post it creating in this way a new subproblem for each disjunct. The strategy LocalConsistency ForEC&DC implements this idea, and again, it is designed by a simple modification to an already existing strategy. The implementation of the rule PostDisjunct

```
[] LocalConsistencyForEC =>
repeat* (ArcConsistency ; first one (Instantiation ;
                                     first one (Elimination , id) ,
                                     Falsity ,
                                     id))


[] FullLookAheadForEC =>
LocalConsistencyForEC ; repeat* (dk (InstantiateFirstValueOfDomain ;
                                     first one (Elimination , id) ;
                                     LocalConsistencyForEC ,
                                     EliminateFirstValueOfDomain))
```

and its use in this strategy shows the elegance of being able to use associative-commutative operators. Indeed, calling PostDisjunct with the *don't know operator* will create just one choice point improving in this way memory management and, as a consequence, performance in terms of time. Now we can easily design a strategy implementing the heuristics *Full Lookahead* considering also disjunctive constraints, the results is the strategy FullLookAheadForEC&DC.

```
[] LocalConsistencyForEC&DC =>
LocalConsistencyForEC; repeat*(dk (PostDisjunct); LocalConsistencyForEC)

[] FullLookAheadForEC&DC =>
LocalConsistencyForEC&DC ; repeat* (dk (InstantiateFirstValueOfDomain ;
                                        first one (Elimination , id) ;
                                        LocalConsistencyForEC ,
                                        EliminateFirstValueOfDomain))
```

4.5   Strategies with Parameters

The art of constraint solving consists in interleaving local consistency verification and domain splitting in a way that a small part of the search tree is explored before reaching a solution. A criteria for guiding the search can be easily incorporated in the strategy FullLookAheadForEC&DC: before splitting a domain we choose the variable that has the minimum set of remaining values.

```
FullLookAheadForEC&DC =>
LocalConsistencyForEC&DC ; repeat* (GetVarWithMinimumDomain ;
                                    dk (InstantiateValue ;
                                    first (Elimination , id) ;
                                    LocalConsistencyForEC ,
                                    EliminateValue))
```

Obviously, the criteria presented here, *Minimum Domain*, an instance of a general principle known as *First Fail Principle*, is not the only one that we can imagine. *Minimal Width Ordering* and *Maximum Cardinality Ordering* [18], are examples of other principles used to choose the splitting variable. This shows the importance of being able to tune strategies in a very flexible way [3]. The following strategy applied with the parameter `GetVarWithMinimumDomain` is equivalent to the previous one:

```
stratop
    global  FullLookAhead(@): (<csp -> csp>)  <csp -> csp>;
end
strategies for csp
    RuleChoiceVar: <csp -> csp>;
[] FullLookAhead(RuleChoiceVar) =>
LocalConsistencyForEC ; repeat* (RuleChoiceVar ;
                                dk (InstantiateValue ;
                                first (Elimination , id) ;
                                LocalConsistencyForEC ,
                                EliminateValue))
```

4.6   Interface with Other Solvers

As pointed out in [17], sometimes one would like to use solvers already designed and implemented for solving specific constraints. As a very simple example we could consider the following case: suppose that a CSP can be decomposed in several subproblems with disjoint sets of variables. In that case, we could deal with each subproblem independently using, for example, the strategy FullLookAheadForEC&DC that we have already presented. Once each subproblem has been solved we could built a global solution as the union of the solutions of each subproblem. Next, we briefly present how we can built such a solver using a rule based approach.

A 5-tuple CSP[P,luCSP,lsCSP,laPID,liPID] is the object rewritten by the main solver, where P is the input CSP, luCSP is a list containing all unsolved CSPs, lsCSP is a list containing all solved CSPs, laPID is a list containing the identification of active processes (running solvers) and liPID is a list containing the idle process. The main solver is defined by the following strategy:

```
[] ParallelResolution =>
DecomposeCSP ;
CreateSolvers ;
repeat* (SendToSolver) ;
repeat* (GetFirstOutputOfSolver) ;
repeat* (ComposeCSP) ;
repeat* (CloseProcess)
```

The rule DecomposeCSP creates luCSP, a list of variable-disjoint unsolved subproblems obtained from the original CSP P. The rule CreateSolvers creates n idle processes, each one containing a solver as specified by the strategy FullLookAheadForEC&DC, where n is the size of the list luCSP. The rule SendToSolver takes an unsolved problem from the list luCSP and an idle solver from the list liPID creating an active solver in the list laPID. The rule GetFirst

OutputOfSolver reads a solution from the list of active solvers laPID and put it in the list of solved CSPs lsCSP. The rules ComposeCSP and CloseProcess built a solution to the original problem and close all process, respectively.

```
[CreateSolvers]
CSPII[P,luCsp,lsCsp,laPid,liPid]
=>
CSPII[P,luCsp,lsCsp,laPid,append(liPid,newliPid)]
where   n       := ()size(luCsp)
where   newliPid := ()CreateProcess(n)
```

```
[SendToSolver]                          [GetFirstOutputOfSolver]
CSPII[P,csp.luCsp,lsCsp,laPid,pid.liPid]  CSPII[P,luCsp,lsCsp,
                                                  pid.laPid,liPid]
=>                                      =>
CSPII[P,luCsp,lsCsp,pid.laPid,liPid]
CSPII[P,luCsp,csp.lsCsp,laPid,pid.liPid]
where   csp1    := ()write(pid,csp)     where   csp := ()read(pid)
```

5 Examples

The popularity of constraint solving techniques is mainly due to their succesful application for dealing with real-life problems, like industrial scheduling. In this section, we just concentrate on some classical benchmarks widely used by the CSP community. Given that our goal is to introduce a general framework to deal with CSPs, the analysis of different models for representing the problems and tuning of heuristics for solving specific problems are beyond the scope of this paper. It is well known that models considering redundant and global constraints (like *alldiff*), can improve constraint propagation, however all the examples presented here were solved with a very naive strategy: *Full Lookahead*, choosing the enumeration variable based on the *Minimum Domain* criteria, and enumerating the values *from first to last*. In figure 2, columns I to VI contain, for each problem, the number of variables, the number of constraints, the time (in seconds) to get the first solution, the number of rewrite steps for the first solution, the time to get all the solutions (and also the total number of solutions), and the number of steps for all solutions, respectively.

With respect to the performances presented in [19], where specific elaborated models and heuristics are used to solve these problems, we require between 60 and 70 times its computation times. Taking into account that we use very naive models and search strategies, we think that these performances are indeed very good. We have also solved random instances of job-shop problems: using very simple models and optimization strategies we are able to solve up to 6×6 problems. We can obtain in few seconds the first solution of 10×10 problems, however, optimization is not possible in a reasonable time. For this kind of problems the need for more ellaborated models and smart search strategies is really important. Considering the average performance of about 500,000 rewrite steps per second, and the stability of the computation process (this average is

Problem	I	II	III	IV	V	VI
CROSS+ROADS=DANGER	14	50	0.10	27,092	0.28 (1)	169,649
DONALD+GERALD= ROBERT	15	59	0.20	90,250	0.20 (1)	92,408
LIONNE+TIGRE=TIGRON	13	41	0.09	45,687	0.14 (1)	82,151
SEND+MORE=MONEY	12	39	0.08	36,934	0.09 (1)	40,478
8 queens	8	84	1.00	480,226	12.00 (92)	7,607,120
9 queens	9	108	1.06	223,275	59.00 (352)	34,279,650
10 queens	10	135	3.10	100,557	317.84 (724)	154,172,782
11 queens	11	165	6.21	1,261,591	1,525.77 (2680)	746,934,337
12 queens	12	198	5.22	865,741	9,713.19 (14200)	3,913,842,503

Fig. 2. Some classical benchmarks

the same for problems requiring from some seconds up to some hours) we think that we could go towards more complex problems.

6 Conclusions

We have presented COLETTE, a computational system for solving CSPs. In this framework, a constraint solver is viewed as a strategy that specifies the order of application of a set of rewrite rules in order to reach a set of constraints in a solved form. Transformation rules are a natural way to express operations on constraints, and ELAN indeed allows one to control the application of the rules using a flexible strategy language. Standard algorithms based on *Chronological Backtracking* can be naturally implemented, and one can realize how easy is to prototype new heuristics using the powerful of strategies. However, it is difficult to implement heuristics like *Intelligent Backtracking* [5] and *Conflict-Directed Backjumping* [16]: the basic strategy operator, *don't know choose*, does not allow to jump in the search tree. We have also shown how, in the same framework, we can built constraint solvers that collaborate in order to solve a global problem. Using the terminology of the constraint solving community, this is an attempt to propose a general framework for collaboration of solvers [14], however, much more work has to be done in that direction.

References

1. *ILOG Solver 4.0, User's Manual*, May 1997.
2. K. R. Apt. A Proof Theoretic View of Constraint Programming. *Fundamenta Informaticae*, 1998. To appear.
3. P. Borovanský, C. Kirchner, and H. Kirchner. Rewriting as a Unified Specification Tool for Logic and Control: The ELAN Language. In *Proceedings of The International Workshop on Theory and Practice of Algebraic Specifications, ASF+SDF'97*, Workshops in Computing, Amsterdam, September 1997. Springer-Verlag.
4. P. Borovanský, C. Kirchner, H. Kirchner, P.-E. Moreau, and M. Vittek. *ELAN Version 3.00 User Manual*. CRIN & INRIA Lorraine, Nancy, France, first edition, January 1998.

5. M. Bruynooghe. Solving Combinatorial Search Problems by Intelligent Backtracking. *Information Processing Letters*, 12(1):36–39, 1981.

6. C. Castro. Solving Binary CSP Using Computational Systems. In J. Meseguer, editor, *Proceedings of The First International Workshop on Rewriting Logic and its Applications, RWLW'96*, volume 4, pages 245–264, Asilomar, Pacific Grove, CA, USA, September 1996. Electronic Notes in Theoretical Computer Science.

7. C. Castro. Building Constraint Satisfaction Problem Solvers Using Rewrite Rules and Strategies. *Fundamenta Informaticae*, 1998. To appear.

8. T. Frühwirth. Constraint Handling Rules. In A. Podelski, editor, *Constraint Programming: Basic and Trends. Selected Papers of the 22nd Spring School in Theoretical Computer Sciences*, volume 910 of *Lecture Notes in Computer Science*, pages 90–107. Springer-Verlag, Châtillon-sur-Seine, France, May 1994.

9. J. Jaffar and M. J. Maher. Constraint Logic Programming: A Survey. *Journal of Logic Programming*, 19(20):503–581, 1994.

10. C. Kirchner, H. Kirchner, and M. Vittek. Designing Constraint Logic Programming Languages using Computational Systems. In P. V. Hentenryck and V. Saraswat, editors, *Principles and Practice of Constraint Programming. The Newport Papers*, pages 131–158. The MIT press, 1995.

11. A. K. Mackworth. Consistency in Networks of Relations. *Artificial Intelligence*, 8:99–118, 1977.

12. A. K. Mackworth. Constraint Satisfaction. In S. C. Shapiro, editor, *Encyclopedia of Artificial Intelligence*, volume 1. Addison-Wesley Publishing Company, 1992. Second Edition.

13. J. Meseguer. Conditional rewriting logic as a unified model of concurrency. *Theoretical Computer Science*, 96(1):73–155, 1992.

14. E. Monfroy. *Collaboration de solveurs pour la programmation logique à contraintes.* PhD thesis, Université Henri Poincaré - Nancy 1, November 1996. Also available in english.

15. P.-E. Moreau and H. Kirchner. Compilation Techniques for Associative-Commutative Normalisation. In *Proceedings of The International Workshop on Theory and Practice of Algebraic Specifications, ASF+SDF'97*, Amsterdam, September 1997. Technical report CRIN 97-R-129.

16. P. Prosser. Domain filtering can degrade intelligent backtracking search. In *Proceedings of The 13th International Joint Conference on Artificial Intelligence, IJCAI'93, Chambéry, France*, pages 262–267, August 29 - September 3 1993.

17. G. Smolka. Problem Solving with Constraints and Programming. *ACM Computing Surveys*, 28(4es), December 1996. Electronic Section.

18. E. Tsang. *Foundations of Constraint Satisfaction.* Academic Press, 1993.

19. J. Zhou. *Calcul de plus petits produits cartésiens d'intervalles: application au problème d'ordonnancement d'atelier.* PhD thesis, Université de la Mediterranée, Aix-Marseille II, March 1997.

An Evolutionary Algorithm for Welding Task Sequence Ordering

Martin Damsbo[1] and Peder Thusgaard Ruhoff[2]

[1] AMROSE A/S, Forskerparken 10, DK-5230 Odense M, Denmark
emailmez@amrose.spo.dk
[2] The Maersk Mc-Kinney Moller Institute for Production Technology,
Odense University, Campusvej 55, DK-5230 Odense M, Denmark
ptr@mip.ou.dk

Abstract. In this paper, we present some of the results of an ongoing research project, which aims at investigating the use of the evolutionary computation paradigm for real world problem solving in an industrial environment. One of the problems targeted in the investigation is that of job sequence optimization for welding robots operating in a shipyard. This is an NP-hard combinatorial optimization problem with constraints. To solve the problem, we propose a hybrid genetic algorithm incorporating domain-specific knowledge. We demonstrate how the method is successful in solving the job sequencing problem. The effectiveness and usefulness of the algorithm is further exemplified by the fact, that it has been implemented in the RoboCopp application program, which is currently used as the task sequence scheduler in a commercially available robot programming environment.

1   Introduction

The welding of large structures using welding robots is a complicated problem. An example of such a large structure is the ship section from Odense Steel Shipyard[1] shown in Fig. 1. The primary reason for these complications is, that propagation of expanses in the structure can lead to unacceptable inaccuracies in the final product. In order to avoid these problems, one must be very careful in executing the individual welding jobs and ensure that they are only executed in accordance with what we call welding expert knowledge. Basically, this means that the welding jobs must be done in a sequence, which adhere to a set of constraints, where the constraints are derived from the welding expert knowledge. To find such a sequence is not a difficult task. However, the problem of finding a sequence, which minimize the time needed to complete the welding operations is an altogether different matter. It is not difficult to realize that finding this sequence is similar to solving a constrained sequential ordering problem; an NP-hard combinatorial optimization problem. Finding such a minimal sequence is clearly important in optimizing the overall performance of the construction yard.

---

[1] Odense Steel Shipyard is mainly producing huge containerships and supertankers. It is recognized throughout the world as one of the most modern and efficient shipyards.

Jacques Calmet and Jan Plaza (Eds.): AISC'98, LNAI 1476, pp. 120–131, 1998.
© Springer-Verlag Berlin Heidelberg 1998

Fig. 1. Overview of a typical shipsection to be welded using welding robots.

Several methods have been developed to solve this type of combinatorial problems, e.g. simulated annealing, tabu search and genetic algorithms. For an overview of these and other methods in combinatorial optimization consult the book by Reeves [13]. Of course, there is no guarantee that these methods are able to determine the optimal soultion. Nevertheless, we expect the methods to be able to find good solutions at a reasonable computational cost.

The purpose of the present contribution is not to compare the various methods. Instead we will show how one can improve a standard genetic algorithm by incorporating problem-specific information and use it in an industrial application. The application of genetic algorithms to problems in industry is also discussed by Stidsen [16].

The remainder of the paper is organized as follows. First, we present the evolutionary computation paradigm. Then, in Sect. 3, we give a detailed description of the welding task sequence ordering problem. In Sect. 4, we outline the evolutionary algorithm. Sect. 5 is devoted to applications of the developed algorithm. In Sect. 6 we show how one can easily modify the algorithm to give a task room division. Finally, we make some concluding remarks and give some directions for future work.

2   The Evolutionary Computation Paradigm

One has only to observe the progress and diversity of nature to fully appreciate the incredible force of evolution. It is not just a careful pruning of weaker individuals in populations and 'survival of the fittest'. Co-evolving species, biological arms races, host versus parasite wars, symbiotic relationships and emerging cooperative and altruistic behavior are but a few of the facets that make up our natural world. All interacting in extremely complex ways to create, sustain and diversify the entire biosphere on earth. Natural evolution and selection operates on many levels, from genetic effects to full phenotype interaction, involving emerging properties on innumerable levels of abstraction. On a macroscopic scale it is *the* most important force of nature; molding life and the environment through time.

Motivated by the apparent success and progression of the evolutionary paradigm in nature, a number of computational methods using the concepts of evolution have emerged over time. They are collectively known as evolutionary algortihms and include genetic algorithms, evolutionary programming, evolutionary strategies. The term evolutionary computation describes any computer-based problem solving system, which uses computational models of the mechanisms of the evolutionary process, as key elements in their design and implementation. These mechanisms include maintaining a pool of candidate solution structures, which evolve according to genetic operators and selection. Evolutionary computation methods operate on populations of several solution representations simultaneously. Through generations of evolving solutions, individual representations are modified (mutated), and subjected to fitness evaluation and selection pressure. Most of these methods also employ recombination of existing candidates, to encourage development and combination of partial solutions. Evolutionary computation methods rely on these mutation and recombination operators to gradually produce more and more optimal solutions, when subjected to the desired fitness selection criteria.

John Holland was one of the first to experiment with evolutionary algorithms during the 1960s, and he established the genetic algorithm theory formally in 1975 [10]. This included the renowned *schema theorem*, which stated, that partial solution building blocks (schemas) would propagate according to how well they performed in a genetic algorithm framework. Until recently virtually all subsequent theoretical work on genetic algorithms has been based on this notion of schemas. Some new approaches to establishing a theoretical foundation for evolutionary algortihms have emerged, with the statistical mechanics view by Prügel-Bennett and Shapiro [14] as one of the most promising.

The initial motivation was to create an artificial model to study adaptation phenomena occurring in nature, but genetic algortihms were soon put to use as general optimization algorithms. The original design has since been developed and extended by many researchers. Good introductions to the field are Goldberg [9] and Mitchell [12].

Evolutionary programming in a sense takes genetic algorithms a step further, by not operating directly on solution representations. This approach was introduced by Fogel *et al.* [7], and has since been applied to a broad range of problems. Evolutionary programming methods distinguish between the genotype and phenotype of a particular solution, and evaluate genotypes only indirectly, by rating the phenotypes. Genetic algorithms on the other hand modify and evaluate the solution representations directly. The direct encoding approach has so far proven the most successful, but evolutionary programming methods are credited with the potential to do at least equally well, denoting a concept that is closer to natural design.

In recent years the boundaries between evolutionary algorithm categories have broken down to some extent. In fact, the way evolutionary algorithms have evolved and developed into specialized sub-disciplines, in many ways resemble the very natural concepts, that the algorithms seek to mimic. Hybridization of known algorithms to achieve superior performance has fueled this development. As already mentioned, the work presented here focuses mainly on the use of a hybrid genetic algorithm.

For a comprehensive introduction to the ongoing research within the field of genetic algorithms, one can consult the conference proceedings [15,18,19,3,2,8,6].

3   The Welding Task Sequence Ordering Problem

The basic representation of a welding project is a CAD 3D-object model and accompanying tasks with interelated dependencies. These data are processed and transformed into the resulting robot control programs, through the use of several specialized application modules. Each takes the output from previous modules and creates the input for subsequent processing in a batch execution as illustrated in the data processing flow diagram of Fig. 2.

A welding task consists of a number of jobs which are described by their spatial extension. The jobs must all be executed in sequence and in accordance with a set of specified dependency conditions.

There are two different types of jobs in the welding tasks considered here

1. *Regular welding jobs* - defined as lines in space with two fixed end points.
2. *Sense jobs* - defined as a point in space, and used to dynamically measure and correct any inconsistencies between the real world and the internal model of the environment used during the execution of the welding robot program.

The primary objective is to minimize the path between welding jobs in a route, where all jobs are executed. Minimum path means minimum time to complete the welding operations. The actual welding length is not included in the path as it remains constant regardless of different job sequences.

The optimization problem is further complicated by the fact that various types of dependencies or constraints are imposed on the welding task. They come in three varieties

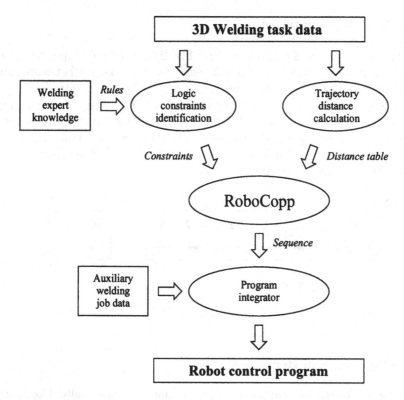

Fig. 2. Schematic overview of the robot programming process. RoboCopp is the welding task scheduler application program developed on the basis of the evolutionary algorithm outlined in Sect. 4.

1. *Order* - some welding jobs must be performed before others to prevent expanses from being displaced. Likewise, some sense jobs must precede certain weld jobs to ensure a sufficient accuracy of the welding process.
2. *Directional* - some jobs (primarily vertical welding seams) need to be welded in a certain direction to ensure the quality of the welding process.
3. *Adjoining* - some welding jobs must be executed in immediate succession of each other. Again, this constraint is necessary in order to avoid gaps and inaccuracies in the finished product.

The above mentioned dependencies are *hard* constrains. Thus, a solution which do not adhere to them is unsuable and is termed illegal.

4   A Hybrid Genetic Algorithm for Task Sequence Ordering

In the follwoing we give a detailed account of the hybrid genetic algorithm which form the basis for the RoboCopp application program shown in Fig. 2.

4.1 Data Representation

The distance tables and constraints (see, Fig. 2) are the only input the genetic algorithm uses to realize the problem space. It makes no further assumptions regarding model or type of problem, except for the heuristics introduced.

The genotype consists of an ordered list of job identifiers (task no.) representing the sequence in which jobs should be executed. Information about the direction of each weld job is also included. The representation is shown in Fig. 3. Every genotype started out as a random permutation of job numbers, i.e. each job number appeared only once in the list.

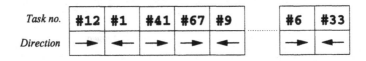

Fig. 3. Representation of candidate solution.

4.2 Crossover and Heuristics

The crossover operation is applied to new specimens stochastically. The operator intends to combine characteristics from both parents in an effort to produce even better solutions from previous partially good solutions. The conventional binary encoded crossover operator does not work well for the type of constrained combinatorial optimization problems considered here. Several alternative approaches have been proposed. One of the most promising is the genetic edge recombination method introduced by Whitley *et al.* [17]. In the algorithm presented here an enhanced version of their method is developed. The crossover algorithm reads

1. Create an edge recombination matrix with information about job connections and directions in the parents.
2. Choose the first job: Either the first job of parent 1, the first job of parent 2 or randomly chosen in a manner depending upon directional and adjoining constraints (see, discussion of heuristics below).
3. Remove the selected job from the edge recombination matrix.
4. If both parents have the same job connected to the previously selected job, then pick this as the next job in the sequence.
5. Otherwise select the job connected to the previously selected job which has the least further edge connections. Ties are broken randomly. If there are no jobs left connected to the newly selected, then pick one at random from the remaining jobs.
6. If the whole sequence has been constructed then stop else go to 3.

The following heuristics are used to enhance the quality of solutions

1. Jobs with adjoining dependencies are treated as one job cluster, i.e. if an adjoining constrained job is selected, then the rest of the adjoining jobs will be selected to follow. This will always be the optimum choice, as no constraints are broken and the path length between them is zero.
2. Jobs with a directional constraint are always inserted in that direction. If unconstrained, they are inserted according to their parents.
3. If adjoining jobs are directionallly constrained, they are inserted in the only legal way.

The outlined crossover algorithm ensures that created specimens contain information about the sequential structure of their parents, propagating favorable sub-paths to the next generations.

4.3   Mutation

Following the crossover operation the specimens in the population can be mutated. The algorithm distinguish between two different forms of mutation. A minor mutation which modify the arrangement of two jobs in the sequence and a major mutation which can involve any number of jobs. Following is a more detailed description of the mutation operators and their relative frequencies

1. Minor mutation
   - 50%: two single random jos are exchanged.
   - 25%: two single adjacent jobs are interchanged.
   - 25%: reversal of execution direction for a single job.
2. Major mutation
   - 80%: a selected sub-path is inserted elsewhere in the sequence.
   - 20%: a selected sup-path is reversed.

The mutational operators were designed to maximize the number of possible favorable sequence modifications, that is creating shorter paths and obeying constraints. The values of the relative frequencies are determined experimentally.

4.4   Fitness Criteria and Selection

The evolving candidate solutions are allowed to break the three types of dependency constraints introduced in Sect. 3. Of course, any legal solution found in the end must not contain any dependency errors. To achieve this we penalize dependency errors through a penalty term in the objectve function. Illegal candidate solutions are then expected to die out. Including the penalty term, the fitness or objective function is

$$F(s) = -\left(L(s) + N_e(s)L_{\max}\right),$$

where s is a given sequence, $L(s)$ is the path length of s, $N_e(s)$ is the number of dependency errors in s and L_{max} is a penalty length. $N_e(s)L_{max}$ is the penalty term.

Another approach to avoid the above problem is to construct crossover and mutation operators that only produce solutions without dependency errors. However, not only are they much harder to implement, they also impose restrictions in the state space search.

To find the appropriate parents for the next generation, individuals are ranked according to fitness. The expected number of offspring is calculated by assinging each specimen an expected fertility value. This value is a linear function of the rank. Selection is implemented primarily by stochastic universal sampling [1]. It should be pointed out that 90 % of the entire new generation of parents were selected using this method. The remaining 10 % were chosen by generic roulette wheel selection. This resembles Boltzmann selection and gives every specimen an appropiate chance of reproducing itself. It was introduced to further diversify the population.

5   An Example

The ship section shown in Fig. 1 is a typical example of a welding project. It consists of 288 weld jobs and 140 sense jobs, giving a total of 428 jobs constrained by 849 dependencies. Of these, the 550 is order dependencies, 168 specified adjoining weld jobs, and 131 restricted the direction in which a job should be welded. For a problem this size, initially randomly created solutions had an average of 510 unsatisfied dependencies and a path length of 1430 for the population average, and 430 errors and a path length of 1390 for the best specimen.

In Fig. 4 and Fig. 5 we see the typical development scenario for a genetic algorithm population. Note that Fig. 5 is a magnification of the lower part of Fig. 4, showing more clearly the details of the path length and dependency error graphs. Curves for population average and best specimen converge almost indentically, with the population average displaced some percentages above the best specimen. Solutions with no dependency errors (best specimen) are found relatively fast, but the average population specimens retain a certain amount of unfulfilled dependencies. The best total fitness graph converges as a displacement of the population average fitness, until it coincides with the best path length curve, when legal solutions are found.

It should be stressed that the parameter settings used in the hybrid genetic algorithm was determined through massive test runs on the departments SGI Onyx with 24 MIPS R4400 processors. In fact, the accumulated computing power expended in test and research, amounted to more than 5 years of CPU time on a single MIPS R4400 processor. However, we will not in this paper go into a detailed treatment of the parameter optimization. A detailed account of this process can be found in Damsbo [4].

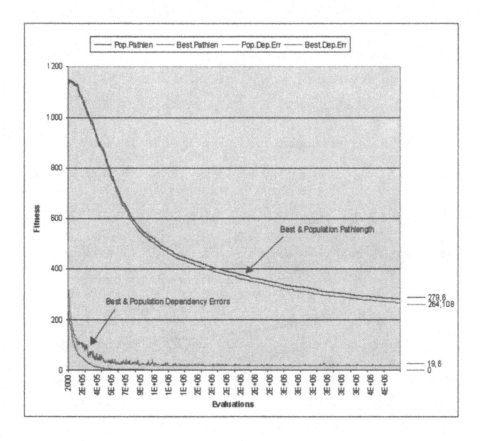

Fig. 4. Typical development of a genetic algorithm population in terms of path length, dependency errors and total fitness.

6   Task Room Division

The sequential ordering of welding jobs, is just one phase in the development of robot control programs. One of the other tasks that participates in this process (see, Fig. 2) is a division of jobs into separate working rooms. Jobs in the same room were supposed to be appropiately related, i.e. rooms reflected the most feasible grouping of the welding sequence in terms of execution flexibility. Furthermore, no dependencies could exist between jobs inhabiting different rooms, making them constraint independent clusters of jobs as well. With rooms divided this way, operators could easily reschedule and rearrange job sections, if the the original welding sequence ran into problems during execution. The task of grouping jobs were originally carried out using heuristics, which considered the location and orientation of jobs.

The room division ensuing from the heuristics approach is not always satisfactory, making alternative implementations interesting. A straightforward extension to the evolutionary algortihm allows one to incorporate the room division

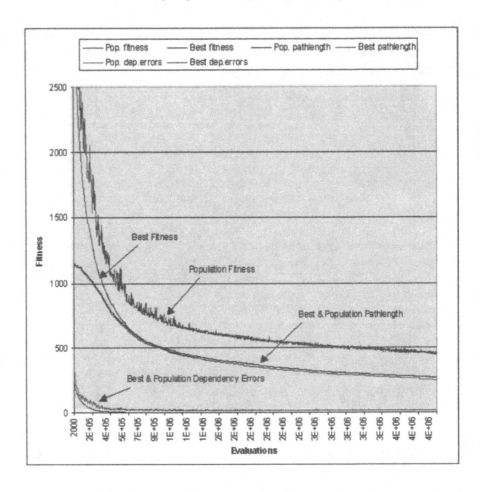

Fig. 5. Typical development of a genetic algorithm population in terms of path length, dependency errors for average and best specimen.

task in the sequence ordering optimization. Analogous to the penality term, a room reward is introduced in the fitness evaluation. Rather than reducing fitness due to unfulfilled constraints, it adds a certain fitness bonus proportionally to the number of rooms in the solutions. Rooms are identified as constraint independent clusters of jobs in each objective function evaluation.

One incentive to the above addition is to determine whether a fitness reward has the same effect as a punishment. The two optimization objectives skew the distance based fitness landscape in different ways, and they are associated with different characteristics in solutions. Clearly, there is no guarantee that the two can operate effectively in parallel and extract their respective solution traits, without disturbing each other. This is a recurring problem in multi-objective optimization, involving the extent to which two or more objectives can be independently achieved. In this particular case, dependency errors must be weeded out before the room division can begin. This is due to the fact, that the definition of rooms assume error-free candidate solutions.

Usiong the above approach, the algortihm can produce well-functioning room divisions. However, for large problems some manual editing may be needed afterwards. Nevertheless, it is considered an effective tool in the robot programming process.

7   Conclusion

Evolutionary algorithms appears fit to apply to combinatorial optimization problems. They are relatively easy to implement, fine-tune and deploy. It is also possible to utilize the same algorithm core dynamics for different applications of genetic algorithm optimization. Only solution representation, objective function and recombination and mutation operators need to be customized for each problem type. Furthermore, problem specific information can be incorporated into the optimization process, to take advantage of the characteristics of the problem space and enhance algorithm performance.

The successful results of the hybrid evolutionary algorithm presented in this paper seems to support the general notions above. The genetic algorithm consistently finds feasible, error-free solutions with competitively short sequence paths. Moreover, the task room division described in Sect. 6 clearly demonstrates the flexibilty of the algorithm. Of course, this flexibilty is very important in real world industrial applications where sudden changes in the production environment may lead to changes in the problem-specific assumptions. The effectiveness and usefulness of the outlined approach which has been implemented in the RoboCopp application program is further exemplified by the fact that it is currently used as the task sequence scheduler in a commercially available robot programming environment.

Encouraged by the immediate success of the evolutionary approach, we have also developed a distributed parallel version of the algorithm displaying superior performance in accordance with [5,11]. Details about the parallel version can be found in Damsbo [4]. In the future, we will try to further enhance the performance of the algorithm by incorporating new strategies for exchange of information between sub-populations.

References

1. Baker, J. E.: Reducing Bias and Inefficiency in the Selection Algorithm. In Grefenstette, J. J. (Ed.): *Proceedings of the 2nd International Conference on Genetic Algorithms.* 14-21, Erlbaum, 1987.
2. Belew, R. K., Booker, L. B., (Eds.): *Proceedings of the Fourth International Conference on Genetic Algorithms.* Morgan Kaufmann, 1991.
3. Belew, R. K., Vose, M. D., (Eds.): *Foundations of Genetic Algorithms 4.* Morgan Kaufman, 1997.
4. Damsbo, M.: *Evolutionary Algorithms in Constrained Sequence Optimization.* M.Sc. thesis, Odense University, 1998.

5. Davidor, Y., Yamada, T., Nakano, R.: The ECOlogical Framework II: Improving GA Performance at Virtually Zero Cost. In Forrest, S., (Ed.): *Proceedings of the Fifth International Conference on Genetic Algorithms.* 171-176, Morgan Kaufmann, 1993.

6. Eshelman, L. J., (Ed.): *Proceedings of the Sixth International Conference on Genetic Algorithms.* Morgan Kaufmann, 1995.

7. Fogel, L. J., Owens, A. J., Walsh, M. J.: *Artificial Intelligence through Simulated Evolution.* Wiley, 1966.

8. Forrest, S., (Ed.): *Proceedings of the Fifth International Conference on Genetic Algorithms.* Morgan Kaufmann, 1993.

9. Goldberg, D. E.: *Genetic Algorithms in Search, Optimization and Machine Learning.* Addison-Wesley, 1989.

10. Holland, J. H.: *Adaption in Natural and Artificial Systems.* Second edition. MIT Press, 1992.

11. Lin S., Goodman, E. D., Punch, W. F.: Investigating Parallel Genetic Algorithms on Job Shop Scheduling Problems. In Angeline, P. J., Reynolds, R. G., McDonnell, J. R., Eberhardt, R., (Eds.): *Evolutionary Programming VI. 6th International Conference, EP97.* 383-393, Springer, 1997.

12. Mitchell, M.: *An Introduction to Genetic Algorithms.* MIT Press, 1996.

13. Reeves, C. R. (Ed.): *Modern Heuristic Techniques for Combinatorial Problems.* McGraw-Hill, 1995.

14. Prügel-Bennett, A., Shapiro, J. L.: An Analysis of Genetic Algorithms Using Statistical Mechanics. Physical Review Letters, 72, 1305-1309, 1994.

15. Rawlins, G. J. E., (Ed.): *Foundations of Genetic Algorithms.* Morgan Kaufmann, 1991.

16. Stidsen, T.: Genetic Algorithms for Industrial Planning. Presented at *Emerging Technologies Workshop,* University College London, 1997. Electronically available at http://www.daimi.aau.dk/~evalia/

17. Whitley, D., Starkweather, T., Shaner, D.: The Travelling Salesman and Sequence Scheduling: Quality Solutions using Genetic Edge Recombination. In Davis, L., (Ed.): *Handbook of Genetic Algorithms.* 350-372, Van Nostrand Reinhold, 1991.

18. Whitley, L. D., (Ed.): *Foundations of Genetic Algorithms 2.* Morgan Kaufmann, 1993.

19. Whitley, L. D., Vose, M. D., (Eds.): *Foundations of Genetic Algorithms 3.* Morgan Kaufmann, 1993.

Intuitionistic Proof Transformations and Their Application to Constructive Program Synthesis

Uwe Egly[1] and Stephan Schmitt[2*]

[1] Abt. Wissensbasierte Systeme 184/3
TU Wien, Treitlstr. 3, A–1040 Wien
uwe@kr.tuwien.ac.at
[2] Department of Computer Science
Cornell University, Ithaca, NY
steph@cs.cornell.edu

Abstract. We present a translation of intuitionistic sequent proofs from a multi-succedent calculus $\mathcal{LJ}_{mc}$ into a single-succedent calculus $\mathcal{LJ}$. The former gives a basis for automated proof search whereas the latter is better suited for proof presentation and program construction from proofs in a system for constructive program synthesis. Well-known translations from the literature have a severe drawback; they use cuts in order to establish the transformation with the undesired consequence that the resulting program term is not intuitive. We establish a transformation based on permutation of inferences and discuss the relevant properties with respect to proof complexity and program terms.

1 Introduction

Constructive program synthesis relies on the parallel process of program *construction* and program *verification*. Using constructive program development systems, for example the NuPRL-system [3], this process can be divided into two steps (see top of Fig. 1). Assume that we have a logical specification of a program within a constructive logic, i.e., Intuitionistic Type Theory (ITT) [9]. In the first step, this specification "formula" will be proven valid using an interactive proof editor based on a sequent calculus for ITT. More precisely, one finds a *constructive* proof for the existence of a function f which maps input elements to output elements of the specified program. In the second step, f will be *extracted* from the computational content of the proof according to the "proofs-as-programs" paradigm [1]. Hence, f forms a correctly verified *program term* with respect to the given specification.

Since the interactive nature of the proof process stands in contrast to an efficient development of programs, every effort has been made in order to support automated proof search in fragments of ITT. This approach turns NuPRL into a coherent program synthesis system which integrates a variety of interactively

* The research report is supported by the German Academic Exchange Service DAAD with a fellowship to the second author.

Jacques Calmet and Jan Plaza (Eds.): AISC'98, LNAI 1476, pp. 132–144, 1998.
© Springer-Verlag Berlin Heidelberg 1998

controlled and automated techniques from theorem proving and algorithm design at different levels of abstraction [2].

In this paper we focus on the automated construction of the purely logical parts of a NuPRL-proof, i.e., subproofs formalized in first-order intuitionistic logic $\mathcal{J}$. The first-order fragment in NuPRL's calculus corresponds to Gentzen's sequent calculus $\mathcal{LJ}_{\mathsf{cut}}$ [6] (including the *cut rule*). The integration of an automated theorem prover for $\mathcal{J}$ into NuPRL is depicted at the bottom of Fig. 1: given a subgoal in $\mathcal{J}$, a separation process constructs a $\mathcal{J}$-formula which serves as input for an intuitionistic *matrix prover*[1]. The resulting matrix proof $\mathcal{MJ}$ has to be integrated into the actual context of the NuPRL-system in order to provide global program extraction from the whole proof. Thus, the $\mathcal{MJ}$-proof has to be transformed back into an $\mathcal{LJ}$-proof which can be integrated as a proof plan for solving the original $\mathcal{J}$-goal.

Since the matrix characterization for $\mathcal{J}$ [15] is based on a multiple-succedent sequent calculus $\mathcal{LJ}_{mc}$, proof reconstruction has to be done in two steps (T1 and T2 in Fig. 1). For realization of the presented concept, efficient proof search procedures [10] as well as reconstruction procedures for efficient generation of $\mathcal{LJ}_{mc}$-proofs from machine-generated proofs have been developed [12]. The construction of $\mathcal{LJ}_{\mathsf{cut}}$-proofs from $\mathcal{LJ}_{mc}$-proofs (step T2) is presented for example in [11]. This transformation T2 is based on a simulation of the multiple succedent in an $\mathcal{LJ}_{mc}$-proof by a disjunction and using the cut rule (see also [8,4]). For this reason, the extracted program term from the resulting $\mathcal{LJ}_{\mathsf{cut}}$-proof to a large extent differs from the original specification since the cut rule is applied *in each* proof step when a multiple succedent is involved. An important impact on the program term results: each operation in the succedent has to be prepared by a selection function which identifies the subformula to be reduced. This was, of course, not intended when specifying the original problem.

In this paper we focus on an alternative transformation step T2 which emphasizes its relation to the original goal, i.e., constructive program synthesis. We present a permutation-based transformation from $\mathcal{LJ}_{mc}$-proofs to $\mathcal{LJ}$-proofs (without using the cut rule) and investigate the complexity of these proofs. On the one hand, we show that there exists no transformation which yields in every case an $\mathcal{LJ}$-proof with polynomial length (with respect to the $\mathcal{LJ}_{mc}$-proof). On the other hand, the resulting $\mathcal{LJ}$-proofs preserve the intended (sub)specification of the program to be synthesized since introduction of additional connectives and inference rules will be avoided. We emphasize that an exponential increase of proof length by our transformation occurs rather seldom, but *every* program term benefits from our construction. The key aspect of our approach is given by a construction of a *normal form* for $\mathcal{LJ}_{mc}$-proofs. This will be achieved by applying permutation schemata *locally* within a given $\mathcal{LJ}_{mc}$-proof which will be pre-structured using so-called *layers*. From this we obtain a proof transformation procedure which is based on a hierarchical system of permutation steps and hence, can easily be implemented into the environment of the NuPRL-system.

---

[1] Its use has historical reasons. There are only a few theorem provers for $\mathcal{J}$ (e.g., in [13]), but there is nearly no work on comparing implementations.

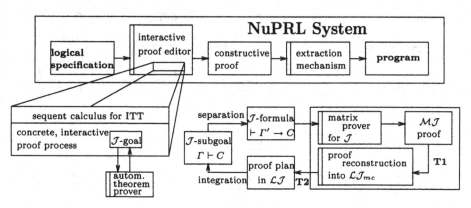

Fig. 1. Application to Constructive Program Synthesis.

2   Preliminaries

Throughout this paper we use a first-order language consisting of *variables*, *constants*, *function symbols*, *predicate symbols*, *logical connectives*, *quantifiers* and *punctuation symbols*. *Terms* and *formulae* are defined according to the usual formation rules.

A *sequent* S is an ordered tuple of the form $\Gamma \vdash \Delta$, where Γ, Δ are finite sets of first-order formulae. Γ is the *antecedent* of S, and Δ is the *succedent* of S. The semantical meaning of a sequent $A_1, \ldots, A_n \vdash B_1, \ldots, B_m$ is the same as the semantical meaning of the formula $(\bigwedge_{i=1}^{n} A_i) \to (\bigvee_{i=1}^{m} B_i)$. We write A, Δ (A, Γ) instead of $\{A\} \cup \Delta$ $(\{A\} \cup \Gamma)$. Furthermore, we denote $\Delta = \emptyset$ $(\Gamma = \emptyset)$ in sequents by $\Gamma \vdash$ $(\vdash \Delta)$ with the intended meaning $\Gamma \vdash \bot$ $(\top \vdash \Delta)$. The empty sequent '$\vdash$' is interpreted by $\top \vdash \bot$.

As proof system we use the (cut-free) sequent calculi $\mathcal{LJ}_{mc}$ and $\mathcal{LJ}$ shown in Fig. 2. In order to reduce the number of cases in our proofs, we consider *negation* as defined; e.g., $\neg A$ is $A \to \bot$, and use an additional $\bot$-*axiom*. Sometimes, we will use the above calculi with an explicit negation rule (and without the $\bot$-axiom) in order to improve readability. These calculi are denoted by $\mathcal{LJ}_{mc}^{\neg}$ and $\mathcal{LJ}^{\neg}$, respectively. We will also consider $\mathcal{LJ}$ extended by the *cut* rule; this calculus is denoted by $\mathcal{LJ}_{\text{cut}}$.

There exist two "directions" in which an inference rule can be read. The first one, the "synthetic direction" is from the rule's premise(s) to its conclusion. In this case, we say that the *principal formula* (i.e., the formula which is new in the conclusion) is *introduced* by the inference. The second one, the "analytic direction", is from the rule's conclusion to its premise(s). In this case, we say that the principal formula is *reduced* to its immediate subformula(e). These subformulae are called the *active* formulae of the inference.

The main difference between $\mathcal{LJ}_{mc}$ and $\mathcal{LJ}$ is given by the fact that $\mathcal{LJ}_{mc}$-sequents may contain several succedent formulae whereas in $\mathcal{LJ}$ at most *one* succedent formula is allowed; sequents of the latter kind are called *intuitionistic*.

An inference rule R of $\mathcal{LJ}_{mc}$ is called *generative* if $R \in \{\to l, \vee r, \exists r\}$; it is called *critical* if $R \in \{\to r, \forall r\}$. For a generative rule R, at least one premise

$\mathcal{LJ}_{mc}$:

$$\frac{}{\Gamma, A \vdash A, \Delta} \; ax.$$

$$\frac{}{\Gamma, \bot \vdash A, \Delta} \; \bot ax.$$

$$\frac{\Gamma \vdash A, B, \Delta}{\Gamma \vdash A \vee B, \Delta} \; \vee r$$

$$\frac{\Gamma \vdash A, \Delta \quad \Gamma \vdash B, \Delta}{\Gamma \vdash A \wedge B, \Delta} \; \wedge r$$

$$\frac{\Gamma, A \vdash B}{\Gamma \vdash A \to B, \Delta} \; \to r$$

$$\frac{\Gamma \vdash A[x \backslash a]}{\Gamma \vdash \forall x. A, \Delta} \; \forall r \; a^*$$

$$\frac{\Gamma \vdash A[x \backslash t], \exists x. A, \Delta}{\Gamma \vdash \exists x. A, \Delta} \; \exists r \; t$$

$$\frac{\Gamma, A \vdash \Delta \quad \Gamma, B \vdash \Delta}{\Gamma, A \vee B \vdash \Delta} \; \vee l$$

$$\frac{\Gamma, A, B \vdash \Delta}{\Gamma, A \wedge B \vdash \Delta} \; \wedge l$$

$$\frac{\Gamma, A \to B \vdash A, \Delta \quad \Gamma, B \vdash \Delta}{\Gamma, A \to B \vdash \Delta} \; \to l$$

$$\frac{\Gamma, \forall x. A, A[x \backslash t] \vdash \Delta}{\Gamma, \forall x. A \vdash \Delta} \; \forall l \; t$$

$$\frac{\Gamma, A[x \backslash a] \vdash \Delta}{\Gamma, \exists x. A \vdash \Delta} \; \exists l \; a^*$$

$\mathcal{LJ}$:

$$\frac{}{\Gamma, A \vdash A} \; ax.$$

$$\frac{}{\Gamma, \bot \vdash A} \; \bot ax.$$

$$\frac{\Gamma \vdash A}{\Gamma \vdash A \vee B} \; \vee r \; 1$$

$$\frac{\Gamma \vdash B}{\Gamma \vdash A \vee B} \; \vee r \; 2$$

$$\frac{\Gamma \vdash A \quad \Gamma \vdash B}{\Gamma \vdash A \wedge B} \; \wedge r$$

$$\frac{\Gamma, A \vdash B}{\Gamma \vdash A \to B} \; \to r$$

$$\frac{\Gamma \vdash A[x \backslash a]}{\Gamma \vdash \forall x. A} \; \forall r \; a^*$$

$$\frac{\Gamma \vdash A[x \backslash t]}{\Gamma \vdash \exists x. A} \; \exists r \; t$$

$$\frac{\Gamma, A \vdash C \quad \Gamma, B \vdash C}{\Gamma, A \vee B \vdash C} \; \vee l$$

$$\frac{\Gamma, A, B \vdash C}{\Gamma, A \wedge B \vdash C} \; \wedge l$$

$$\frac{\Gamma, A \to B \vdash A \quad \Gamma, B \vdash C}{\Gamma, A \to B \vdash C} \; \to l$$

$$\frac{\Gamma, \forall x. A, A[x \backslash t] \vdash C}{\Gamma, \forall x. A \vdash C} \; \forall l \; t$$

$$\frac{\Gamma, A[x \backslash a] \vdash C}{\Gamma, \exists x. A \vdash C} \; \exists l \; a^*$$

$\mathcal{LJ}^{\neg}_{mc}$:

$$\frac{\Gamma, A \vdash}{\Gamma \vdash \neg A, \Delta} \; \neg r$$

$$\frac{\Gamma, \neg A \vdash A, \Delta}{\Gamma, \neg A \vdash \Delta} \; \neg l$$

$\mathcal{LJ}^{\neg}$:

$$\frac{\Gamma, A \vdash}{\Gamma \vdash \neg A} \; \neg r$$

$$\frac{\Gamma, \neg A \vdash A}{\Gamma, \neg A \vdash C} \; \neg l$$

$\mathcal{LJ}_{cut} \; / \; \mathcal{LJ}^{\neg}_{cut}$:

$$\frac{\Gamma \vdash A \quad \Gamma, A \vdash C}{\Gamma \vdash C} \; cut$$

* a must satisfy the eigenvariable condition for $\forall r$ and $\exists l$.

Fig. 2. The calculi $\mathcal{LJ}_{mc}$, $\mathcal{LJ}$, and their extensions.

contains more succedent formulae than the conclusion of R does. An application of an inference r *depends* on an application of an inference r' iff r' reduces a superformula of the formula reduced by r in an analytic sequent proof. An application of an inference r is called *dependent generative* (*d-generative*), iff r is generative or $r = \wedge r$ and r depends on a generative rule. The *generative branches* of a d-generative rule r are the branches above the premises of r except for $r = \to l$, where only the branches above the *left* premise are generative.

3   Complexity of Proof Transformations

Our goal is to construct an $\mathcal{LJ}$-proof of a sequent S for the NuPRL-system from an $\mathcal{LJ}_{mc}$-proof of S which itself is a reconstruction of a proof in some matrix calculus. The question which we consider in this section is whether such a construction (named T2 in Fig. 1) yields an $\mathcal{LJ}$-proof whose length is "approximately" the length of the corresponding $\mathcal{LJ}_{mc}$-proof. More precisely, we consider minimal (or shortest) $\mathcal{LJ}$-proofs and $\mathcal{LJ}_{mc}$-proofs for some classes of formulae and compare their length. As the *length* of a proof, we use the number of its axioms. "Approximately" means that the length of the $\mathcal{LJ}$-proof is bounded by a *polynomial* function in the length of the $\mathcal{LJ}_{mc}$-proof. We give a negative answer to the question stated above by providing a class of formulae for which there are short $\mathcal{LJ}_{mc}$-proofs but *any* $\mathcal{LJ}$-proof is long (i.e., its length *cannot* be bounded by a polynomial in the length of the former proof).

The following definition of a polynomial simulation is adapted from [5] and restricted to the case that the connectives in both calculi are identical. A calculus P_1 can *polynomially simulate* ($\mathcal{P}$-simulate) a calculus P_2 if there is a polynomial p such that the following holds. For every proof of a formula (or sequent) F in P_2 of length n, there is a proof of F in P_1, whose length is not greater than $p(n)$.

We restrict our attention to *cut-free* $\mathcal{LJ}$ and $\mathcal{LJ}_{mc}$; if both calculi are extended by the cut rule then they $\mathcal{P}$-simulate each other [8,4,11]. The next lemma can be proved easily.

Lemma 1. $\mathcal{LJ}_{mc}$ $\mathcal{P}$-*simulates* $\mathcal{LJ}$.

We will show in the remainder of this section that the reverse simulation is not polynomial. We present a class of formulae for which any $\mathcal{LJ}$-proof is exponential. For arbitrary $n \in \mathbb{IN}$ we consider the formulae

$$F(n) \equiv \forall w.(A_n w) \wedge [\bigwedge_{i=0}^{n-1} O_i \wedge N_i] \rightarrow \forall z.A_0 z$$

where $O_i \equiv \forall x.((B_i \vee A_i x) \vee B_i)$ and $N_i \equiv \neg(B_i \wedge \forall y.A_{i+1} y)$. In order to simplify the discussion, we use the calculi $\mathcal{LJ}^{\neg}$ and $\mathcal{LJ}_{mc}^{\neg}$ with negation rules instead of $\mathcal{LJ}$ and $\mathcal{LJ}_{mc}$ where $\neg A$ is an abbreviation for $A \rightarrow \bot$. The result can be lifted to the latter calculi by adding axioms of the form $\bot a x$.

Lemma 2. $\mathcal{LJ}$ *does* not $\mathcal{P}$-*simulate* $\mathcal{LJ}_{mc}$.

Proof (sketch). By counterexample $F(n)$. We show:

  (i) There exists an $\mathcal{LJ}_{mc}^{\neg}$-proof with $3n + 1$ axioms.

  (ii) *Each* $\mathcal{LJ}^{\neg}$-proof requires $2^{n+2} - 3$ axioms.

(i) By induction on n. (ii) First we prove that the eigenvariable condition forces a unique direction for deriving the axioms in a cut-free $\mathcal{LJ}^{\neg}$-proof: namely, from A_0 up to A_n. Let G, H be subformulae of $F(n)$. We define a *reduction ordering* $G \mapsto H$ which means that G has to be *reduced* before H. Each $\mathcal{LJ}^{\neg}$-proof for $F(n)$ has to respect the reduction ordering $O_j \mapsto N_j \mapsto O_{j+1} \mapsto N_{j+1}$ for $0 \leq j \leq n-2$. This can be shown by induction on j, separately for $O_j \mapsto N_j$ and $N_j \mapsto O_{j+1}$. Then, by induction on n, we prove (ii).   (q.e.d.)

The above result can be understood as follows. An $\mathcal{LJ}_{mc}^{\neg}$-proof of $F(n)$ consists of n subproofs each of them uses atoms with index i. Subproof i needs 3 axioms,

$i = 0, \ldots, n - 1$, whereas subproof n needs only one axiom. This results in $(\sum_{i=0}^{n-1} 3) + 1 = 3n + 1$ axioms. In contrast, in an $\mathcal{LJ}^-$-proof, each subproof i is duplicated 2^i times. Thus, we have $(\sum_{i=0}^{n-1} 3 \cdot 2^i) + 2^n = 3(2^n - 1) + 2^n = 2^{n+2} - 3$ axioms. From a complexity theoretical viewpoint, the result can easily be generalized to the calculi $\mathcal{LJ}$ and $\mathcal{LJ}_{mc}$ since avoiding the $\neg$-rules yields a polynomial (with respect to the length of the $\mathcal{LJ}$-proof) increase of proof length.

4 Permutation-Based Transformations

Let us start with an overview of our procedure which constructs an $\mathcal{LJ}$-proof from an $\mathcal{LJ}_{mc}$-proof. Given an $\mathcal{LJ}_{mc}$-proof α^{mc} of a sequent S (essentially from our matrix prover for $\mathcal{J}$), we first structure α^{mc} into *layers*. A layer of α^{mc} is a maximal subtree L of α^{mc} such that (i) the leaf nodes of L are either axioms or intuitionistic sequents, and (ii) no intuitionistic sequents occur below the leaves. The *excluded root* S_L of a layer L is the topmost intuitionistic sequent not contained in L. α^{mc} can be considered to consist of a couple of layers where the boundaries between layers are intuitionistic sequents and distinct layers do not overlap. Applications of critical inferences c form boundaries between layers because the premise of c is an intuitionistic sequent. The basic idea is to stepwise transform each layer L into an $\mathcal{LJ}$-proof with endsequent S_L using (essentially) the leaf sequents of L as "given axioms". The stepwise transformation is based on two key observations, namely (i) that each relevant application of a *d-generative inference rule* can be permuted above all "highest" $\vee l$-inferences in L, and (ii) that an $\mathcal{LJ}_{mc}$-proof β^{mc} of $\Gamma \vdash \Delta$ can be transformed easily into an $\mathcal{LJ}$-proof β of $\Gamma \vdash D$ (for some $D \in \Delta$) if β^{mc} does not contain any $\vee l$-inference. After the application of all permutations of (i), the resulting $\mathcal{LJ}_{mc}$-proof is in a specific normal form.

4.1 Permutation Schemata

We start our detailed description with an explanation of permutable inferences (rules) and necessary permutation schemata.

Rule R' is *permutable over* rule R (towards the axioms), if, for all applications r of R, r' of R', r immediately above r' such that

1. the principal formula F of r is not active in r'; hence, the two principal formulae do not *overlap*.
2. $\mathcal{A}$ is the set of all premises of r, S the conclusion of r, r' takes premises from $\mathcal{B} \cup \{S\}$ ($\mathcal{B}$ possibly empty) and yields the conclusion S',

there is a proof of S' from $\mathcal{A} \cup \mathcal{B}'$ in which an application of R' occur immediately above an application of R, or one of the two applications disappear. If F is an α-formula, then the set $\mathcal{B}'$ is obtained by replacing some sequent $s \in \mathcal{B}$ by a sequent s' such that s' is the premise of the inference which introduces F. If F is a β-formula, then $\mathcal{B}'$ is obtained by replacing some sequent $s \in \mathcal{B}$ by two sequents s_1, s_2 such that s_1, s_2 are the premises of the inference which introduces F.

For a discussion of non-permutabilities of inference rules in Gentzen's calculus $\mathcal{LJ}$ see [7,14]. Before we prove the *local permutation lemma for* $\mathcal{LJ}_{mc}$, we first introduce some notational conventions. Let $S := A_1,..,A_n \vdash B_1,..,B_m$ be a sequent. We consider S as a set of *signed formulae* $(A_1)^a,..,(A_n)^a,(B_1)^s,..,(B_m)^s$ where a is chosen for antecedent formulae and s is chosen for succedent formulae. For sequents, we write $S, T, U, \ldots$, possibly subscripted. If $S = \Gamma \vdash \Delta$ and $T = \Gamma' \vdash \Delta'$, then ST denotes $\Gamma\Gamma' \vdash \Delta\Delta'$. Arbitrary signs are denoted by p, q.

Lemma 3. *In* $\mathcal{LJ}_{mc}$-*proofs, rule* R' *is always permutable over* R *(towards the axioms) except in the following cases:*

$$
\begin{array}{c||c|c|c|c|c}
R & \forall l & \forall l, \exists r & \to l & \to r, \forall r & \to r, \forall r \\
R' & \forall r & \exists l & \to r, \forall r & \vee l & \to l
\end{array}
$$

The first two non-permutabilities concerning quantifiers are identical for $\mathcal{LJ}$ and $\mathcal{LJ}_{mc}$. The last non-permutabilities occur only if $\to r$ or $\forall r$ occurs in the right premise of the $\to l$-inference. The non-permutabilities except the first two categories do not affect the transformation of a layer into an $\mathcal{LJ}$-proof because a critical rule forms a boundary (the premise of a critical rule is intuitionistic). Important for our approach is the permutability of any generative rule and of the $\wedge r$-rule over any non-generative rule (except critical rules) and over the $\to l$-rule. Permuting the $\wedge r$-rule is important if a generative formula contains a conjunction as a subformula which eventually appears as the principal formula of an $\wedge r$-inference; this inference depends on the generative inference and has to be permuted over all highest $\vee l$ in the layer before the generative rule can be permuted. After these explanations, we prove Lemma 3.

Proof. The following sequents justify the non-permutabilities (R, R') from Lemma 3.

$(\forall l, \forall r)$	$\forall x Ax \vdash \forall x(Ax \vee B)$		$(\forall l, \exists l)$	$\forall x Ax, \exists x(Ax \to \bot) \vdash$
$(\exists r, \exists l)$	$\exists x(Ax \wedge B) \vdash \exists x Ax$		$(\to l, \to r)$	$A \to (A \to \bot) \vdash A \to B$
$(\to l, \forall r)$	$C, C \to B \vdash \forall x Ax, B$		$(\to r, \vee l)$	$A \vee B \vdash C \to B, A$
$(\forall r, \vee l)$	$A \vee \forall x Bx \vdash \forall y B(y), A$		$(\to r, \to l)$	$A \vee E, A \to B \vdash C \to B, E$
$(\forall r, \to l)$	$A \vee E, A \to \forall x Bx \vdash \forall x Bx, E$			

Due to the critical rules, some permutations need not to be considered (for instance, the case when R and R' are critical). Moreover, we only consider $\mathcal{LJ}_{mc}$-deductions without locally superfluous applications of inference rules. For instance, consider the sequent $A \to B, \Gamma \vdash C \to D, \Delta$. If the (critical) inference $\to r$ with principal formula $C \to D$ occurs immediately above the $\to l$-inference with principal formula $A \to B$ (in the left branch), then the latter inference is locally superfluous and can be deleted.

One-premise rule over one-premise rule. If R is a critical rule and $R' \in \{\vee r, \exists r\}$ then the second rule is superfluous. ψ^p, ϕ^q are sequences of signed formulae which contain the principal formula and additional formulae introduced by the critical rule.

$$
\cfrac{\cfrac{STU}{(\phi)^p TU}\ R}{(\phi)^p (\psi)^q U}\ R' \quad\Longrightarrow\quad \cfrac{\cfrac{STU}{(\psi)^q SU}\ R'}{(\phi)^p (\psi)^q U}\ R
$$

One-premise rule over two-premise rule. If R' is a critical rule then $R \neq \wedge r$. Moreover, the indicated non-permutabilities forbid the case when $R = \to l$ and R' is critical.

$$\frac{\dfrac{S_1TU \quad S_2TU}{(F)^pTU}R}{(F)^p(\psi)^qU}R' \quad \Longrightarrow \quad \frac{\dfrac{S_1TU}{(\psi)^qS_1U}R' \quad \dfrac{S_2TU}{(\psi)^qS_2U}R'}{(F)^p(\psi)^qU}R$$

Two-premise rule over one-premise rule. If R is critical then $R' \notin \{\to l, \wedge r\}$. Moreover, the indicated non-permutabilities forbid the case when R is critical and $R' = \vee l$. Observe that $(F)^p$ (occurring in the end sequent of α) is replaced by S resulting in ST_2U. No problem can arise in the modified deduction since $(F)^p$ is replaced by S and S consists of the immediate subformulae of $(F)^p$. If $(F)^p$ is of the form $(\forall x A)^a$ then $S = \{(F)^a, (A\{x\backslash t\})^a\}$; eigenvariable conflicts with the new free variables in t can be avoided by renaming eigenvariables in α'. Moreover, since $(F)^a \in S$, appropriate "instances" can be generated in α' if necessary. A similar argument applies to the case if $(F)^p = (\exists x A)^s$ and $S = \{(F)^s, (A\{x\backslash t\})^s\}$. The "right permutation" case is similar.

$$\frac{\dfrac{ST_1U}{(F)^pT_1U}R \quad \overset{\alpha}{(F)^pT_2U}}{(F)^p(G)^qU}R' \quad \Longrightarrow \quad \frac{ST_1U \quad \overset{\alpha'}{ST_2U}}{\dfrac{(G)^qSU}{(F)^p(G)^qU}R}R'$$

Two-premise rule over two-premise rule. Observe that $(F)^p$ is introduced by $\vee l$, $\to l$, or $\wedge r$, and that replacing $(F)^pT_2U$ by S_1T_2U and S_2T_2U results in two deductions which are simpler than the deduction of $(F)^pT_2U$ (some branches of an inference with principal formula F in polarity p become obsolete; we will call the deletion of these branches *branch modification*). The "right permutation" case is similar.

$$\frac{\dfrac{S_1T_1U \quad S_2T_1U}{(F)^pT_1U}R \quad (F)^pT_2U}{(F)^p(G)^qU}R' \quad \Longrightarrow \quad \frac{\dfrac{S_1T_1U \quad S_1T_2U}{(G)^qS_1U}R' \quad \dfrac{S_2T_1U \quad S_2T_2U}{(G)^qS_2U}R'}{(F)^p(G)^qU}R$$

(q.e.d.)

The following example illustrates the "two-premise over two-premise" case.

Example 1. Consider the following $\mathcal{LJ}_{mc}$-proof with two $\to l$-inferences.

$$\frac{\dfrac{\overset{\alpha}{A \to B, C \to D, \Gamma \vdash C, A, \Delta} \quad \overset{\beta}{A \to B, D, \Gamma \vdash A, \Delta}}{A \to B, C \to D, \Gamma \vdash A, \Delta} \to l \quad \overset{\gamma}{B, C \to D, \Gamma \vdash \Delta}}{A \to B, C \to D, \Gamma \vdash \Delta} \to l$$

This proof is transformed into the following one (all inferences are $\to l$-inferences).

$$\frac{\dfrac{\overset{\alpha}{A \to B, C \to D, \Gamma \vdash C, A, \Delta} \quad \overset{\gamma'}{B, C \to D, \Gamma \vdash C, \Delta}}{A \to B, C \to D, \Gamma \vdash C, \Delta} \quad \dfrac{\overset{\beta}{A \to B, D, \Gamma \vdash A, \Delta} \quad \overset{\gamma''}{D, B, \Gamma \vdash \Delta}}{A \to B, D, \Gamma \vdash \Delta}}{A \to B, C \to D, \Gamma \vdash \Delta}$$

This yields the following instantiations: $S_1 := (C \to D)^a, (C)^s$, $T_1 = (A \to B)^a, (A)^s$, $S_2 := (D)^a$, $T_2 := (B)^a$, $U := (\Gamma)^a, (\Delta)^s$, $(F)^p := (C \to D)^a$, and $(G)^q := (A \to B)^a$.

4.2   Constructing $\mathcal{LJ}$-Proofs via Normal Form Proofs in $\mathcal{LJ}_{mc}$

In this subsection, we describe a transformation of $\mathcal{LJ}_{mc}$-proofs into $\mathcal{LJ}$-proofs. The following lemma provides a transformation for $\mathcal{LJ}_{mc}$-proofs without $\vee l$-inferences.

Lemma 4. *Let α^{mc} be an $\mathcal{LJ}_{mc}$-proof of $S := \Gamma \vdash \Delta$, without an application of $\vee l$ and Δ is non-empty[2]. Then there exists an $\mathcal{LJ}$-proof α of $\Gamma \vdash D$ for some $D \in \Delta$. Moreover, the length of α is polynomial in the length of α^{mc}.*

Proof. By induction on the depth of α^{mc}.

Base. α^{mc} consists of an axiom $\Gamma, A \vdash A, \Delta$ or $\Gamma, \bot \vdash \Delta$. Then $\Gamma, A \vdash A$ or $\Gamma, \bot \vdash A$ are the corresponding axioms in $\mathcal{LJ}$.

Step. (IH) Assume that the lemma holds for all $\mathcal{LJ}_{mc}$-proofs of depth $< n$. We consider an $\mathcal{LJ}_{mc}$-proof α^{mc} of depth n and proceed by case analysis with respect to the last inference in α^{mc}. In the different cases, the $\mathcal{LJ}$-proof(s) β (or β_1 and β_2 in the binary case) are given by IH.

cases $\forall l, \exists l, \wedge l, \forall r, \rightarrow r$ can be transformed directly into $\mathcal{LJ}$-inferences.

case $\rightarrow l$. Consider the left deduction below. IH provides an $\mathcal{LJ}$-proof of $B \rightarrow C, \Gamma \vdash D$.

$$
\frac{\begin{array}{cc} \beta_1^{mc} & \beta_2^{mc} \\ B \rightarrow C, \Gamma \vdash B, \Delta & C, \Gamma \vdash \Delta \end{array}}{B \rightarrow C, \Gamma \vdash \Delta} \quad \overset{(ii)}{\Longrightarrow} \quad \frac{\begin{array}{cc} \beta_1 & \beta_2 \\ B \rightarrow C, \Gamma \vdash B & C, \Gamma \vdash A \end{array}}{B \rightarrow C, \Gamma \vdash A}
$$

Now, we have the following two subcases: (i) $D \in \Delta$. Then the indicated occurrence of B is not relevant in β_1^{mc} and we can replace α^{mc} by an $\mathcal{LJ}_{mc}$-proof of $B \rightarrow C, \Gamma \vdash \Delta$ of depth $n - 1$. IH provides an $\mathcal{LJ}$-proof α in this case. (ii) $D \notin \Delta$. By IH, we obtain $\mathcal{LJ}$-proofs β_1 and β_2 of $B \rightarrow C, \Gamma \vdash B$ and $C, \Gamma \vdash A$, respectively. A final application of $\rightarrow l$ results in α.

case $\exists r$. IH provides an $\mathcal{LJ}$-proof β of $\Gamma \vdash D$. There are the following two sub-

$$
\frac{\begin{array}{c} \beta^{mc} \\ \Gamma \vdash B\{x \backslash t\}, \exists x B, \Delta \end{array}}{\Gamma \vdash \exists x B, \Delta}
$$

cases: (i) $D = B\{x \backslash t\}$. Extending β by an $\exists r$-inference yields α. (ii) Otherwise, take β as α.

case $\wedge r$. IH provides two $\mathcal{LJ}$-proofs β_1 and β_2 of $\Gamma \vdash D$ and $\Gamma \vdash E$, respec-

$$
\frac{\begin{array}{cc} \beta_1^{mc} & \beta_2^{mc} \\ \Gamma \vdash B, \Delta & \Gamma \vdash C, \Delta \end{array}}{\Gamma \vdash B \wedge C, \Delta}
$$

tively. (i) $D = B$ and $E = C$ or $D = C$ and $E = B$. Then an additional $\wedge r$-inference yields α. (ii) $D \in \Delta$. Set $\alpha = \beta_1$. (iii) $E \in \Delta$. Set $\alpha = \beta_2$.

case $\vee r$. IH provides an $\mathcal{LJ}$-proof β of $\Gamma \vdash D$.

$$
\frac{\begin{array}{c} \beta^{mc} \\ \Gamma \vdash A, B, \Delta \end{array}}{\Gamma \vdash A \vee B, \Delta}
$$

(i) $D = A$ or $D = B$. Then an additional $\vee r$ 1 or $\vee r$ 2 yields α. (ii) Otherwise, set $\alpha = \beta$.

(q.e.d.)

Let us return to the general case, i.e., a transformation of $\mathcal{LJ}_{mc}$-proofs *with* $\vee l$-inferences. Lemma 3 provides a "micro-step" for this transformation. The lemma is called the *local permutation lemma* for $\mathcal{LJ}_{mc}$ because only two adjacent inferences with non-overlapping principal formulae can be permuted. The following lemma provides a "macro-step" relying on the micro-steps and requiring the concept of an *admissible branch*.

Let α^{mc} be an $\mathcal{LJ}_{mc}$-proof of a sequent S, L be a layer in α^{mc}. A branch b in L is called *admissible* if it contains inferences r, $\circ$ such that the following conditions are satisfied.

---

[2] Δ can be considered to contain at least $\bot$.

1. r is a topmost $\vee l$-inference in L.
2. $\circ$ is the topmost d-generative rule below r such that (i) b is a generative branch of $\circ$, and (ii) each $\to l$-inference *between* r and $\circ$ does not contain any $\vee l$-inferences in its generative branches.

Lemma 5. *Let α^{mc} be an $\mathcal{L}\mathcal{J}_{mc}$-proof of a sequent S, L be a layer in α^{mc}. Moreover, let b be an admissible branch in L containing r and $\circ$. Let r' be the lowmost $\vee l$-inference between r and $\circ$ on b. Then α^{mc} can be transformed into an $\mathcal{L}\mathcal{J}_{mc}$-proof β^{mc} of S such that $\circ$ occurs above r'.*

Proof (sketch). We proceed by induction on the number d of inferences between $\circ$ and r'. It is important that b is a branch in L, i.e., no critical rules occur on b. Moreover, no eigenvariable problems can occur when we permute $\exists r$ towards the axioms.

Base $d = 0$. The inferences r' and $\circ$ are adjacent, r' is above $\circ$, and $\circ$ is a d-generative rule. Since $\circ$ can be (locally) permuted above $\vee l$, we obtain β^{mc} by Lemma 3.

Step. We assume that the lemma holds for all generative branches between r' and $\circ$ (containing r, r', $\circ$ as described above) and $d < n$. Consider a generative branch b with $d = n$ and let p, $\circ$ the two inferences corresponding to the last two elements of b where p is above $\circ$. First, we show that the principal formulae of p and $\circ$ cannot overlap. We first consider the subbranch b_1 ending at p. Observe that p is neither a critical rule (because these rules do not occur on b) nor the d-generative rule satisfying the admissibility conditions for b (because $\circ$ is the topmost occurrence of such a rule). p is either $\to l$ without $\vee l$-inferences in its generative branches, another left rule or a non-dependent $\wedge r$. Moreover, $\circ$ affects only the succedent of its (left) premise. In all of these cases, no overlap can occur between principal formulae of p and $\circ$ for which reason permutation cannot fail.

Second, we have to guarantee termination. For this, consider the case where $\circ = p = \to l$ and reconsider Example 1 for the proof pattern. Observe that α does not contain any $\vee l$-inferences since b has to be admissible. Apply Lemma 4 to α in order to obtain the (single) relevant succedent formula F. If $F \equiv A$ then delete p. If $F \in \Delta$ then delete p and $\circ$. If $F \equiv C$ then permute $\circ$ over p. Since $F \equiv C$, the $\to l$-inference $\circ$ below α and the subdeduction γ' have to be deleted. Hence, p remains $\vee l$-free on its generative branches which yields termination of the transformation. A similar analysis can be performed for $\circ = \wedge r$ and $p = \to l$ occurring in the left or right premise of $\circ$. For all other cases, permuting $\circ$ with its predecessor p also reduces the number of inferences between r' and $\circ$.   (q.e.d.)

The following lemma provides the construction of normalized $\mathcal{L}\mathcal{J}_{mc}$-proofs.

Lemma 6. *Let $R_{I,L}$ be the set of all topmost $\vee l$-inferences in the layer L which occur above the inference I. Let α^{mc}, L, $\circ$ be given as in Lemma 5 such that all generative branches of $\circ$ containing an $r \in R_{\circ,L}$ are admissible. Then α^{mc} can be transformed into β^{mc} such that, for all $r \in R_{\circ,L}$, $\circ$ occurs above r.*

Let n_r be the number of $\vee l$-inferences between $\circ$ and r, and let $n = \max\{n_r \mid r \in R_{\circ,L}\}$. The proof is based on nested induction: (1) on n (outer induction); (2) on the maximal number of not permuted $\wedge r$-inferences and $\to l$-inferences with $\vee l$-free generative branches between $\circ$ and a lowmost $\vee l$-inference r' above $\circ$ (inner induction).

A normal form for an $\mathcal{LJ}_{mc}$-proof α^{mc} is defined by a normal form for each layer L of α^{mc}. The *excluded root* of a layer L is the topmost intuitionistic sequent S_L not contained in L. Let $N = \{S_1, \ldots, S_n\}$ be the leaves of L where each S_i is either of the form $\Gamma_i \vdash C_i$ or an axiom. The *layer normal form* of L is a subdeduction M_L which consists of several layers and is structured as follows:

(1) S_L remains the endsequent of M_L.

(2) M_L has the leaves $N' = \{S_1', \ldots, S_m'\}$, $m \geq n$, such that for each S_j' there exists a $S_i \in N$ with either $S_j' = S_i$, or S_j' and S_i differ only in eigenvariables renaming and/or branch modification according to the proof of Lemma 3.

(3) Let R' be the set of topmost $\vee l$-inferences in M_L. Each premise of $r' \in R'$ is either a leaf $S_j' \in N'$, or it is an excluded root of a layer l from M_L. l itself is either a single layer (possibly $l \in N'$), or it has a subset of N' as leaves.

(4) Each sequent between S_L and the premises of all $r' \in R'$ is intuitionistic and forms a single layer in M_L.

The normal form α_N^{mc} of an $\mathcal{LJ}_{mc}$-proof α^{mc} is defined by all its layer normal forms. The normal form can be constructed by repeated application of Lemma 6 which locally transforms each layer L of α^{mc} into a layer normal form M_L.

Theorem 1. (normal form) *Each $\mathcal{LJ}_{mc}$-proof α^{mc} can be transformed into a normal form proof α_N^{mc} via permutation of inferences.*

By definition of the layer normal form, no $\vee l$-inferences occur within the topmost layers of M_L. In order to construct an $\mathcal{LJ}$-proof α from a normalized $\mathcal{LJ}_{mc}$-proof α_N^{mc} we have to eliminate redundant formulae and inferences in the topmost (non-single) layers of each subdeduction M_L. Such a construction is accomplished by a procedure extracted from the constructive proof of Lemma 4.

Example 2. Consider the formula $F \equiv (\forall x.Ax \vee Bx) \wedge (\exists y.Ay \rightarrow \exists z.\neg Az) \rightarrow \exists x.Bx$ with its $\mathcal{LJ}_{mc}^-$-proof α^{mc} shown in Fig. 3. [3] We have two layers where permutations take place, namely (i) L_1 in subgoal 1, between the premise of the $\exists l$-inference as its excluded root and the axioms, and (ii) L_2 containing the left branch of the $\rightarrow l$-inference with the premise of $\wedge l$ as its excluded root. First, we permute in L_1 the $\neg l$-inference (i.e., an $\rightarrow l$-inference) above the (only) $\vee l$-inference. Second, in L_2 permute the $\rightarrow l$-inference above the $\vee l$-inference in its generative (i.e., left) branch.

The resulting normal form α_N^{mc} is depicted in Fig. 4. The layer normal form M_{L_1} of L_1 is given within *subgoal 1'* above the excluded root of L_1 (the antecedent formula X will be ignored). M_{L_2} is given by the deduction above the excluded root of L_2, containing eigenvariable renaming b for a, and the duplication of *subgoal 1*, which results in *subgoal 1'* and *subgoal 1''* due to the permutation schemata of Lemma 3. Renaming requires a second instance of $\forall x.Ax \vee Bx$ and $\exists x.Bx$, whereas $X = Aa$ and $X = Ba$ reflects branch modification in these two subgoals.

In order to obtain an $\mathcal{LJ}$-proof α of $\vdash F$, we have to delete inferences according to the proof of Lemma 4. In *subgoal 2*, the $\rightarrow l$-inference can be deleted since the additional succedent formula $\exists y.Ay$ in its left premise does not contribute to the axiom $Ba \vdash \exists y.Ay, Ba$. Hence, *subgoal 1''* can be deleted as well. For a similar reason, the $\neg l$-inference in the right branch of the remaining *subgoal 1'* is deleted. Putting together the results yields an $\mathcal{LJ}$-proof α of $\vdash F$.

---

[3] If possible, we omit explicit contractions in $\forall l, \rightarrow l, \neg l$, and $\exists r$.

$$\cfrac{\cfrac{\cfrac{\cfrac{\cfrac{\cfrac{Aa \vdash Aa, \exists x.Bx}{Aa \vdash \exists y.Ay, \exists x.Bx}\ \exists r\ a \quad \cfrac{Ba \vdash \exists y.Ay, Ba}{Ba \vdash \exists y.Ay, \exists x.Bx}\ \exists r\ a}{Aa \vee Ba \vdash \exists y.Ay, \exists x.Bx}\ \vee l}{\forall x.Ax \vee Bx \vdash \exists y.Ay, \exists x.Bx}\ \forall l\ a \quad subg.\ 1}{\forall x.Ax \vee Bx, \exists y.Ay \to \exists z.\neg Az \vdash \exists x.Bx}\ \to l}{(\forall x.Ax \vee Bx) \wedge (\exists y.Ay \to \exists z.\neg Az) \vdash \exists x.Bx}\ \wedge l}{\vdash (\forall x.Ax \vee Bx) \wedge (\exists y.Ay \to \exists z\neg Az) \to \exists x.Bx}\ \to r$$

subgoal 1:

$$\cfrac{\cfrac{\cfrac{\cfrac{Aa \vdash Aa, \exists x.Bx \quad \cfrac{Ba \vdash Aa, Ba}{Ba \vdash Aa, \exists x.Bx}\ \exists r\ a}{Aa \vee Ba \vdash Aa, \exists x.Bx}\ \vee l}{\forall x.Ax \vee Bx \vdash Aa, \exists x.Bx}\ \forall l\ a}{\forall x.Ax \vee Bx, \neg Aa \vdash \exists x.Bx}\ \neg l}{\forall x.Ax \vee Bx, \exists z.\neg Az \vdash \exists x.Bx}\ \exists l\ a$$

Fig. 3. An $\mathcal{LJ}_{mc}^{-}$ proof of $\vdash F$ from Example 2.

$$\cfrac{\cfrac{\cfrac{\cfrac{\cfrac{\cfrac{Aa \vdash Aa, \exists x.Bx}{Aa \vdash \exists y.Ay, \exists x.Bx}\ \exists r\ a \quad subg.\ 1'}{Aa, \exists y.Ay \to \exists z.\neg Az \vdash \exists x.Bx}\ \to l \quad subg.\ 2}{Aa \vee Ba, \exists y.Ay \to \exists z.\neg Az \vdash \exists x.Bx}\ \vee l}{\forall x.Ax \vee Bx, \exists y.Ay \to \exists z.\neg Az \vdash \exists x.Bx}\ \forall l\ a}{(\forall x.Ax \vee Bx) \wedge (\exists y.Ay \to \exists z.\neg Az) \vdash \exists x.Bx}\ \wedge l}{\vdash (\forall x.Ax \vee Bx) \wedge (\exists y.Ay \to \exists z\neg Az) \to \exists x.Bx}\ \to r$$

subgoal 1' $X = Aa$:

$$\cfrac{\cfrac{X, Ab \vdash Ab, \exists x.Bx}{X, Ab, \neg Ab \vdash \exists x.Bx}\ \neg l}{X, Ab \vee Bb, \neg Ab \vdash \exists x.Bx}$$

subgoal 1'' $X = Ba$:

$$\cfrac{\cfrac{\cfrac{X, Bb \vdash Ab, Bb}{X, Bb \vdash Ab, \exists x.Bx}\ \exists r\ b}{X, Bb, \neg Ab \vdash \exists x.Bx}\ \neg l}{X, Ab \vee Bb, \neg Ab \vdash \exists x.Bx}\ \vee l$$

$$\cfrac{\cfrac{X, Ab \vee Bb, \neg Ab \vdash \exists x.Bx}{X, \forall x.Ax \vee Bx, \neg Ab \vdash \exists x.Bx}\ \forall l\ b}{X, \forall x.Ax \vee Bx, \exists z.\neg Az \vdash \exists x.Bx}\ \exists l\ b$$

subgoal 2:

$$\cfrac{\cfrac{Ba \vdash \exists y.Ay, Ba}{Ba \vdash \exists y.Ay, \exists x.Bx}\ \exists r\ a \quad subgoal\ 1''}{Ba, \exists y.Ay \to \exists z.\neg Az \vdash \exists x.Bx}\ \to l$$

Fig. 4. The normal form α_N^{mc} from Example 2.

5 Conclusion and Future Work

We have presented a permutation-based proof transformation from $\mathcal{LJ}_{mc}$-proofs into $\mathcal{LJ}$-proofs. It relies on a layer-oriented construction of normal form proofs in $\mathcal{LJ}_{mc}$. Furthermore, we have shown that, in general, no polynomial simulation exists between $\mathcal{LJ}_{mc}$ and $\mathcal{LJ}$ (both without cut). Our approach will be integrated into NuPRL as an improved transformation step T2 (see Fig. 1) since, in contrast to existing approaches, it preserves the original logical specifications when extracting the program terms from the resulting $\mathcal{LJ}$-proofs. Again, we stress that, in practice, there are only few examples where an exponential increase of proof length occurs but *every* program term benefits from our construction. In future work, we have to combine a *controlled* introduction of cut with our permutation approach in order to obtain $\mathcal{LJ}_{cut}$-proofs which preserve the intended specifications *and* provide an at most polynomial increase of proof length for *all* formulae. We will investigate the computational correspondence between structuring $\mathcal{LJ}$-proofs with the cut rule and procedural programming concepts.

References

1. J. L. BATES AND R. L. CONSTABLE. Proofs as programs. *ACM Transactions on Programming Languages and Systems*, 7(1):113–136, January 1985.
2. W. BIBEL, D. KORN, C. KREITZ, F. KURUCZ, J. OTTEN, S. SCHMITT, AND G. STOLPMANN. A Multi-Level Approach to Program Synthesis. In 7th LoPSTr Workshop, LNCS, 1998.
3. R. L. CONSTABLE, S. F. ALLEN, AND H. M. BROMLEY. *Implementing Mathematics with the NuPRL proof development system*. Prentice Hall, 1986.

4. H. B. CURRY. *Foundations of Mathematical Logic.* Dover, Dover edition, 1977.
5. E. EDER. *Relative Complexities of First Order Calculi.* Vieweg, 1992.
6. G. GENTZEN. Untersuchungen über das logische Schließen. *Mathematische Zeitschrift*, 39:176–210, 405–431, 1935.
7. S. C. KLEENE. Permutability of Inferences in Gentzen's Calculi LK and LJ. *Memoirs of the AMS*, 10:1–26, 1952.
8. S. MAEHARA. Eine Darstellung der intuitionistischen Logik in der klassischen. *Nagoya Mathematical Journal*, 7:45–64, 1954.
9. P. MARTIN-LÖF. *Intuitionistic Type Theory*, volume 1 of *Studies in Proof Theory Lecture Notes.* Bibliopolis, Napoli, 1984.
10. J. OTTEN AND C. KREITZ. A Uniform Proof Procedure for Classical and Nonclassical Logics. In *20th German Annual Conference on AI*, LNAI 1137, pp. 307–319, 1996.
11. S. SCHMITT AND C. KREITZ. On transforming intuitionistic matrix proofs into standard-sequent proofs. In *4th TABLEAUX Workshop*, LNAI 918, pp. 106–121, 1995.
12. S. SCHMITT AND C. KREITZ. Converting non-classical matrix proofs into sequent-style systems. In *CADE-13*, LNAI 1104, pp. 418–432, 1996.
13. T. TAMMET A Resolution Theorem Prover for Intuitionistic Logic. In *CADE-13*, LNAI 1104, pp. 2–16, 1996.
14. A. S. TROELSTRA AND H. SCHWICHTENBERG. *Basic Proof Theory.* Cambridge Univ. Press, 1996.
15. L. WALLEN. *Automated deduction in nonclassical logics.* MIT Press, 1990.

Combining Algebraic Computing and Term-Rewriting for Geometry Theorem Proving*

Stéphane Fèvre and Dongming Wang

Laboratoire LEIBNIZ – Institut IMAG
46, avenue Félix Viallet, 38031 Grenoble Cedex, France

Abstract. This note reports some of our investigations on combining algebraic computing and term-rewriting techniques for automated geometry theorem proving. A general approach is proposed that requires both Clifford algebraic reduction and term-rewriting. Preliminary experiments for some concrete cases have been carried out by combining routines implemented in Maple V and Objective Caml. The experiments together with several examples illustrate the suitability and performance of our approach.

1 Motivation

This work is motivated by our investigations on automated geometry theorem proving (GTP) using Clifford algebra and rewrite rules [4,12]. Recent research by Li [7,8] and us has demonstrated the power and capability of Clifford algebra in expressing geometric problems for the purpose of automated reasoning. Using the proposed methods, one needs to deal with Clifford algebraic expressions (or *Clifford expressions* for short), for which computer algebra (CA) systems such as Maple and Mathematica have shown to be appropriate. Since Clifford expressions usually involve operators (e.g., outer and inner products) other than sum and product as in the polynomial case, an expression that is identically equal to 0 does not simply evaluate to 0. We found that effective evaluation of such expressions may be achieved by developing a term-rewriting system (TRS) with suitably chosen rules.

Implementing a geometry theorem prover based on our approach thus requires tools for both symbolic algebraic computation and term-rewriting. The former are available typically in CA systems, while the latter have been developed largely in the community of rewriting techniques. There is no satisfactory common environment in which the two kinds of tools are integrated with desired performance. Although some CA systems like Mathematica provide rewriting functionality in their programming languages, the power of rewriting techniques is not fully implemented therein; whereas the existing TRS and tools do not contain functions for advanced algebraic computation. Given the nature and design

---

* This work is supported partially by CEC under Reactive LTR Project 21914 (CUMULI) and a project of LIAMA.

of such systems, it is clearly inefficient to implement algorithms for algebraic computation in a TRS, and *vice versa*. Take for example Maple, in which algebraic computations such as simplification and factorization that we need work pretty well. For experimental purpose, we wrote a set of rewriting routines in Maple to evaluate Clifford expressions to 0. It turns out that the performance of these routines is rather poor. The reason seems quite simple: Maple is not suitable for implementing term-rewriting techniques.

For these and other technical reasons, we chose to implement functions for Clifford algebraic computation in Maple and those for term-rewriting in Objective Caml; these functions are combined by building an interface between the two systems. Implementation issues and other considerations will be described in Sect. 3 of this paper. In the following section, we shall explain a general approach that combines Clifford algebraic computation and term-rewriting for GTP. The suitability and performance of this approach will be illustrated by some examples and our preliminary experiments in Sect. 4.

2   A Combinational Approach

In this section we sketch a general approach for GTP, based on a combination of calculation with Clifford expressions and evaluation of such expressions to 0 by term-rewriting.

2.1   Clifford Algebra

We shall work with a ground field $\mathbb{K}$ of characteristic $\neq 2$, an n-dimensional vector space $\mathcal{V}$, and a Clifford algebra $\mathcal{C}$ associated to some fixed quadratic form over $\mathbb{K}$. The reader is referred to [5,8,12] for some basics of Clifford algebra. The elements of $\mathcal{C}$, sometimes called *Clifford numbers*, are multivectors. Among Clifford numbers, various operators may be introduced. Typical operators include *geometric sum, geometric product, inner product, outer product, cross product, meet,* and *dual operator*. They will be referred to as *Clifford operators*.

For example, let $\mathbf{a}$ and $\mathbf{b}$ be any two vectors in $\mathcal{V}$. The geometric sum of $\mathbf{a}$ and $\mathbf{b}$, $\mathbf{a} + \mathbf{b}$, is also a vector of $\mathcal{V}$; the inner product $\mathbf{a} \cdot \mathbf{b}$ of $\mathbf{a}$ and $\mathbf{b}$ is an element of $\mathbb{K}$; the outer product $\mathbf{a} \wedge \mathbf{b}$ of $\mathbf{a}$ and $\mathbf{b}$ is a bivector in $\mathcal{C}$. All these Clifford operators have clear geometric meanings, see [5] for details. They obey certain basic calculational laws and are related to each other. For instance,

$$\mathbf{a} \wedge \mathbf{b} = \mathbf{a}\mathbf{b} - \mathbf{a} \cdot \mathbf{b},$$

where $\mathbf{a}\mathbf{b}$ denotes the geometric product of $\mathbf{a}, \mathbf{b} \in \mathcal{V}$. Some of the laws/relations are selected and grouped as rewrite rules in [12]. The rich variety of operators together with such rules makes Clifford algebra a powerful language for expressing geometric notions and relations.

2.2   Formulation of Geometric Theorems

Now, let us take a concrete geometry of fixed dimension. Examples of such geometries are plane Euclidean geometry, solid geometry, non-Euclidean geometry, and differential geometry. By a *geometric entity* we restrict it to be a point, or a geometric scalar such as the length of a segment or the area of a triangle. Consider those theorems in the geometry, each of which may be formulated constructively as follows.

Starting with finitely many given free geometric entities $\mu_1, \ldots, \mu_e$, construct new geometric entities one by one. Some of the constructed entities may be completely free, while the others are constrained by the geometric hypotheses. Let the former be denoted $\mu_{e+1}, \ldots, \mu_d$ ($d \geq e$) and the latter $\chi_1, \ldots, \chi_r$. For convenience, we shall call the free geometric entities $\mu_1, \ldots, \mu_d$ *parameters*, abbreviated to μ, and the constrained geometric entities $\chi_1, \ldots, \chi_r$ *dependents*.

Take a proper Clifford algebra so that the geometric relations for the constructed dependents χ_i can be given algebraically in the following form

$$
\begin{aligned}
h_1(\mu, \chi_1) &= 0, \\
h_2(\mu, \chi_1, \chi_2) &= 0, \\
&\cdots\cdots \\
h_r(\mu, \chi_1, \chi_2, \ldots, \chi_r) &= 0,
\end{aligned}
\tag{H}
$$

where each h_i is composed by means of Clifford operators. Then, (H) constitutes the hypothesis of the geometric theorem under discussion. The reader may consult [4,7,8,12] for how to represent geometric relations using Clifford algebra.

Or more generally, one may consider any geometric theorem whose hypothesis can be first Clifford-algebraized and then transformed into the form (H). One such triangularization procedure as proposed by Li and Cheng [7,9] is an analogy to Ritt-Wu's well-ordering principle [13] for polynomials.

It is a special and simple case when each dependent χ_i can be solved in terms of μ and $\chi_1, \ldots, \chi_{i-1}$ using Clifford operators; namely,

$$
\chi_i = f_i(\mu, \chi_1, \ldots, \chi_{i-1}), \quad 1 \leq i \leq r.
\tag{H*}
$$

Suppose that the conclusion of the theorem to be proved is given as another Clifford expression

$$
g = g(\mu, \chi_1, \ldots, \chi_r) = 0.
\tag{C}
$$

Proving the theorem amounts to verifying whether (C) follows from (H), possibly under some subsidiary conditions.

2.3   Reduction and Rewriting

In order to show that (H) implies (C), one proceeds to reduce the conclusion-expression g by the hypothesis-expressions $h_r, \ldots, h_1$ successively. During the reduction, some conditions of the form $d_i \neq 0$ may have to be imposed; such conditions are usually taken as the *non-degeneracy conditions* for the theorem to be true. After the reduction is completed, a Clifford expression r will be obtained.

In the special case when (H) is replaced by (H*), the reduction is performed by simply substituting the expressions f_i of the dependents into the conclusion-expression g; the obtained r involves μ only, i.e., $r = r(\mu)$. This has been detailed in [4,12].

Next, we want to verify whether r is identically equal to 0, i.e., whether $r = 0$ is an identity. Recall that r may comprise several Clifford operators. It does not automatically evaluate to 0 even if it is identically 0. How to evaluate r or a factor of r to 0 is a crucial issue that was not addressed in the work of Li and his collaborators. An easy way that was taken by Li is to use Wu's coordinate-based method when r cannot be simplified to 0. Note that the main advantage of using Clifford algebra for GTP is to work with points and invariants, so we apparently wish to avoid introducing coordinates if possible. This is one of the major motivations for us to advocate using term-rewriting techniques explained below.

The Clifford operators involved in the expression r are related by a set of computational laws. It is under these laws that $r = 0$ may become an identity. The laws can be written as equational relations and thus can be represented as rewrite rules. So now the problem is how to rewrite r to 0 using the rules. For this, one can develop a TRS on the basis of various efficient techniques that have been proposed. In [4] we have presented one such rewriting system for the special case in which $r = r(\mu)$ and only geometric sum, inner/outer/geometric products, and plane dual operator are involved.

2.4 Combination

What we have suggested above is a general approach for GTP that requires the manipulation and computation with Clifford expressions and the proof of identities by term-rewriting. It is known that algebraic computation and term-rewriting are two related yet somewhat separate areas of research. The existing CA systems do not have rich rewrite functionality, nor provide ideal features for easy and efficient implementation of up-to-date rewrite techniques.

On the other hand, the well-developed rewrite libraries and tools do not support effective algebraic computation. In our case, some rather advanced algebraic computations such as simplification and factorization are needed. It is certainly expensive and unrealistic to develop routines for such computations in a rewriting environment. As a consequence, we propose to combine routines and tools developed in CA and rewriting systems. Advantages and general methodologies of combining algebraic computing and theorem proving have been discussed, for instance, in [2,3,6]. Following some of the suggested strategies, we have been experimenting with our combinational approach for some restricted cases, in particular, proving theorems in plane Euclidean geometry.

3   Implementation in Maple and Caml

To experiment with the combinational approach suggested in the preceding section, we have started an implementation in Maple V and Objective Caml for a large enough class of theorems in plane Euclidean geometry, whose hypotheses can be expressed in the form (H*). A technical account of our method for proving this class of theorems is given in [4,12]. This section provides details about implementation issues and combination strategies. The objective of this work is twofold: on one hand, we want to develop a new and high-performance geometry theorem prover based on coordinate-free methods, in particular, those using Clifford algebra; the current program will be part of this prover. On the other hand, we are interested in recent research on combining algebraic computing and theorem proving (see [2,6]). Our investigation may be considered as a practical exercise on this subject for the case of geometry.

3.1   Why Maple and Caml

In order to achieve a fine combination of term-rewriting and algebraic computing, we need polynomial formulas coded into the rewriting steps. Most popular rewriting packages do not supply any evaluation mechanism for such formulas and thus are not capable of handling our problem appropriately. So it is necessary to develop a new sample package for such a very special kind of rewriting. We decided to do it in Objective Caml, a functional polymorphic language dialect of ML. Note that ML is widely used in the automated theorem proving community and recognized for being well-suited to develop prototypes. Its type inference mechanism is well-adapted to programming, and polymorphism allows to program once an algorithm used in several contexts. Our representation of Clifford expressions uses terms both for scalar expressions and purely geometric expressions. Although being of different types, the two kinds of terms are mostly treated by the same polymorphic functions; only few of them had to be adapted. Moreover functionality provides a good solution for combining strategies. Last, the interpreter provides a good interactive system.

On the other hand, it is very time consuming and difficult to program efficient algebraic computation procedures (in a rewriting environment) as mentioned before. It is desirable to take one of the existing CA systems in which powerful algebraic computation routines have already been well implemented. Despite that Mathematica and other CA systems are good candidates, our long-time experience influenced our choice of Maple. The main features we need are substitution, simplification, factorization, collection and sorting, which exist in Maple as well as many other CA systems. Previous experiments with Maple have showed us the importance of designing sophisticated strategies and efficient rewriting procedures. However, the Maple language tends to make large programs very tricky and makes difficult the modification and test of new strategies — it is not adequate for programming rewriting techniques. The above considerations give us the major motivations for pursuing a combination of Caml and Maple.

3.2 Rewriting

In [4] we have presented a technique allowing to perform algebraic computations while rewriting a term into another by replacing a variable with a constant standing for the result of these computations. This idea is presented as an operational semantic of normalized rewriting [10].

In our implementation, we make a distinction between constant symbols of the signature of the term algebra and constants, also called *atoms*, representing algebraic objects (here, polynomials). Normalized rewriting is presented by its authors as an attempt to unifying the Knuth-Bendix completion, Buchberger critical-pair and many other similar algorithms. In some previous work, for example [2], external computations are modeled by mappings from variables or constants to external objects. Actually the notion of *extended term rewriting system* is more general than what is presented here as we do not consider rules with variables in their left-hand side matching external objects. In our scheme they only match ground terms whose constants denote algebraic laws or elements in an algebraic domain. Moreover, as suggested in [4], these objects are normalized before a rule is applied. This normalization concretely consists in sending an algebraic expression to a CA system and then reading the result. The simple interface may be done either by using Unix pipes as we did or by using the Maple's facilities for calling kernel functions. The normalization strategy is called *immediate computation* in [2]. Following the notations in [10], this could be defined as follows. If $s \downarrow_S$ denotes the S-normal form of the term s, the rewrite relation by a rule r at position p is defined by: $s \rightarrow^p_{r/S} t$ iff there is a term u such that $s \rightarrow^p_r u$ and $t = u \downarrow_S$. The two previous approaches to including external computations in rewriting are different, while ours may be situated in their intersection. Also, it seems that none of the former has been validated by combining a CA system with a TRS.

3.3 Design

The architecture of our prototype is simple and allows one to use the user interface of GEOTHER [11], a geometry theorem prover developed by the second author using several methods (Wu's characteristic sets [13], Buchberger's Gröbner bases [1], etc.).

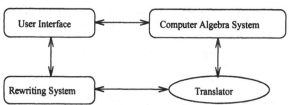

High-level functions of GEOTHER are used to send the Clifford expressions to be reduced to the rewriting module. The latter then uses the CA capabilities to simplify some expressions involving polynomials for instance. When the expression is reduced, a function is called with the result as argument. Thus both the CA system and the TRS cooperate.

3.4   Term Representation

Clifford expressions are represented by terms based on the signature

$$\Sigma = \{+, -, *, /, 1, 0, **, \wedge, \cdot, v\},$$

where every symbol is binary except the unary "$-$", the ternary v, and the constants "$1, 0$". Multivectors are either stated and thus represented by a new constant or built up from others. The term $v(g, f, e)$ stands for the geometric product of f and g, where f is any expression in the basic field (e.g., a fraction), e is a Clifford expression and g is the grade of the expression e when e is not equal to 0. If e equals 0, g does not mean anything. For instance, $v(2, +(-(t), 1)), \wedge(x, y))$ represents $(1 - t)\, x \wedge y$, where t is a scalar parameter and x, y are multivectors. As the multivector $\wedge$ is assumed to be of grade 2, x and y here should necessarily represent vectors (see [4] for an overview of the rules for computing a grade). In the previous expression, the two first arguments are what we call *atoms*. Atoms are considered as constant symbols by the rewrite system, except that they actually are subtrees and assumed to be in canonical form. This form is computed by the CA system. An atom can only be matched by another atom structurally identical. Rules have four parts: a conditional part, the left-hand side, the right-hand side and a definition part. The last defines what computation should be done for atoms. For instance, the following rule, written in infix form, is part of the system:

$$a > b \,:\, v(g_1, f_1, v(g_2, f_2, a) \wedge v(g_3, f_3, b)) \rightarrow v(g_1, f_4, v(g_3, 1, b) \wedge v(g_2, 1, a))$$
$$\{\, f_4 = (-1)^{(g_2 + g_3)} f_1 f_2 f_3 \,\}.$$

In general, the fourth part of a rule defines a substitution to apply after computation of the terms to substitute. Any variable occurring in this part denotes an atom. Thus variables are replaced by atoms.

The key point in designing a strategy for reducing Clifford expressions is to avoid a dramatic combinatorial explosion. This is due to the definition of inner and outer products of two vectors:

$$x \cdot y = \frac{xy + yx}{2}, \quad x \wedge y = \frac{xy - yx}{2}$$

which are bilinear. That is why any derived term is further reduced by another simplification system. The whole strategy is defined by composition of several elementary strategies (see [4] for more details).

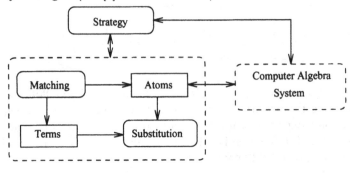

This kind of implementation is made very easy by the use of a special-purpose rewriting system named **EZTerm** we have developed. Moreover, pattern-matching and substitution are performed by linear time complexity algorithms. The use of a CA system not only makes easy the computation of algebraic expressions but also helps in normalizing expressions. For instance, to improve efficiency of our system for disproving conjectures, summands are efficiently grouped by grade and sorted by Maple. Then the system attempts to reduce each homogeneous expression to zero starting from the (supposedly) highest grade expression to the lowest (see the above figure). This reduces the size of each treated expression and improves the performance. Thus combining systems can also contribute to use new strategies.

4   Examples and Performance

As shown in the table below, a number of non-trivial geometric theorems may be proved effectively by combining algebraic computing and term-rewriting. The following three examples serve to illustrate the suitability of this combinational approach.

Example 1. Let ABC_1, BCA_1 and CAB_1 be three equilateral triangles drawn all inward or all outward on the three sides of an arbitrary $\triangle ABC$. It is proved as Example 3 (b) in [4] that the circumcircles of $\triangle ABC_1, \triangle BCA_1, \triangle CAB_1$ are concurrent. Now denote the centroids of $\triangle ABC_1, \triangle BCA_1, \triangle CAB_1$ by C_0, A_0, B_0 respectively. Then the circumcircles of $\triangle ABC_0, \triangle BCA_0, \triangle CAB_0$ are also concurrent.

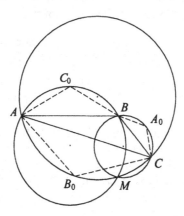

Let the vertex C be located at the origin, i.e., $C = 0$. Then the points A_0, B_0, C_0 may be represented in terms of A and B as follows

$$A_0 = \frac{B}{2} \pm \frac{\sqrt{3}}{6} B^{\sim}, \quad B_0 = \frac{A}{2} \mp \frac{\sqrt{3}}{6} A^{\sim}, \quad C_0 = \frac{A+B}{2} \pm \frac{\sqrt{3}}{6}(A^{\sim} - B^{\sim}),$$

where $\sim$ is the dual operator (see [12] for example). Denote the circumcenters of $\triangle BCA_0, \triangle CAB_0, \triangle ABC_0$ by O_A, O_B, O_C respectively. We have

$$O_A = \frac{B \cdot B\, A_0^\sim - A_0 \cdot A_0\, B^\sim}{2\, B \cdot A_0^\sim},$$

$$O_B = \frac{A \cdot A\, B_0^\sim - B_0 \cdot B_0\, A^\sim}{2\, A \cdot B_0^\sim},$$

$$O_C = \frac{(A \cdot A - B \cdot B)\, C_0^\sim + (B \cdot B - C_0 \cdot C_0)\, A^\sim + (C_0 \cdot C_0 - A \cdot A)\, B^\sim}{2\,(A \cdot C_0^\sim - B \cdot C_0^\sim - A \cdot B^\sim)}.$$

Let M be the reflection of C with respect to $O_A O_B$; then

$$M = 2\,\frac{(O_B \cdot O_B - O_A \cdot O_B)\, O_A + (O_A \cdot O_A - O_A \cdot O_B)\, O_B}{(O_B - O_A) \cdot (O_B - O_A)}.$$

We want to prove that

$$g = (M + B - 2\,O_A) \wedge (O_C - O_A) = 0,$$

i.e., M is the reflection of B with respect to $O_A O_C$. For this purpose, substitute the expressions of $M, O_C, O_B, \ldots$ into g. This is a typical problem of algebraic computation with functions and radicals. For instance, after the substitution done in Maple V the numerator of g is of the form

$$c_1 \sqrt{3} + c_0,$$

where the coefficients c_1 and c_0 of $\sqrt{3}$ are two Clifford polynomials consisting of 17 and 16 terms, respectively. The collection of c_1 and c_0 is also typical in CA systems. Obviously, $g \equiv 0$ under the conditions

$B \cdot A_0^\sim \neq 0,$	% not col(A_0, B, C)
$A \cdot B_0^\sim \neq 0,$	% not col(B_0, A, C)
$A \cdot C_0^\sim - B \cdot C_0^\sim - A \cdot B^\sim \neq 0,$	% not col(C_0, A, B)
$(O_B - O_A) \cdot (O_B - O_A) \neq 0$	% $O_A O_B$ is non-isotropic

iff both $c_1 \equiv 0$ and $c_0 \equiv 0$ (here "not col$(A, B, C, \ldots)$" means that $A, B, C, \ldots$ are not collinear). To show the latter we need rewrite c_0 and c_1 to 0 by applying the rules relating the Clifford operators.

Application of our rewriting package EZTerm in Objective Caml to both c_0 and c_1 yields 0, so the theorem is proved to be true under the above-mentioned non-degeneracy conditions.

It is not so simple to prove this theorem using Wu's and other coordinate-based methods.

Example 2. Let ABC be an arbitrary triangle with orthocenter H and circumcenter O. Denote the circumcenters of $\triangle BCH$ and $\triangle ACH$ by A_1 and B_1 respectively. Then the three lines AA_1, BB_1 and OH are concurrent.

Using the constructions listed in
[4,12], the hypothesis of this theorem may be given as follows

$$H = \text{ort\_ctr}(A, B, C),$$
$$O = \text{cir\_ctr}(A, B, C),$$
$$A_1 = \text{cir\_ctr}(B, C, H),$$
$$B_1 = \text{cir\_ctr}(A, C, H).$$

To simplify calculation, one may
take C as the origin: $C = 0$. Let

$$I = \text{int}(A, A_1, B, B_1).$$

The explicit expressions of O,
A_1, B_1 and I can be written out
according to the formulas in [4,12]. The conclusion of the theorem to be proved
is

$$g = (I - O) \wedge (H - O) = 0.   \quad \text{% col}(I, O, H)$$

Substituting the Clifford expressions of I, B_1, etc. into g, we get a fraction of
Clifford polynomials in A and B. The numerator consists of 42 terms and can
be proved to be identically 0 by our TRS, wherefore the theorem is true under
some non-degeneracy conditions.

Example 3. Using our combinational approach to prove the butterfly theorem,
we need to rewrite a large Clifford expression r (consisting of 588 terms) to 0.
Application of EZTerm without using Maple takes 58.9 seconds. When r is sent
to and factored by Maple, we get six factors, one of which is easily rewritten
to 0. So the theorem may be proved in about 44 seconds of total CPU time as
shown in the table.

Factorization is one of the most difficult tasks in computer algebra, for which
modern research has been conducted for several decades. The above example ex-
hibits an obvious advantage of combination, as implementing another (efficient)
factorizer in Caml is neither easy nor reasonable.

We provide below a table of experimental results for 15 significant well-known
theorems proved automatically; some of them cannot be easily proved by using
other methods like Wu's [13]. These experiments were made on a Sparcserver
400 with 128MB of memory using Maple V $R3$ and Objective Caml v1.6. Maple,
Comm and Caml in the heading entries of the table indicate the computing times
in Maple, for communication, and in Caml, given in CPU seconds. The size of
the expressions (to be rewritten to 0) measured by the Maple function length
is shown in the Size column. As usual, times give only an indication and may
slightly vary from one session to another. It should be noticed that these times
also depend on the initial statements of the theorems. Some of them are stated
using high-level predicates further translated into Clifford expressions, while the
others are stated directly using Clifford expressions.

Theorem	Maple	Comm	Caml	Size
Orthocenter	1.4	0.01	0.01	74
Centroid	0.74	0.01	1.18	297
Euler line	2.09	0.02	0.34	429
Desargues	0.92	0.02	1.26	457
Pappus	3.5	0.04	0.01	1665
Gauss line	2.12	0.02	0.75	489
Ex 3 (a) [4]	2.09	0.43	1.79	797
Ex 3 (b) [4]	6.08	1.27	30.5	1741
Ex 3 (c) [4]	0.51	0.22	0.51	154
Ex 3 (d) [4]	0.34	0.02	0.57	112
Example 1	2.6	1.01	9.17	1892
Example 2	0.03	0.01	0.01	529
Pivot	1.02	0.34	49.24	274
Simson	2.24	0.51	15.63	3591
Butterfly	12.85	8.17	22.7	36828

This table shows that the time needed for communication between Maple and EZTerm is negligible compared with that for computation. As opposed to the approach described in [4] (where expressions were extracted from the initial statements by hand), Maple is used here more intensively. For instance, every summation is now simplified by Maple. Moreover, we try to factor large expressions to find a null factor. Last, a better strategy (compared with [4]) used for computing the statements allows to reduce the problem size. It is interesting to note that for larger theorems the rewriting side becomes less efficient with respect to algebraic computing.

Comparing the proofs and timings produced by our approach with those by others is beyond our current intention and interest. The comparison and how to do it fairly and significantly still remain for further investigations.

5 Conclusion and Remarks

This paper studies the combinational aspects of algebraic computing and term-rewriting, and respectively two strongly interfaced systems, Maple V and Objective Caml. A significant extension has been implemented that makes possible to work on base fields other than $\mathbb{Q}$ (the field of rationals). This also allows to deduce necessary and sufficient algebraic conditions from a reduced Clifford expression involving scalar parameters. Preliminary experiments have shown that combination in our case is a realistic and promising approach, by which difficult geometric theorems can be proved efficiently in a coordinate-free environment.

Although this work is incremental with respect to [4,12], the approach proposed here is more general, the TRS is extended and refined, the combinational strategies have not been described before, and the implementation of modules combining Caml and Maple is new. Our implementation is still in progress. For the Maple part, the Clifford operators like inner/outer products and the dual operator are defined as Maple procedures. When used in computation, they often have some uneven behavior. We continue observing such behavior and believe that a lot of improvements can be obtained by optimizing the Maple code. We also try to mechanize the analysis of produced Clifford expressions to get a minimal set of conditions (for the theorem to be true) and the interpretation of such conditions in a high-level language involving the usual predicates (collinearity, parallelism and so on). For the rewriting part, we still study the properties of the equational system to improve the application of rules. For the combined system, we are introducing new strategies to limit the size of computed expressions: some experiments by hand seem to show that lots of improvements are possible. Studying new communication strategies is the most promising possibility for proving still more complicated theorems. We also plan to extend the application domain of our implementation for proving theorems in other geometries including solid geometry and differential geometry in the near future.

References

1. Buchberger, B.: Gröbner bases: An algorithmic method in polynomial ideal theory. In: Multidimensional systems theory (N. K. Bose, ed.), D. Reidel, Dordrecht Boston, pp. 184–232 (1985).
2. Bündgen, R.: Combining computer algebra and rule based reasoning. In: Proc. AISMC-2 (Cambridge, UK, August 3–5, 1994), LNCS **958**, pp. 209–223 (1995).
3. Fèvre, S.: Integration of reasoning and algebraic calculus in geometry. In: Automated deduction in geometry (D. Wang et al., eds.), LNAI **1360**, pp. 218–234 (1998).
4. Fèvre, S., Wang, D.: Proving geometric theorems using Clifford algebra and rewrite rules. In: Proc. CADE-15 (Lindau, Germany, July 5–10, 1998), LNAI (to appear).
5. Hestenes, D., Sobczyk, G.: Clifford algebra to geometric calculus. D. Reidel, Dordrecht Boston (1984).
6. Homann, K., Calmet, J.: Combining theorem proving and symbolic mathematical computing. In: Proc. AISMC-2 (Cambridge, UK, August 3–5, 1994), LNCS **958**, pp. 18–29 (1995).
7. Li, H.: New explorations on mechanical theorem proving of geometries. Ph.D thesis, Beijing University, China (1994).
8. Li, H.: Vectorial equations-solving for mechanical geometry theorem proving. J. Automat. Reason. (to appear).
9. Li, H., Cheng, M.-t.: Proving theorems in elementary geometry with Clifford algebraic method. Chinese Math. Progress **26**: 357–371 (1997).
10. Marché, C.: Normalized rewriting: An alternative to rewriting modulo a set of equations. J. Symb. Comput. **3**: 253–288 (1996).
11. Wang, D.: GEOTHER: A geometry theorem prover. In: Proc. CADE-13 (New Brunswick, USA, July 30 – August 3, 1996), LNAI **1104**, pp. 166–170 (1996).
12. Wang, D.: Clifford algebraic calculus for geometric reasoning with application to computer vision. In: Automated deduction in geometry (D. Wang et al., eds.), LNAI **1360**, pp. 115–140 (1998).
13. Wu, W.-t.: Mechanical theorem proving in geometries: Basic principles. Springer, Wien New York (1994).

Cooperation Between Top-Down and Bottom-Up Theorem Provers by Subgoal Clause Transfer

Dirk Fuchs

FB Informatik
Universität Kaiserslautern
67663 Kaiserslautern, Germany
dfuchs@informatik.uni-kl.de

Abstract. Top-down and bottom-up theorem proving approaches have each specific advantages and disadvantages. Bottom-up provers profit from strong redundancy control and suffer from the lack of goal-orientation, whereas top-down provers are goal-oriented but have weak calculi when their proof lengths are considered. In order to integrate both approaches our method is to achieve cooperation between a top-down and a bottom-up prover: the top-down prover generates subgoal clauses, then they are processed by a bottom-up prover. We discuss theoretic aspects of this methodology and we introduce techniques for a relevancy-based filtering of generated subgoal clauses. Experiments with a model elimination and a superposition prover reveal the high potential of our approach.

1   Introduction

Automated deduction is—at its lowest level—a search problem that spans huge search spaces. In the past, many different calculi have hence been developed in order to cope with problems from the area of automated theorem proving. Essentially, for first-order theorem proving two main different paradigms for calculi are in use: *Top-down calculi* like *model elimination* (ME, see [Lov68]) attempt to recursively break down and transform a goal into subgoals that can finally be proven immediately with the axioms or with assumptions made during the proof. *Bottom-up calculi* like *superposition* (see [BG94]) go the other way by producing consequences from the initial clauses until the empty clause is derived.

When comparing results of various provers it is obvious that provers based on different paradigms often have quite a different behavior. There are problems where bottom-up theorem provers perform considerably well, but top-down provers perform poorly, and vice versa. The main reason for this is that many bottom-up provers suffer from the lack of goal-orientation of their search, but profit from their strong redundancy control mechanisms. In contrast, top-down provers profit from their goal-orientation and suffer from insufficient redundancy control thus entailing long proofs for many problems. Therefore, a topic that has come into focus of research is the integration of both approaches. In particular, cooperation between theorem provers based on top-down and bottom-up principles (see, e.g., [Sut92, Sch94, Fuc98b]) appears to be a promising way because by

Jacques Calmet and Jan Plaza (Eds.): AISC'98, LNAI 1476, pp. 157–169, 1998.
© Springer-Verlag Berlin Heidelberg 1998

exchanging information each approach can profit from the other. Note that it is also possible to modify calculi or provers which work according to one paradigm so as to introduce aspects of the other paradigm into it. This, however, requires a lot of implementational effort to modify the provers, whereas our approach does not require changes of the provers but only changes of their input (see section 3). As a consequence, we can employ arbitrary state-of-the-art provers.

Information well-suited for top-down provers are lemmas deduced by bottom-up provers. These lemmas are added to the input of a top-down prover and can help to shorten the proof length by immediately solving subgoals. Since the arising problem to filter relevant lemmas has already been discussed (see, e.g., [AS92], [AL97], [Fuc98b]) we are not going to deal with this aspect here.

Instead, we want to consider top-down/bottom-up integration by transferring information from a top-down prover to a bottom-up prover. By transferring top-down generated *subgoal clauses* we are able to introduce a goal-oriented component into a bottom-up prover which enables it to solve proof problems considerably faster (see section 5). However, since similar to the use of lemmas an unbounded transfer of subgoal clauses is not sensible, techniques for *filtering relevant subgoal clauses* must be developed. Note that the method proposed here is a possible solution to the problem of extracting information suitable for cooperation from top-down theorem provers. This problem remained unsolved in [DF98] where a general framework for cooperation of different kinds of theorem provers was introduced.

In order to examine our kind of top-down/bottom-up integration we start by giving a brief overview about superposition and model elimination theorem proving in section 2. After that, in section 3 we introduce subgoal clauses and discuss effects of the integration of ME subgoal clauses into the search state of a superposition-based prover. In section 4, we point out two variants of a relevancy-based filtering of subgoal clauses. An experimental study conducted with the theorem provers SETHEO [MIL+97] and SPASS [WGR96] reveals the potential of our techniques. Finally, a discussion and an outlook at possible future work conclude the paper.

2 Automated Theorem Proving with Superposition and Model Elimination

The general problem in first-order theorem proving is to show the inconsistency of a given set C of clauses. Clauses are sets of literals. As already discussed, theorem provers utilize either top-down or bottom-up calculi to accomplish this task. In the following, we want to introduce the calculi that we employ for our theoretic and experimental study.

Typically, a bottom-up calculus contains several inference rules which can be applied to a set of clauses that constitute the search state. The superposition calculus (e.g., [BG94]) contains the inference rules superposition, equality resolution, and equality factoring. It is to be emphasized that we employ the version of the superposition calculus introduced in [BG94]. Specifically this entails that factoring is only applied to positive literals.

Usually, a bottom-up theorem prover maintains a set $\mathcal{F}^P$ of so-called *potential* or *passive clauses* from which it selects and removes one clause C at a time. This clause is put into the set $\mathcal{F}^A$ of *activated clauses*. Activated clauses are, unlike potential clauses, allowed to produce new clauses via the application of some inference rules. The inferred new clauses are put into $\mathcal{F}^P$. Initially, $\mathcal{F}^A = \emptyset$ and $\mathcal{F}^P = \mathcal{C}$. The indeterministic selection or *activation step* is realized by heuristic means. To this end, a heuristic $\mathcal{H}$ associates a natural number $\varpi_C \in \mathbb{N}$ with each $C \in \mathcal{F}^P$, and the $C \in \mathcal{F}^P$ with the smallest weight ϖ_C is selected.

A typic top-down calculus is model elimination, a restricted version of the *connection tableau calculus* (CTC) [LMG94]. This calculus works on *connected tableaux* for the initial clause set $\mathcal{C}$. A tableau for a set of clauses $\mathcal{C}$ is a tree whose non-root nodes are marked with literals and where it holds: if the immediate successor nodes $n_1, \ldots, n_m$ of a node n are marked with literals $l_1, \ldots, l_m$, then the clause $l_1 \vee \ldots \vee l_m$ *(tableau clause)* is an instance of a clause from $\mathcal{C}$. A tableau is called connected if it holds: if a non-leaf node n is marked with literal l then there is an immediate successor leaf-node n' marked with literal l' such that $l = \sim l'$. (Note that $\sim l = \neg l$, if l is positive, $\sim l = \tilde{l}$, if $l = \neg \tilde{l}$.)

The connection tableau calculus contains inference rules for transforming tableaux into others. The rules are *start, extension,* and *reduction*. The start rule can only be applied to the trivial tableau consisting of the unmarked root. It selects a clause from $\mathcal{C}$ and attaches its literals to the unmarked root. The step to select a clause from $\mathcal{C}$ and to attach its literals to an *open* leaf node (the branch leading to this node does not contain complementary literals) is also called *expansion*. Note that the start rule can be restricted to so-called *start relevant* clauses without causing incompleteness. Start relevancy of a clause is defined as follows. If $\mathcal{C}$ is an unsatisfiable set of clauses we call $S \in \mathcal{C}$ start relevant if there is a satisfiable subset $\mathcal{C}' \subset \mathcal{C}$ such that $\mathcal{C}' \cup \{S\}$ is unsatisfiable. Since the set of negative clauses contains at least one start relevant clause, in the following we also consider a restricted calculus which employs only negative clauses for the start expansion (CTC_{neg}). Tableau reduction closes a branch by unifying the literal which is the mark of an open leaf node (also called *subgoal*) with the complement of a literal on the same branch and applying the substitution to the whole tableau. Extension is performed by expanding a subgoal s and immediately performing a reduction step involving s and one of the newly introduced subgoals. We write $T \vdash T'$ if the tableau T' can be derived from T by the application of one inference rule. A *search tree* $\mathcal{T}^C$ defined by a set of clauses $\mathcal{C}$ is given by a tree, whose root is labeled with the trivial tableau. Every inner node in $\mathcal{T}^C$ labeled with tableau S has as immediate successors the maximal set of nodes $\{v_1, \ldots, v_n\}$, where v_i is labeled with S_i and $S \vdash S_i$, $1 \leq i \leq n$.

A theorem prover based on the connection tableau calculus tries to solve a proof problem $\mathcal{C}$ by deriving from the trivial tableau a closed tableau, i.e. a tableau with no open branch. A common way to perform derivations is to employ iterative deepening procedures ([Kor85]): in this approach iteratively larger finite segments of $\mathcal{T}^C$ are explored in a depth-first manner. These segments are defined by a so-called *bound* (which poses structural restrictions on the tableaux which are allowed in the current segment) and a fixed natural number, a so-called *resource*. Iterative deepening is performed by starting with a basic resource value $n \in \mathbb{N}$ and iteratively increasing n until a closed tableau is found in the current segment. A prominent example for a bound is the *inference bound* (see [Sti88]) which limits the number of inferences needed to infer a certain tableau. Note that the connection tableau calculus has no specific rules for handling equality. Hence, the common equality axioms must be added to the initial clause set when equality is involved in a problem.

3   Subgoal Clauses for Top-Down/Bottom-Up Integration

3.1   Transferring Top-Down Generated Clauses to a Bottom-Up Prover

Because of the fact that connection tableau-based provers have a search state which contains deductions (tableaux) instead of clauses, it is at first sight not obvious how to extract valid clauses from such a search state which are well-suited for a superposition-based prover. A common method in order to extract valid clauses is to employ lemma mechanisms of ME provers: Assume that a literal s is a label of the root node of a closed subtableau T^s. Let $l_1, \ldots, l_n$ be the literals that are used in reduction steps and that are outside of T^s. Then, the clause $\sim s \vee \sim l_1 \vee \ldots \vee \sim l_n$ may be derived as a new lemma (since it is a logical consequence of the tableau clauses in T^s). Then, such a lemma could be transferred to a bottom-up prover. As appealing as this idea sounds, it has some grave restrictions: Usually, such lemmas are—due to instantiations needed to close other branches before—not as general as they could be. Hence, they often cannot be used in inferences, especially not in *contracting inferences* (subsumption, rewriting) which are very important for bottom-up provers. Moreover, since these clauses are generated during the proof run in a rather unsystematic way they do not really introduce much goal-orientation and hence do not make use of the advantages of the search scheme typic for ME.

The concept of subgoal clauses, however, allows for the generation of clauses derived by inferences involving a proof goal: A *subgoal clause* S_T regarding a tableau T is the clause $S_T \equiv l_1 \vee \ldots \vee l_m$, where the literals l_i are the subgoals of the tableau T. The *subgoal clause set* $\mathcal{S}^{\mathcal{B},n,\mathcal{C}}$ w.r.t. a bound $\mathcal{B}$, a resource n, and a clause set $\mathcal{C}$, is defined by $\mathcal{S}^{\mathcal{B},n,\mathcal{C}} = (\cup S_T) \setminus \mathcal{C}$, T is a tableau which is a label of a node in the initial segment of the search tree for $\mathcal{C}$ that is defined by bound $\mathcal{B}$ and resource n. Note that subgoal clauses are valid clauses, i.e. logical consequences of the initial clause set. In order to make our method more goal-directed it is wise to only consider subgoal clauses which are derived from "real" proof goals, i.e. which are derived from start relevant clauses. E.g., it might be sensible to restrict the set of subgoal clauses to descendants of the set of negative clauses, i.e. to only consider such subgoal clauses where the start expansion was performed with a negative clause. We call the set of these subgoal clauses $\mathcal{S}^{\mathcal{B},n,\mathcal{C}}_{neg}$.

Example 1. Let $\mathcal{C} = \{\neg g, \neg p_1 \vee \ldots \vee \neg p_n \vee g, \neg q_1 \vee \ldots \vee \neg q_m \vee g\}$. Then, $\neg p_1 \vee \ldots \vee \neg p_n$ is the subgoal clause S_T belonging to the tableau obtained when extending the goal $\neg g$ with the clause $\neg p_1 \vee \ldots \vee \neg p_n \vee g$. If we employ $\mathcal{B} = $ inference bound (Inf) and resource $k = 2$, then $S^{\mathcal{B},k,\mathcal{C}} = S_{neg}^{\mathcal{B},k,\mathcal{C}} = \{\neg p_1 \vee \ldots \vee \neg p_n, \neg q_1 \vee \ldots \vee \neg q_m\}$.

A subgoal clause S_T represents a transformation of an original goal clause (which is the start clause of the tableau T) into a new subgoal clause realized by the deduction which led to the tableau T. The set $S^{Inf,k,\mathcal{C}}$ is the set of all possible goal transformations into subgoal clauses within k inferences if we consider *all input clauses* to be goal clauses, the set $S_{neg}^{Inf,k,\mathcal{C}}$ is the set of all possible goal transformations into subgoal clauses within k inferences if we only consider *the negative clauses* to be goal clauses.

Now, in order to couple a ME and a superposition prover, we generate with the inference bound and a fixed resource $k > 1$ either the set $S^{Inf,k,\mathcal{C}}$ or the set $S_{neg}^{Inf,k,\mathcal{C}}$, depending on the fact whether CTC or CTC_{neg} is used. A superposition-based prover obtains then $\mathcal{C} \cup S^{Inf,k,\mathcal{C}}$ $(\mathcal{C} \cup S_{neg}^{Inf,k,\mathcal{C}})$ as input.

3.2   Reduction of Proof Length and Search Through Subgoal Clauses

The introduced method gives rise to the question whether a *proof length reduction* is possible, i.e. whether there are shorter superposition proofs of the inconsistency of $\mathcal{C} \cup S^{Inf,k,\mathcal{C}}$ or $\mathcal{C} \cup S_{neg}^{Inf,k,\mathcal{C}}$ than of the inconsistency of $\mathcal{C}$. Note that we measure the length of a proof by counting the number of inference steps *needed in it*. This question is mainly of theoretical interest. It is more important whether a bottom-up prover can really profit from a possible proof length reduction in form of a *proof search reduction*, i.e. a reduction of the number of inferences the prover *needs in order to find* a proof. Specifically, it is interesting to find out in which cases the proof search reduction is high.

First, we assume that no equality is involved in the problem, i.e. superposition corresponds to (ordered) resolution.

Theorem 1.

1. *Let $\mathcal{C}$ be a set of ground clauses not containing equality, let $\square \notin \mathcal{C}$, and let $k > 1$ be a natural number. Let P_1 and P_2 be minimal (w.r.t. the number of inference steps $|P_1|$ and $|P_2|$) resolution refutation proofs for $\mathcal{C}$ and $\mathcal{C} \cup S^{Inf,k,\mathcal{C}}$, respectively. Then, it holds: $|P_1| > |P_2|$.*
   But there is a set of ground clauses $\mathcal{C}$ not containing equality $(\square \notin \mathcal{C})$ where no minimal resolution refutation proof of the inconsistency of $\mathcal{C} \cup S_{neg}^{Inf,2,\mathcal{C}}$ has a shorter length than a minimal proof of the inconsistency of $\mathcal{C}$.

2. *For each $k > 1$ there is a set of (non-ground) clauses $\mathcal{C}_k$ not containing equality $(\square \notin \mathcal{C}_k)$, such that no minimal resolution refutation proof for $\mathcal{C}_k \cup S^{Inf,k,\mathcal{C}_k}$ or $\mathcal{C}_k \cup S_{neg}^{Inf,k,\mathcal{C}_k}$ is shorter than a minimal resolution refutation proof for $\mathcal{C}_k$.*

Proof:

1. Due to lack of space we only prove the first part. (The proof of the second part can be found in [Fuc98a].) Note that no factorization steps are needed in the case of ground clauses (recall that clauses are sets of literals). Then, the claim is trivial since the result of the first resolution step of each minimal proof is an element of $S^{Inf,k,C}$.

2. Let $k > 1$. Let C_k be defined by $C_k = \{\neg p(x_1) \vee \ldots \vee \neg p(x_k), p(y_1) \vee \ldots \vee p(y_k)\}$. Let $>= \emptyset$ be the ordering used for superposition. Then, a minimal resolution refutation proof for C_k requires $k-1$ factorization steps (resulting in the clause $p(y_1)$) and k resolution steps. Furthermore, in S^{Inf,k,C_k} are only clauses which contain at least one positive and one negative literal. Thus, none of these clauses can lead to a refutation proof for $C_k \cup S^{Inf,k,C_k}$ in less than $2k - 1$ inferences. Since S_{neg}^{Inf,k,C_k} is a subset of S^{Inf,k,C_k} we obtain the same result in this case.

□

Hence, a reduction of the proof length is at least for ground clauses possible. However, the (heuristic) proof search of a superposition-based prover need not always profit from the proof length reduction obtained. E.g., it is possible that all clauses of a minimal refutation proof for C have smaller heuristic weights than the clauses from $S^{Inf,k,C}$ ($S_{neg}^{Inf,k,C}$) and will hence be activated before them:

Example 2. Let $>= \emptyset$ be the ordering used for superposition. Let the clause set C be given by $C = \{\neg a \vee \neg b \vee c, \neg g \vee b, a, g, \neg c\}$. The heuristic $\mathcal{H}$ corresponds to the FIFO heuristic. Further, resolvents of the two most recently activated clauses are preferred by $\mathcal{H}$. Then, following clauses are activated by the prover (in this order): $\neg a \vee \neg b \vee c, \neg g \vee b, \neg a \vee \neg g \vee c, a, \neg g \vee c, g, c, \neg c, \square$. Furthermore, if the subgoal clauses of $S^{Inf,k,C}$ ($S_{neg}^{Inf,k,C}$) are inserted behind the original axioms the prover will find the same refutation proof as before and the proof search can hence not profit from a possible proof length reduction.

However, since the example (especially the chosen heuristic) is somewhat contrived it can be expected that for many problems clauses from $S^{Inf,k,C}$ ($S_{neg}^{Inf,k,C}$) will be activated and can contribute to a reduction of the search effort.

In the case that equality is involved in the problem, a proof length reduction is not guaranteed even for ground clauses.

Theorem 2. *For each resource $k > 1$ there is a set of ground unit equations C_k ($\square \notin C_k$) where the minimal superposition refutation proofs for $C_k \cup S^{Inf,k,C_k}$ ($C_k \cup S_{neg}^{Inf,k,C_k}$) are not shorter than minimal proofs for C_k.*

Proof: Let $>= \emptyset$ be the ordering used for superposition. Consider the set of unit equations $C_k = \{a = b, f^{k-1}(a) \neq f^{k-1}(b)\}$. We assume that $>= \emptyset$ is used as an ordering for superposition. Then, a minimal superposition refutation proof for C_k requires two inferences, a superposition step into $f^{k-1}(a) \neq f^{k-1}(b)$ resulting in the inequation $f^{k-1}(a) \neq f^{k-1}(a)$, and then an equality resolution step. In the set S^{Inf,k,C_k} are either non-unit clauses whose refutation requires at least 2 inferences or the units $\mathcal{U} = \{f^i(a) \neq f^i(b), f^j(a) = f^j(b) : 0 \leq i < k - 1, 0 < j \leq k - 1\}$. Since also the refutation of $\mathcal{C}_k \cup \mathcal{U}$ requires a superposition and an equality resolution step a proof length reduction is impossible. Since $\mathcal{S}_{neg}^{Inf,k,\mathcal{C}_k}$ is a subset of $\mathcal{S}^{Inf,k,\mathcal{C}_k}$ we obtain the same result in this case. $\square$

Despite this negative result, it is sometimes possible to shorten the proof search if the search space is restructured in a favorable way.

All in all, we obtain that in general the reduction of the heuristic search for a proof cannot be guaranteed although sometimes proof lengths—at least for ground clauses and if no equality is involved in the problems—are shortened. Nevertheless, in practice it might often be the case that a restructuring of the search caused by using subgoal clauses allows for finding proofs faster. The integration of subgoal clauses into the search state of a superposition-based prover promises a strong gain of efficiency in the following cases:

Firstly, it is important that some of the subgoal clauses can be proven quite easily, especially more easily than the original goal(s). In order to estimate this, it is necessary to judge whether they can probably be solved with the help of clauses of the initial clause set. E.g., measuring similarity between a goal and other clauses with the techniques developed in [DF94] may be well-suited.

Secondly, a solution of a newly introduced subgoal clause should not always entail a solution of an original goal within few steps of the prover. If this were the case then the integration of new subgoal clauses would not promise much gain. Criteria in order to estimate this are: on the one hand, the transformation of an original goal clause into a subgoal clause by a ME prover should have been performed by using many inferences, i.e. k should be quite high. Then, it is possible that a solution of a new subgoal clause does not entail a solution of an original goal within few steps because the probability is rather high that a bottom-up prover cannot—due to its heuristic search—quickly reconstruct the inferences needed to infer the original goal. On the other hand, if there is a subgoal clause S_T and some of the tableau clauses of the tableau T have a high heuristic weight regarding the heuristic of the superposition-based prover, a high gain of efficiency may occur if the prover can prove S_T. This is because inferences needed to infer the original goal may not be performed by the prover.

4 Relevancy-Based Selection of Subgoal Clauses

Already when using small resources k the set $\mathcal{S}^{Inf,k,\mathcal{C}}$ ($\mathcal{S}_{neg}^{Inf,k,\mathcal{C}}$) can become quite large. Thus, it is not sensible to integrate all subgoal clauses from $\mathcal{S}^{Inf,k,\mathcal{C}}$ ($\mathcal{S}_{neg}^{Inf,k,\mathcal{C}}$) into the search state of a superposition-based prover: Integrating too many clauses usually does not entail a favorable rearrangement of the search because the heuristic "gets lost" in the huge number of clauses which can be derived from many subgoal clauses. Hence, it is reasonable to develop techniques for filtering subgoal clauses that entail a large gain of efficiency for a superposition prover if they can be proven. I.e. we are interested in filtering *relevant* subgoal clauses. Therefore, our approach is as follows: At first, we generate a *set of subgoal clause candidates* and then we select some subgoal clauses from this set. The

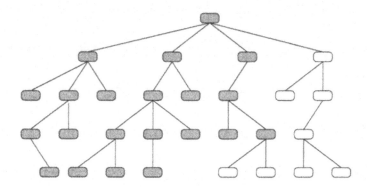

Fig. 1. Inference-based generation of a set of subgoal clause candidates

chosen subgoal clauses are added to the search state of the bottom-up prover. We shall first introduce two techniques for generating a set of subgoal clause candidates. After that, we shall deal with selecting relevant subgoal clauses.

In order to generate a set of interesting subgoal clauses it is important that we employ a large resource for generating subgoal clauses. As we have already discussed, subgoal clauses that are generated with a small number of inferences do not promise much gain because a bottom-up prover may easily reconstruct the inferences needed to infer them. Thus, if the prover is able to prove the subgoal clause it can also prove the original goal clause with few inferences and we do not gain much efficiency. However, it is not possible to generate all subgoal clauses $\mathcal{S}^{Inf,k,\mathcal{C}}$ ($\mathcal{S}_{neg}^{Inf,k,\mathcal{C}}$) for a sufficiently large resource k as subgoal clause candidates because their large number entails too high costs for the generation and additional selection. Hence, we are only able to choose as subgoal clause candidates a subset of $\mathcal{S}^{Inf,k,\mathcal{C}}$ ($\mathcal{S}_{neg}^{Inf,k,\mathcal{C}}$), k sufficiently large (see section 5).

Our first variant, an *inference-based* method, starts by generating subgoal clauses from $\mathcal{S}^{Inf,k,\mathcal{C}}$ ($\mathcal{S}_{neg}^{Inf,k,\mathcal{C}}$) for a rather large resource k and stops when N_{sg} subgoal clause candidates are generated. The advantage of this method is that it is very easy and can efficiently be implemented: Tableaux are enumerated with a fixed strategy for selecting subgoals for inferences (usually left-most/depth-first) and for each tableau its subgoal clause is stored. The main disadvantage of this method is that due to the fixed strategy and the limit of the number of subgoal clauses, we only obtain subgoal clauses which are inferred from goal clauses by expanding certain of their subgoals with a high number of inferences, other subgoals only with a small number of inferences. (See also Figure 1: Ovals are tableaux in a finite segment of the search tree $\mathcal{T}$, the lines represent the ⊢ relation. Grey ovals represent enumerated tableaux, i.e. their subgoal clauses are stored, white ovals represent tableaux which are not enumerated within N_{sg} inferences.) Thus, the method is somewhat unintelligent because no information about the quality of the transformation of an original goal clause into a subgoal clause is used. Certain transformations are favored only due to the uninformed subgoal selection strategy.

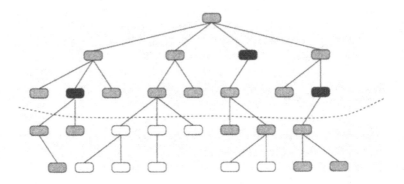

Fig. 2. Adaptive generation of a set of subgoal clause candidates

Our second variant, an *adaptive* method, tries to overcome the disadvantages of the first in the following way: Instead of allowing for more inferences when generating subgoal clauses due to an uninformed subgoal selection strategy, we want to allow for more inferences at certain *interesting positions* of the search tree T^C for a given set of clauses C.

In detail, our approach is as follows: At first, we generate all subgoal clauses $S^{Inf,k_1,C}$ ($S_{neg}^{Inf,k_1,C}$) with a resource k_1 which is smaller compared to the first variant. Then, a fixed number N_{ref} of subgoal clauses is chosen which promise the highest gain of efficiency regarding the criteria from section 3. More exactly, we choose subgoal clauses which are maximal w.r.t. function ψ:

$$\psi(S_T) = \alpha_1 \cdot I(S_T) + \alpha_2 \cdot \max\{\varpi_C : C \text{ is a tableau clause in } T\}$$
$$+ \alpha_3 \cdot \max\{sim(S_T, C) : C \in \mathcal{C}, |C| = 1\}$$

The higher the number of inferences $I(S_T)$ is which are needed to infer S_T the higher $\psi(S_T)$ should be. Hence, α_1 should be positive. Also a rating $\alpha_2 > 0$ is sensible. If there are tableau clauses C in T which have a high heuristic weight ϖ_C regarding the heuristic $\mathcal{H}$ of the superposition-based prover we can perhaps gain a lot of efficiency. The function sim measures whether literals from S_T could probably be solved with unit clauses from $\mathcal{C}$. We utilized the function *occnest* [DF94] for accomplishing this task.

Now, let $M^{N_{ref}} \subseteq S^{Inf,k_1,C}$ ($S_{neg}^{Inf,k_1,C}$) be the set of chosen subgoal clauses. Then, we generate again with a resource k_2 subgoal clauses but employ as start clauses for the tableau (subgoal clause) enumeration the clauses from $M^{N_{ref}}$. We call the set of subgoal clauses generated with this method $S^{Inf,k_2,C,M^{N_{ref}}}$. (Consider also Figure 2: The dotted line shows which subgoal clauses are generated with resource k_1. Then some of them are selected (black ovals) and used as starting points for the generation of new subgoal clauses (grey ovals) with resource k_2.) The resource k_2 should again be not too high in order to allow for a fast enumeration of the subgoal clauses. The set of subgoal clause candidates is then given by $S^{Inf,k_1,C} \cup S^{Inf,k_2,C,M^{N_{ref}}}$ ($S_{neg}^{Inf,k_1,C} \cup S^{Inf,k_2,C,M^{N_{ref}}}$) . Thus, subgoal clause candidates are all subgoal clauses derived with a certain number of inferences which may—at least for ground clauses and if no equality is involved in the problem—guarantee a reduction of the proof length. Further, we have some subgoal clause candidates derived with a higher number of inferences, at most $k_1 + k_2$. These subgoal clauses promise a high gain of efficiency because they are derived from subgoal clauses selected with function ψ.

For selecting subgoal clauses from the set of subgoal clause candidates we employed function φ, defined by $\varphi(S_T) = \psi(S_T) - \theta(S_T)$ and selected clauses with the highest weight. θ simply counts a weighted sum of the number of variables in S_T and two times the number of function or predicate symbols in S_T. Hence, "general" subgoal clauses are preferred because they can more often be used in inferences of a bottom-up prover.

5 Experimental Results

In order to evaluate our concept of integrating top-down/bottom-up provers by cooperation, we coupled two renowned provers: the ME prover SETHEO (which employs CTC_{neg}) and the superposition prover SPASS. Each prover runs on an own processor and obtains the initial clause set C as input. When tackling simple problems it is unnecessary to let the provers cooperate. Therefore, each prover tries to solve the problem independently with a timeout of 4 seconds. If no prover could solve the problem, the top-down prover generates subgoal clauses with one of the two variants. Note that in our context this does not require changes in the top-down prover but can be performed with built-ins of the PROLOG-style input language of SETHEO. Then, these subgoal clauses are filtered, transferred to the bottom-up prover, and integrated into its search state. Finally, the provers proceed to tackle the problem in parallel.

Hence, we can efficiently solve simple problems. Moreover, cooperation can be performed for harder problems after the first timeout. Note that during this cooperation phase it is also possible to add some clauses that the bottom-up prover has generated to the axiomatization of the ME prover. Thus, we can achieve cooperation by exchanging lemmas and subgoal clauses without one concept disturbing the other.

We experimented in the light of problems from the well-known problem library TPTP [SSY94]. In order to obtain a reliable collection of data, we used two domains of TPTP as our test set, the domains CAT (category theory) and LDA (LD-algebras). The CAT domain consists of 58 problems, the LDA domain of 22. From these domains we extracted 22 and 15 non-trivial problems, respectively, i.e. problems none of the provers can solve within 4 seconds. Note that the problems in both domains contain equality. The subgoal clause candidates were generated in the following way: For variant 1 we employed the resource $k = 10$ which performed best in the experiments. The use of higher resources did not entail better results. We limited the set of subgoal clauses by $N_{sg} = 500$. For variant 2 we employed as resources $k_1 = k_2 = 9$. These resources allowed for the efficient generation of all subgoal clauses within the initial segments of the search tree. Usually at most 500 subgoal clauses were generated, i.e. about the same number as when employing variant 1. As start clauses for an adaptive refinement we selected $N_{ref} = 5$ clauses.

Table 1. Integration of top-down/bottom-up approaches by cooperative provers

problem	SPASS	SETH.	inf.	adapt.	problem	SPASS	SETH.	inf.	adapt.
LDA004-1	–	–	–	–	CAT001-1	–	–	–	–
LDA005-1	–	–	–	–	CAT001-3	134s	32s	6s	6s
LDA005-2	279s	–	265s	8s	CAT001-4	33s	11s	5s	5s
LDA006-1	–	–	–	–	CAT002-2	–	–	–	–
LDA006-2	276s	–	304s	10s	CAT003-1	–	–	–	–
LDA007-1	16s	366s	19s	21s	CAT004-3	–	23s	–	9s
LDA007-2	–	50s	7s	7s	CAT008-1	91s	126s	6s	6s
LDA008-1	–	–	–	–	CAT009-1	–	–	10s	10s
LDA008-2	–	–	–	–	CAT009-3	–	–	–	29s
LDA009-1	–	–	–	–	CAT009-4	53s	–	47s	50s
LDA009-2	–	–	–	24s	CAT010-1	–	–	11s	9s
LDA010-1	–	–	–	9s	CAT011-3	17s	–	12s	12s
LDA010-2	–	–	–	26s	CAT014-3	18s	–	11s	11s
LDA011-1	54s	–	58s	9s	CAT018-3	–	–	–	74s
LDA011-2	21s	–	35s	7s	CAT019-4	–	–	–	–

Finally, 100 clauses were selected from the set of candidates and transmitted to SPASS. Table 1 presents results of our experiments (we omitted some problems which could not be solved by any of the alternatives from table 1).

Column 1 of each part of the table displays the name of the problem. Columns 2 and 3 present the runtimes of SPASS and SETHEO (on a SPARCstation 20) when working alone. SPASS was used with its standard setting. SETHEO employed the weighted-depth bound ([MIL+97]) which performed best in the considered domains. Columns 4 and 5 display the runtimes of SPASS when it obtains subgoal clauses from SETHEO which are generated regarding variants 1 and 2, respectively. Note that the runtimes include the 4 seconds before the cooperation, the selection of subgoal clauses, and the transmission to SPASS. The entry "–" means that the problem could not be solved within 1000 seconds.

The results reveal the high potential of our approach to significantly improve on single provers. However, when considering the results of variant 1, they also show that a naive and uninformed generation of subgoal clauses usually does not entail much gain. In the LDA domain, we can solve 9 of 15 hard problems by using variant 2 for subgoal generation, whereas SPASS can only solve 5, SETHEO only 2. A simple competitive prover which employs SPASS and SETHEO in parallel would also only be able to solve 6 problems. Hence, cooperation is really important in order to increase the success rate. We can also in almost all cases significantly decrease the runtimes. E.g. problems LDA005-2 and LDA006-2 which require a runtime of more than 4 minutes when using SPASS can be proven in a few seconds. In the CAT domain the results are similar. By employing subgoal clauses generated regarding variant 2 we can solve 11 of 22 hard problems. SPASS and SETHEO are only able to solve 6 and 4, respectively.

6 Discussion and Future Work

Integration of top-down and bottom-up provers by employing cooperation is very promising in the field of automated deduction. Due to certain strengths and weaknesses of provers following different paradigms, techniques that try to combine the strengths by cooperation can allow for an improvement of the deductive system. Our approach of combining top-down and bottom-up provers by processing top-down generated subgoal clauses in a bottom-up prover achieves this combination by introducing goal-orientation into a bottom-up prover thus combining strong redundancy control mechanisms and goal-directed search.

So far, related approaches mainly aimed at supporting a top-down prover by a bottom-up based lemma component. Results presented, e.g., in [Sch94] or [Fuc98b], reveal that also these approaches are well-suited. However, in some domains, especially if equality is involved, superposition-based provers clearly outperform ME provers. Thus, in such domains it may be more sensible to develop techniques in order to support the more powerful bottom-up prover than the weaker top-down prover. Transmitting information from a top-down to a bottom-up prover was so far—to our knowledge—only discussed in [Sut92]. However, there bottom-up lemmas generated by a ME prover were transferred to resolution-based provers and the results were not satisfactory.

Future work should deal with an extension of the empiric investigation. It would be interesting to detect in which kinds of domains (Horn/non-Horn, equality/no equality) the approach is especially well-suited. Moreover, it might be interesting to develop more complex methods for generating subgoal clause candidates. The results from section 5 suggest that an even more intelligent generation and selection of subgoal clauses leads to further improvements.

References

[AL97]   O.L. Astrachan and D.W. Loveland. The use of Lemmas in the Model Elimination Procedure. *Journal of Automated Reasoning*, 19(1):117–141, 1997.

[AS92]   O.L. Astrachan and M.E. Stickel. Caching and Lemmaizing in Model Elimination Theorem Provers. In *Proceedings of CADE-11*, pages 224–238, Saratoga Springs, USA, 1992. Springer LNAI 607.

[BG94]   L. Bachmair and H. Ganzinger. Rewrite-based equational theorem proving with selection and simplification. *Journal of Logic and Computation*, 4(3):217–247, 1994.

[DF94]   J. Denzinger and M. Fuchs. Goal oriented equational theorem proving. In *Proc. 18th KI-94*, pages 343–354, Saarbrücken, 1994. LNAI 861.

[DF98]   J. Denzinger and D. Fuchs. Enhancing conventional search systems with multi-agent techniques: a case study. In *Proc. Int. Conf. on Multi Agent Systems (ICMAS) 98*, Paris, France, 1998.

[Fuc98a] D. Fuchs. Cooperation between Top-Down and Bottom-Up Theorem Provers by Subgoal Clause Transfer. Technical Report SR-98-01, University of Kaiserslautern, Kaiserslautern, 1998.

[Fuc98b]   M. Fuchs. Similarity-Based Lemma Generation for Model Elimination. In *Proc. CADE-15*, Lindau, Germany, 1998.

[Kor85]   Richard E. Korf. Depth-First Iterative-Deepening: An Optimal Admissible Tree Search. *AI*, 27:97 – 109, 1985. Elsevier Publishers B.V. (North-Holland).

[LMG94]   R. Letz, K. Mayr, and C. Goller. Controlled Integration of the Cut Rule into Connection Tableau Calculi. *Journal of Automated Reasoning*, (13):297–337, 1994.

[Lov68]   D.W. Loveland. Mechanical Theorem-Proving by Model Elimination. *Journal of the ACM*, 15(2), 1968.

[MIL+97]   M. Moser, O. Ibens, R. Letz, J. Steinbach, C. Goller, J. Schumann, and K. Mayr. The Model Elimination Provers SETHEO and E-SETHEO. *special issue of the Journal of Automated Reasoning*, 1997.

[Sch94]   J. Schumann. Delta - a bottom-up preprocessor for top-down theorem provers. system abstract. In *Proceedings of CADE-12*. Springer, 1994.

[SSY94]   G. Sutcliffe, C.B. Suttner, and T. Yemenis. The TPTP Problem Library. In *CADE-12*, pages 252–266, Nancy, 1994. LNAI 814.

[Sti88]   M.E. Stickel. A prolog technology theorem prover: Implementation by an extended prolog compiler. *Journal of Automated Reasoning*, 4:353–380, 1988.

[Sut92]   G. Sutcliffe. A heterogeneous parallel deduction system. In *Proc. FGCS'92 Workshop W3*, 1992.

[WGR96]   C. Weidenbach, B. Gaede, and G. Rock. Spass & Flotter Version 0.42. In *Proc. CADE-13*, pages 141–145, New Brunswick, 1996. LNAI 1104.

Polymorphic Call-by-Value Calculus Based on Classical Proofs
(Extended Abstract)

Ken-etsu Fujita

Kyushu Institute of Technology, Iizuka, 820-8502, Japan
fujiken@dumbo.ai.kyutech.ac.jp

Abstract. We introduce a polymorphic call-by-value calculus, λ_{exc}^v, based on 2nd order classical logic. The call-by-value computation rules are defined based on proof reductions, in which classical proof reductions are regarded as a logical permutative reduction in the sense of Prawitz and a dual permutative reduction. It is shown that the CPS-translation from the core λ_{exc}^v to the intuitionistic fragment, i.e., the Damas-Milner type system is sound. We discuss that the use of the dual permutative reduction is, in general, uncorrected in polymorphic calculi. We also show the Church-Rosser property of λ_{exc}^v, and the soundness and completeness of the type inference algorithm $\mathcal{W}$. From the subject reduction property, it is obtained that a program whose type is inferred by $\mathcal{W}$ never leads to a type-error under the rewriting semantics. Finally, we give a brief comparison with ML plus callcc and some of the existing call-by-value styles.

1 Introduction

Information can be represented as symbols, and symbolic computation is important in artificial intelligence such as problem solving, reasoning, knowledge representation, natural language processing, learning, expert systems and so on [26]. For symbolic computation, it is essential to provide an underlying system or language which describes symbols and algorithm. On the other hand, based on the Curry-Howard-De Bruijn isomorphism [14], types are assigned to formulae and terms to proof trees, and proof reductions can be regarded as computational rules or symbolic rewriting rules. This principle is widely applied to automated theorem proving, constructive programming, analogical reasoning, etc. The computational meaning of proofs has been investigated by many researchers, not only in intuitionistic logic but also in classical logic and modal logic [15]. In the area of classical logic, Griffin [11], Murthy [17], Parigot [21], Berardi&Barbanera [4], Rehof&Sørensen [24], de Groote [6], Ong [19], and so on, are some of the noteworthy investigators. As far as we know, however, polymorphic call-by-value calculus is less studied from the viewpoint of classical logic. As an application of the isomorphism to construct a symbolic computation system with a control operator, we investigate the fragment of second order classical logic, in the sense

Jacques Calmet and Jan Plaza (Eds.): AISC'98, LNAI 1476, pp. 170–182, 1998.
© Springer-Verlag Berlin Heidelberg 1998

of the Damas-Milner type discipline [5], based on the call-by-value strategy. The isomorphism can give a neat guide to define symbolic computation rules, especially in the case of polymorphic call-by-value systems.

First, we introduce a simple system λ_{exc}^v of polymorphic call-by-value language based on classical logic, and the CPS-translation from the core λ_{exc}^v to the intuitionistic fragment, i.e., the Damas-Milner type system. The soundness of the translation is established under a strict notion of values but an extended form. The call-by-value computation rules are defined based on proof reductions, in which classical proof reductions are regarded as a logical permutative reduction in the sense of Prawitz [23] and a dual permutative reduction. Here, continuations can be naturally treated by the logical permutative reductions. Second, we discuss that from the viewpoint of proof reductions, one of the important classical proof reductions, called dual permutative reduction, in general, becomes uncorrected not only in full polymorphic calculi but also in the Damas-Milner style. Our observation reveals some conditions under which the dual permutative reduction can be accepted. Third, it is shown that λ_{exc}^v has the Church-Rosser property. We also show the soundness and completeness of the type inference algorithm $\mathcal{W}$. From the subject reduction property, we found that a program whose type is inferred by $\mathcal{W}$ never leads to a type-error under the rewriting semantics. Finally, we give a brief comparison with ML [16,12] and some of the existing call-by-value styles.

2   Polymorphic Call-by-Value Language Based on Classical Logic

In this section we give the definition of the core language λ_{exc}^v. The types τ and the type schemes σ are defined as usual. The terms are defined by two kinds of variables; x is used as usual, like in λ-calculus. On the other hand, y is called an exceptional variable or a continuation variable, where y is used only for negation types $\neg\tau$ defined as $\tau \to \bot$. The binary relation $\tau \le \forall\alpha_1\cdots\alpha_n.\tau'$ $(n \ge 0)$ is defined such that the type τ is obtained from the type τ' by substituting the type τ_i for α_i $(1 \le i \le n)$, i.e., $\tau = \tau'[\alpha_1 := \tau_1, \cdots, \alpha_n := \tau_n]$. A type assumption is a finite set of declarations with the form $x:\sigma$ or $y:\neg\tau$, all with distinct subjects. The inference rule $(\bot I)$ below is introduced, since we do not consider a term of the form My for some M. Based on the continuation semantics discussed in the next section, the variable y is used for representing a continuation, and the continuation can be regarded as the context in which a term is evaluated. Since the variable y is waiting for the rest of the computation, it is natural to consider an η-expansion form $M(\lambda x.yx)$ instead of My. Here, a negation type $\neg\tau$ plays a role of the type of continuation that expects a term of type τ, which is reflected in the definition of values below. In contrast to a system with the double negation elimination, to establish a classical system we have to introduce both $(\bot E)$ below and the classical rule (exc), which is a variant of the law of the excluded middle.

λ_{exc}^{v}:

Types	Type Schemes	Type Assumptions
$\tau ::= \alpha \mid \bot \mid \tau \to \tau$	$\sigma ::= \tau \mid \forall \alpha.\sigma$	$\Gamma ::= \langle \, \rangle \mid x{:}\sigma, \Gamma \mid y{:}\neg\tau, \Gamma$

Terms

$$M ::= x \mid MM \mid \lambda x.M \mid \text{let } x = M \text{ in } M \mid \text{raise}(M) \mid yM \mid \{y\}M$$

Type Assignment

$$\Gamma \vdash x : \tau \ \ if \ \tau \le \Gamma(x)$$

$$\frac{\Gamma \vdash M_1 : \tau_1 \to \tau_2 \quad \Gamma \vdash M_2 : \tau_1}{\Gamma \vdash M_1 M_2 : \tau_2} \ (\to E) \qquad \frac{\Gamma, x{:}\tau_1 \vdash M : \tau_2}{\Gamma \vdash \lambda x.M : \tau_1 \to \tau_2} \ (\to I)$$

$$\frac{\Gamma \vdash M_1 : \tau_1 \quad \Gamma, x{:}\forall\alpha.\tau_1 \vdash M_2 : \tau_2}{\Gamma \vdash \text{let } x = M_1 \text{ in } M_2 : \tau_2} \ (\text{let})^* \qquad \frac{\Gamma, y{:}\neg\tau \vdash M : \tau}{\Gamma \vdash \{y\}M : \tau} \ (\text{exc})$$

$$\frac{\Gamma \vdash M : \bot}{\Gamma \vdash \text{raise}(M) : \tau} \ (\bot E) \qquad \frac{\Gamma \vdash M : \tau}{\Gamma \vdash yM : \bot} \ (\bot I) \ if \ \Gamma(y) \equiv \neg\tau$$

(let)* denotes the side condition such that the sequence of type variables α, say $\alpha_1 \cdots \alpha_n$, does not appear free in Γ.

The notion of values is defined below as an extended form; the class of values is closed under both value-substitution and term-replacement, as defined later. The notion of values will be discussed based on the CPS-translation. The definition of the reduction rules is given based on the call-by-value strategy. In particular, the classical reduction (5-1) below can be explained as a logical permutative reduction in the sense of Prawitz [23] and Andou [1]. Here, in the reduction of $(\{y\}M)N \rhd \{y\}(M[y \Leftarrow N])N$, since both the type of M and the type of each subterm M' with the form yM' in M can be considered as members of the segments ending with the type of $\{y\}M$, the application of $(\to E)$ is shifted up to the occurrence M and each occurrence M', and then MN and $M[y \Leftarrow N]$ (each yM' is replaced with $y(M'N)$) are obtained. This reduction is also called a structural reduction in Parigot [21]. On the other hand, since a term of the form $\{y\}M$ is not regarded as a value, $(\lambda x.M_1)(\{y\}M_2)$ will not be a β-contractum, but will be a contractum of (5-2) below, which can be considered as a dual permutative reduction. $FV(M)$ stands for the set of free variables in M.

Values $V ::= x \mid \lambda x.M \mid yM$

Term reductions

(1) $(\lambda x.M)V \ \rhd \ M[x := V]$;

(2-1) $(\text{raise } M_1)M_2 \rhd \text{raise } M_1$;   (2-2) $V(\text{raise } M) \rhd \text{raise } M$;

(3-1) $\text{let } x = V \text{ in } M \ \rhd \ M[x := V]$;

(4-1) $\{y\}M \rhd M$ if $y \notin FV(M)$;   (4-2) $\{y\}\text{raise}(yM) \rhd \{y\}M$;

(5-1) $(\{y\}M_1)M_2 \rhd \{y\}(M_1[y \Leftarrow M_2])M_2$;   (5-2) $V(\{y\}M) \rhd \{y\}V(M[V \Rightarrow y])$,

where the term-replacement $M[y \Leftarrow N]$ is used such that an operand of $\{y\}M$, i.e., the right context (continuation) of $\{y\}M$, is replaced on an argument position of y in M. each y The term $M[y \Leftarrow N]$ is defined as a term obtained from M by replacing each subterm of the form yM' in M with $y(M'N)$. Similarly, the dual operation $M[N \Rightarrow y]$ is used such that each subterm of the form yM' in M is replaced with $y(NM')$.

We identify $\{y\}\{y_1\}M$ with $\{y\}M[y_1 := y]$ for technical simplicity. The binary relation $\triangleright^*$ is defined by the reflexive transitive closure of $\triangleright$, and the congruence relation is denoted by $=$. We sometimes use the term $\{y : \neg\tau\}M$ with the type τ of M.

From the definition, λ^v_{exc} is a simple fragment of the second order λ-calculus $\lambda2$ (Girard's F) together with the excluded middle. Moreover, for the finite-type fragment of λ^v_{exc} (i.e., with neither σ nor (let)), there exists a term M such that $\Gamma \vdash M : \tau$ iff τ as a formula is provable from Γ in classical propositional logic.

3   CPS-Translation of λ^v_{exc}-Terms

We provide the CPS-translation from classical logic λ^v_{exc} to the intuitionistic fragment (the Damas-Milner style) ML, which logically induces Kuroda's translation. The system of ML is defined as usual. The reduction rules are defined from (1) and (3-1) without restricting to values. The one step reduction relation and the reflexive transitive closure are denoted by $\triangleright_\beta$ and $\triangleright^*_\beta$, respectively. The congruence relation is denoted by $=_\beta$.
ML:

$$\Gamma \vdash x : \Gamma(x)$$

$$\frac{\Gamma \vdash M_1 : \tau_1 \to \tau_2 \quad \Gamma \vdash M_2 : \tau_1}{\Gamma \vdash M_1 M_2 : \tau_2} \ (\to E) \qquad \frac{\Gamma, x:\tau_1 \vdash M : \tau_2}{\Gamma \vdash \lambda x.M : \tau_1 \to \tau_2} \ (\to I)$$

$$\frac{\Gamma \vdash M : \forall \alpha.\sigma}{\Gamma \vdash M : \sigma[\alpha := \tau]} \ (\text{Inst}) \qquad \frac{\Gamma \vdash M : \sigma}{\Gamma \vdash M : \forall \alpha.\sigma} \ (\text{Gen})^*$$

$$\frac{\Gamma \vdash M_1 : \sigma \quad \Gamma, x:\sigma \vdash M_2 : \tau}{\Gamma \vdash \texttt{let } x = M_1 \texttt{ in } M_2 : \tau} \ (\text{Let})$$

$(\text{Gen})^*$ denotes the side condition such that the type variable α does not appear free in Γ.

As mentioned in Harper&Lillibridge [13], there are difficulties in defining translation rules for polymorphic let-expressions. With respect to λ^v_{exc}, the let-expression, $\texttt{let } x = M \texttt{ in } N$, cannot be interpreted as $\overline{(\lambda x.N)M} = \lambda k.\overline{M}(\lambda x.\overline{N}k)$ in the full polymorphic lambda calculus $\lambda2$ (System F of Girard), since the denotation is not well-typed under our negative translation of Kuroda, which also relates to the use of the dual permutative reduction in polymorphic calculi discussed in the next section. To define the translation, we start with separating the λ-variables x into two categories, depending on a finite or polymorphic use; x is used monomorphically, and x^* is used polymorphically. The syntax of the terms in λ^v_{exc} is rewritten as follows:

$$M ::= x \mid x^* \mid MM \mid \lambda x.M \mid \texttt{let } x^* = M \texttt{ in } M \mid \texttt{raise}(M) \mid yM \mid \{y\}M$$

In an explicit type system, the polymorphic variable x^* is used in the form $x^*\tau'$ by the instantiation rule, where $x^* : \forall \alpha.\tau$ for some τ. This means that even in an implicit type system, the expression x^* cannot be regarded as a value in a strict

sense (a trivial computation may be needed). We also consider a strict class of values, excluding a single occurrence x^*: $V ::= x \mid \lambda x.M \mid yM$
The call-by-value reduction rules are applied for the strict class V in this section. To establish the CPS-translation, the distinction between V and x^* is used only in this section. For programs containing no free variables x^*, the discrimination would not be important.

The translation, with an auxiliary function Ψ for values V, comes from Plotkin [22] and de Groote [6]. It is proved that the translation is sound with respect to conversions.

Definition 1 (CPS-translation from λ_{exc}^v to ML) $\overline{x} = \lambda k.kx$; $\overline{x^*} = \lambda k.x^*k$;
$\overline{\lambda x.M} = \lambda k.k(\lambda x.\overline{M})$; $\overline{yM} = \lambda k.k(\overline{M}y)$;
$\overline{MN} = \lambda k.\overline{M}(\lambda m.\overline{N}(\lambda n.mnk))$; $\overline{\text{let } x^*=M \text{ in } N} = \lambda k.(\text{let } x^*=\overline{M} \text{ in } (\overline{N}k))$;
$\overline{\text{raise}(M)} = \lambda k.\overline{M}(\lambda x.x)$; $\overline{\{y\}M} = \lambda y.\overline{M}y$.
$\Psi(x) = x$; $\Psi(\lambda x.M) = \lambda x.\overline{M}$; $\Psi(yM) = \overline{M}y$.

According to the continuation semantics of Meyer&Wand [18], our definition of the CPS-translation is given as follows, where the type $\bot$ plays the role of the answer type: If we have a variable x, then the value x is sent to the continuation k. In the case of a λ-abstraction, a certain function that will take two arguments is sent to the continuation k. If we have a term with a continuation variable y, then a certain function with the argument y is sent to the continuation k, where the variable y will be substituted by a continuation. Here, it would be natural that a value is regarded as the term that is mapped by Ψ to some term consumed by the continuation k, since the continuation is the context in which a term is evaluated and then to which the value is sent. Our notion of values as an extended form is derived based on this observation.

Lemma 1 i) For any term M where $k \notin FV(M)$, $\lambda k.\overline{M}k \rhd_\beta \overline{M}$.
ii) For any value V, $\overline{V} = \lambda k.k\Psi(V)$.
iii) For any term M and value V, $\overline{M[x := V]} = \overline{M}[x := \Psi(V)]$.
iv) For any terms M and N, $\overline{M}[x^* := \overline{N}] \rhd_\beta^* \overline{M[x^* := N]}$.

The above lemma can be proved by straightforward induction. Based on the CPS-translation, the term-replacements $M[y \Leftarrow N]$ and $M[N \Rightarrow y]$ can be interpreted as the following substitutions for continuation variables, respectively.

Lemma 2 i) For any term M and N, $\overline{M[y \Leftarrow N]} \rhd_\beta^* \overline{M}[y := \lambda m.\overline{N}(\lambda n.mny)]$.
ii) For any term M and value V, $\overline{M[V \Rightarrow y]} \rhd_\beta^* \overline{M}[y := \lambda n.\Psi(V)ny]$.

    Proof. By induction on the structure of M. Only the case of yM is shown:
$\overline{(yM)[V \Rightarrow y]} = \lambda k.k((\lambda k'.\overline{V}(\lambda m.\overline{M[V \Rightarrow y]}(\lambda n.mnk')))y)$
$\rhd_\beta \lambda k.k((\lambda k'.k'\Psi(V))(\lambda m.\overline{M[V \Rightarrow y]}(\lambda n.mny)))$
$\rhd_\beta^* \lambda k.k(\overline{M[V \Rightarrow y]}(\lambda n.\Psi(V)ny)) \rhd_\beta^* \lambda k.k(\overline{M}[y := \lambda n.\Psi(V)ny](\lambda n.\Psi(V)ny))$
$= \lambda k.k(\overline{M}y)[y := \lambda n.\Psi(V)ny] = \overline{yM}[y := \lambda n.\Psi(V)ny]$.   □

Lemma 3 *If $M \triangleright N$, then $\overline{M} =_\beta \overline{N}$.*

Proof. By induction on the derivation of $M \triangleright N$. We show some of the cases:
(3-1) let $x^* = V$ in $M \quad \triangleright \quad M[x^* := V]$:
$\overline{\text{let } x^* = V \text{ in } M} = \lambda k.(\text{let } x^* = \overline{V} \text{ in } (\overline{M}k))$
$\triangleright_\beta \lambda k.\overline{M}[x^* := \overline{V}]k \triangleright_\beta^* \lambda k.\overline{M[x^* := V]}k \triangleright_\beta \overline{M[x^* := V]}$.
(5-2) $V(\{y\}M) \triangleright \{y\}(V(M[V \Rightarrow y]))$:
$\overline{V(\{y\}M)} = \lambda k.\overline{V}(\lambda m.(\lambda y.\overline{M}y)(\lambda n.mnk))$
$\triangleright_\beta \lambda k.(\lambda k'.k'\Psi(V))(\lambda m.\overline{M}[y := \lambda n.mnk](\lambda n.mnk))$
$\triangleright_\beta^* \lambda k.\overline{M}[y := \lambda n.\Psi(V)nk](\lambda n.\Psi(V)nk) = \lambda y.\overline{M}[y := \lambda n.\Psi(V)ny](\lambda n.\Psi(V)ny)$
$=_\beta \lambda y.\overline{M[V \Rightarrow y]}(\lambda n.\Psi(V)ny) =_\beta \lambda y.(\lambda m.\overline{M[V \Rightarrow y]}(\lambda n.mny))\Psi(V)$
$=_\beta \lambda y.(\lambda k.\overline{V}(\lambda m.\overline{M[V \Rightarrow y]}(\lambda n.mnk)))y = \overline{\{y\}V(M[V \Rightarrow y])}$. $\square$

Now, under the strict notion of values, we have confirmed the soundness of the translation in the sense that equivalent λ_{exc}^v-terms are translated into equivalent ML-terms.

Proposition 1 (Soundness of the CPS-Translation) *If we have $M = N$ in λ_{exc}^v, then $\overline{M} =_\beta \overline{N}$ in ML.*

The translation logically establishes the double-negation translation of Kuroda.

Definition 2 (Kuroda's Translation) $\tau^q = \tau$ *where τ is atomic;*
$(\tau_1 \to \tau_2)^q = \tau_1^q \to \neg\neg\tau_2^q; \qquad\qquad (\forall\alpha.\tau)^q = \forall\alpha.\neg\neg\tau^q.$
$(x:\sigma, \Gamma)^q = x:\sigma^q, \Gamma^q; \qquad\qquad (y:\neg\tau, \Gamma)^q = y:\neg\tau^q, \Gamma^q.$

Proposition 2 *If we have $\Gamma \vdash M : \tau$ in λ_{exc}^v, then $\Gamma^q \vdash \overline{M} : \neg\neg\tau^q$ in ML.*

Proof. By induction on the derivation. We show some of the cases.
Case 1-1. $x : \tau$ is derived from $x : \tau$:
$\lambda k.kx : \neg\neg\tau^q$ is derived from $x : \tau^q$ in ML.
Case 1-2. $x^* : \tau[\alpha := \tau_1]$ is derived from $x^* : \forall\alpha.\tau$:
$\lambda k.xk : \neg\neg\tau^q[\alpha := \tau_1^q]$ is derived from $x^* : \forall\alpha.\neg\neg\tau^q$ in ML. One also has that $(\tau[\alpha := \tau_1])^q = \tau^q[\alpha := \tau_1^q]$ for any types τ and τ_1.
Case 2. $\Gamma \vdash$ let $x^* = M_1$ in $M_2 : \tau_2$ from $\Gamma \vdash M_1 : \tau_1$ and $\Gamma, x^*:\forall\alpha.\tau_1 \vdash M_2 : \tau_2$:
By the induction hypotheses, we have $\Gamma^q \vdash \overline{M_1} : \neg\neg\tau_1^q$ where α is not free in types in Γ^q, and $\Gamma^q, x^* : \forall\alpha.\neg\neg\tau_1^q \vdash \overline{M_2} : \neg\neg\tau_2^q$. One also has $\Gamma^q, k : \neg\tau_2^q, x^* : \forall\alpha.\neg\neg\tau_1^q \vdash \overline{M_2}k : \bot$, and $\Gamma^q, k:\neg\tau_2^q \vdash \overline{M_1} : \forall\alpha.\neg\neg\tau_1^q$. Hence, $\Gamma^q \vdash \lambda k.(\text{let } x^* = \overline{M_1} \text{ in } \overline{M_2}k) : \neg\neg\tau_2^q$ is derived in ML. $\square$
From the consistency of ML, it is also derived that λ_{exc}^v is consistent in the sense that there is no closed term M such that $\vdash M : \bot$.

Note that all of the above results are also available for the system with (3-1')
let $x = N$ in $M \quad \triangleright \quad M[x := N]$, without restricting to values. The reason is explained based on the observation in the next section. The polymorphic use of the term $\{y\}M$ cannot treat the left context correctly, i.e., the dual permutative reduction, in general, becomes an uncorrected proof reduction. Here, we do not forbid the polymorphic use of $\{y\}M$, since the polymorphic term can treat the

right context correctly by the permutative reduction. The reduction rule (3-1') supports this computation, and the CPS-translation would work for the 'by name' semantics for let-expressions in [12]. This is one of the reasons why the CPS-translation logically establishes the negative translation of Kuroda without restricting let-bound expressions, M_1 in (let)* to a value, as compared with that of [13]. Next, we introduce new reduction rules that transform the local definition part to a value form, since the system is working with (3-1) under the call-by-value strategy.

4   Dual Permutative Reduction and Polymorphism

In the core language, one problem is that there is no reduction rule for the following polymorphic program:

   let $x = \{y\}(\lambda x.N)$ in M where N is in normal.

As a naive extension of the reduction rule (5-2): $V(\{y\}N) \triangleright \{y\}VN[V \Rightarrow y]$ to the polymorphic case, one may take the rule such that

   (a) let $x = \{y\}N$ in $M \triangleright \{y\}(\lambda x.M)N[\lambda x.M \Rightarrow y]$.

Here, the reduced term cannot be well-typed in the Damas-Milner type discipline. However, the main defect is not the weakness of the underlying type system, but the reduction itself cannot be a correct proof reduction. This situation was discovered by Harper&Lillibridge [13]. Now we observe this from the viewpoint of proof reductions, which reveals some conditions that justify the reduction. Take the following rule in a β-reduced form of the above (a):

   (b) let $x = \{y\}N$ in $M \triangleright \{y\}M[x := N[y(M[x := N'])/yN']]$, where
the term $N[y(M[x := N'])/yN']$ is a term obtained from N, replacing each subterm of the form yN' in N with $y(M[x := N'])$.

Assume that we have the following proof figure in the Damas-Milner type system:

$$
\cfrac{
\cfrac{
\cfrac{
\cfrac{[y:\neg\tau_1] \quad \cfrac{\Pi_1}{N':\tau_1}}{yN':\bot}
}{\cfrac{\Pi_2}{N:\tau_1}}
}{\{y\}N:\tau_1}
\qquad
\cfrac{
\cfrac{[x:\forall\alpha.\tau_1]}{x:\tau_1[\alpha:=\tau_3]} \quad \cfrac{[x:\forall\alpha.\tau_1]}{x:\tau_1[\alpha:=\tau_4]}}{\cfrac{\Pi_3}{M:\tau_2}}
}{\text{let } x = \{y\}N \text{ in } M:\tau_2}
\ (\text{let})^*
$$

First, compute the type of $N[y(M[x := N'])/yN']$, and then one obtains the following type assignment:

$$
\cfrac{
y:\neg\tau_2 S
\qquad
\cfrac{
\cfrac{\Pi_1[\alpha:=\tau_3]\circ S}{N:\tau_1[\alpha:=\tau_3]\circ S} \quad \cfrac{\Pi_1[\alpha:=\tau_4]\circ S}{N:\tau_1[\alpha:=\tau_4]\circ S}}{\cfrac{\Pi_3 S}{M[x:=N']:\tau_2 S}}
}{
\cfrac{\cfrac{y(M[x:=N']):\bot}{\Pi_2[\alpha:=\tau_3]\circ S}}{N[y(M[x:=N'])/yN']:\tau_1[\alpha:=\tau_3]\circ S}}
,
$$

if τ_3 and τ_4 are unifiable under some substitution S, considering the case such that the assumption whose type contains a free variable α in Π_1 is discharged by $(\to I)$ in Π_2. Second, compute the type of $\{y\}M[x := N[y(M[x := N'])/yN']]$, and then finally the following assignment is obtained:

$$
[y:\neg\tau_2 S] \qquad\qquad\qquad\qquad\qquad [y:\neg\tau_2 S]
$$
$$
\vdots \qquad\qquad\qquad\qquad\qquad\qquad\qquad \vdots
$$

$$
\cfrac{\cfrac{N[y(M[x := N'])/yN'] : \tau_1[\alpha := \tau_3]\circ S \quad N[y(M[x := N'])/yN'] : \tau_1[\alpha := \tau_4]\circ S}{\cfrac{M[x := N[y(M[x := N'])/yN']] : \tau_2 S}{\{y\}M[x := N[y(M[x := N'])/yN']] : \tau_2 S}}}{\Pi_3 S}
$$

Following the above observation, we obtain that (b) represents a correct proof reduction only if all types of x in M can be unified. Here, the merit of polymorphism is lost. Moreover, the type of the reduced term becomes a substitution instance of τ_2. It can also be observed that, in the above proof figure, if Π_2 contains no $(\to I)$ that discharges the type containing free α, then there is no need to unify each type of x in M, and (b) becomes correct. For instance, in the case of let $x = \{y\}\lambda x'.\mathtt{raise}(y(\lambda v.x'))$ in M, one may have to unify each type of x in M, if one uses (b). On the other hand, (b) is a correct reduction for the case of let $x = \{y\}\lambda v.\mathtt{raise}(y(\lambda x'.x'))$ in M.

In general, even with $\lambda 2+(\mathrm{exc})$, in which one can obtain a sound type, the reduction rule (5-2) is still uncorrected in the sense that the reduced term does not present a correct proof. Following the similar observation (to be skipped here), there exists a case where the side condition of polymorphic generalization is not satisfied. The observation means that the polymorphic term of the form $\{y\}M$ cannot manage the left context correctly. This also shows that one cannot adopt the dual structural reduction in Parigot's $\lambda\mu$-calculus (2nd order classical logic) [21]. Otherwise, we can obtain a self application for any value. For instance, the following example is derived from that of Harper&Lillibridge [13]:

$$\lambda g.\lambda x.g((\lambda f.(\lambda v.\lambda x.x)(fx)(f(\lambda x.x)))\ (\{y\}\lambda x.\mathtt{raise}(y(\lambda v.x)))\ x)$$
with type $(\alpha \to \alpha) \to \alpha \to \alpha$

is reduced to $\lambda g.\lambda x.g(xx)$ by the use of (5-2), where, in the reduction process with the type, the side condition of polymorphic generalization is not satisfied. This observation is not in conflict with that of Harper&Lillibridge.

Our solution is to take the reduction rule $\{y\}(\lambda x.M) \triangleright \lambda x.\{y\}M[y \Leftarrow x]$. The two terms are extensionally equivalent in the sense that for any V, $(\{y\}(\lambda x.M))V = (\lambda x.\{y\}M[y \Leftarrow x])V$, where both are reduced to $\{y\}M[y \Leftarrow V][x := V]$. However, in general, adding the reduction rule breaks down the Church-Rosser property. Hence, we use the reduction rule in the local definition of (let):

(3-2) let $x = \{y\}(\lambda x'.M_1)$ in $M_2 \triangleright$ let $x = \lambda x'.\{y\}M_1[y \Leftarrow x']$ in M_2.

For example,

$HL \equiv$ let $f = \{y\}\lambda x.\mathtt{raise}(y(\lambda v.x))$ in $(\lambda v.\lambda x.x)(f1)(f\mathtt{true})$ with type bool

is now reduced to true. On the other hand,

$(\lambda f.(\lambda v.\lambda x.x)(f1)(f2))(\{y\}\lambda x.\mathtt{raise}(y(\lambda v.x)))$ with type int is reduced to 1. Relating to the CSP-translation, if one naively interpreted HL as $\overline{(\lambda f.(\lambda v.\lambda x.x)(f1)(f\mathtt{true}))(\{y\}\lambda x.\mathtt{raise}(y(\lambda v.x)))}$, then $\lambda k.k1$ was derived simulating (5-2) by β-reductions. In our turn, $\overline{HL} =_\beta \lambda k.k\mathtt{true}$ is obtained following the CPS-translation.

One may also consider a more general rule, such that

let $x = \{y : \neg(\tau_1 \to \tau_2)\}N$ in M ▷ let $x = \lambda x'.\{y : \neg\tau_2\}Nx'[y \Leftarrow x']$ in M.

However, the reduction rule also breaks down the confluence property under the call-by-value strategy.

Following the above observation, we add the reduction rule (3-2), and similarly   (3-3) let $x = \mathtt{raise}_{\tau_1 \to \tau_2}(M_1)$ in M_2 ▷ let $x = \lambda v.\mathtt{raise}_{\tau_2}(M_1)$ in M_2
where v is fresh.

The idea is that the reduction rules transform the local definition part to a value form, and we treat only the right context with respect to $\{y\}(\lambda x.M)$ for some M in the polymorphic declaration.

With respect to all of the reductions, including (3-2) and (3-3), we can prove that λ_{exc}^v has the Church-Rosser property by the well-known method of parallel reductions [2,22] and the Lemma of Hindley-Rosen, see [2].

Proposition 3 (Church-Rosser Theorem) *If we have $M \triangleright^* M_1$ and $M \triangleright^* M_2$, then $M_1 \triangleright^* N$ and $M_2 \triangleright^* N$ for some N.*

5   Type Inference Algorithm

From a practical point of view, we give the type inference algorithm $\mathcal{W}$ to λ_{exc}^v extended with constants, a recursion operator, and so on.

Types   Type Schemes   Type Assumptions

$\tau ::= b \mid \alpha \mid \tau \to \tau$ $\sigma ::= \tau \mid \forall \alpha.\sigma$ $\Gamma ::= \langle\ \rangle \mid x : \sigma, \Gamma \mid c :$
$\sigma, \Gamma \mid y : \neg\tau, \Gamma$

Terms

$M ::= c \mid x \mid MM \mid \lambda x.M \mid \mathtt{if}\ M\ \mathtt{then}\ M\ \mathtt{else}\ M \mid \mathtt{fix}\ x.M \mid \mathtt{let}\ x = M\ \mathtt{in}\ M$
    $\mid \mathtt{raise}(M) \mid yM \mid \{y\}M$

Type Assignment λ_{exc}^v plus

$$\Gamma \vdash c : \tau\quad if\ \tau \leq \Gamma(c)$$

$$\frac{\Gamma \vdash M : \mathtt{bool}\quad \Gamma \vdash M_1 : \tau\quad \Gamma \vdash M_2 : \tau}{\Gamma \vdash \mathtt{if}\ M\ \mathtt{then}\ M_1\ \mathtt{else}\ M_2 : \tau}\ (\mathrm{if})\qquad \frac{\Gamma, x : \tau \vdash M : \tau}{\Gamma \vdash \mathtt{fix}\ x.M : \tau}\ (\mathrm{fix})$$

Reduction rules: (1) $\sim$ (5) together with

(6-1) if true then M_1 else M_2 ▷ M_1;   (6-2) if false then M_1 else M_2 ▷ M_2;

(7-1) fix $x.M$ ▷ $M[x := \mathtt{fix}\ x.M]$.

Proposition 4 (Subject Reduction) *If we have $\Gamma \vdash M : \tau$ and $M \triangleright N$, then $\Gamma \vdash N : \tau$.*

Proof. By induction on the derivation of $M \triangleright N$.   □

We use the Robinson's unification algorithm $\mathcal{U}$ that computes the most general unification $\mathcal{U}(\tau_1, \tau_2) = S$ for any types τ_1 and τ_2. For any substitutions S_1 and S_2, the composition $S_1 \circ S_2$ is defined as usual.

Following Milner [16] and Damas&Milner [5], we give the type inference algorithm that computes the principal type under the given type assignment.

$\mathcal{W}(\Gamma; M) = (S, \tau)$, where

1) If M is x or c and $\Gamma(M) = \forall \alpha_1 \cdots \alpha_n.\tau'$ ($0 \leq n$), then $S = Id$ (identity substitution) and $\tau = \tau'[\alpha_1 := \beta_1, \cdots, \alpha_n := \beta_n]$ where β_i is a new type variable ($0 \leq i \leq n$).

2), 3), 4), 5), 6) The case M of $\lambda x.M_1$, $M_1 M_2$, if M_0 then M_1 else M_2, fix $x.M_1$, or let $x = M_1$ in M_2 is defined as usual.

7) If M is yM_1, then:
let $\neg \tau_1 = \Gamma(y)$, $(S_1, \tau_2) = \mathcal{W}(\Gamma; M_1)$, and $U = \mathcal{U}(\tau_1, \tau_2)$;
then $S = S_1 \circ U$ and $\tau = \bot$.

8) If M is $\mathbf{raise}(M_1)$, then:
let $(S_1, \tau_1) = \mathcal{W}(\Gamma; M_1)$, and $U = \mathcal{U}(\tau_1, \bot)$;
then $S = S_1 \circ U$ and $\tau = \alpha$ where α is fresh.

9) If M is $\{y\}M_1$, then:
let $(S_1, \tau_1) = \mathcal{W}(\Gamma, y: \neg\alpha; M_1)$ where α is fresh, and $U = \mathcal{U}(\tau_1, \alpha S_1)$;
then $S = S_1 \circ U$ and $\tau = \tau_1$.

Proposition 5 (Soundness and Completeness) i) *For any Γ and M, if $\mathcal{W}(\Gamma, M) = (S, \tau)$, then we have $\Gamma S \vdash M : \tau$.*
ii) For any Γ and M, if we have $\Gamma S \vdash M : \tau$ for some S and τ, then $\mathcal{W}(\Gamma, M) = (S_1, \tau_1)$, and $\Gamma S = \Gamma S_1 \circ S_2$ and $\tau = \tau_1 S_2$ for some S_2.

Proof. By induction on the structure of M. □

From the soundness and completeness of $\mathcal{W}$, it is confirmed that $\mathcal{W}$ computes the principal type under the given type assignment. It is also obtained that from the subject reduction property and the soundness of $\mathcal{W}$, if a type of M is inferred by $\mathcal{W}$, then the program M never gives a type error under the rewriting semantics.

We give a simple example due to Friedman&Felleisen, page 57 in [7]. Given $a : \alpha$ and $l : \alpha$ list, compute a sublist of l, which is the tail part of l from the last occurrence a. Let **remberuptolast** be

$\lambda a.\lambda l'.\{y\}$ (fix $f.\lambda l.$ if l=nil then nil else if car(l)=a
        then $\mathbf{raise}(y(f(\mathrm{cdr}(l))))$ else cons (car(l)) ($f(\mathrm{cdr}(l))$)l'

with type $\alpha \to \alpha$ list $\to \alpha$ list.

Then we have let f=remberuptolast in f 1 $[1, 2, 1, 3, 1, 4, 5]$ $\rhd^*$ $[4, 5]$.
One can also define **remberuptolast** $a\ l = \mathcal{P}(\mathbf{aux}\ a\ l)$, where
aux $a\ l'$ is

$\lambda exit.$ (fix $f.\lambda l.$ if l=nil then nil else if car(l)=a
        then $exit(f(\mathrm{cdr}(l)))$ else cons (car(l)) ($f(\mathrm{cdr}(l))$)l', and
$\mathcal{P}$ is a proof of Peirce's law $((\alpha \to \beta) \to \alpha) \to \alpha$, i.e., $\lambda x_1.\{y\}x_1(\lambda x_2.\mathbf{raise}(yx_2))$.
Along this line, together with the proof $\mathcal{F} = \lambda x_1 gf.\{y\}g(x_1(\lambda x_2.\mathbf{raise}(y(fx_2))))$
with type $((\alpha \to \beta) \to \gamma) \to (\gamma \to \delta) \to (\alpha \to \delta) \to \delta$, one can write a program with a case analysis: **remberuptolast'** $a\ l\ g\ f = \mathcal{F}(\mathbf{aux}\ a\ l)\ g\ f$.

When l contains a, $\texttt{remberuptolast}' \ a \ l \ g \ f = f(\texttt{remberuptolast} \ a \ l)$; otherwise $\texttt{remberuptolast}' \ a \ l \ g \ f = g(\texttt{remberuptolast} \ a \ l)$.

6   Comparison with Related Work and Concluding Remarks

We briefly compare λ^v_{exc} with ML [16] together with $\texttt{callcc}$ [12]. In ML, the class of type variables is partitioned into two subclasses, i.e., the applicative and the imperative type variables, where the imperative type variable is introduced for polymorphic references in Tofte [25]. The type of $\texttt{callcc}$ is declared with imperative type variables to guarantee the soundness of the type inference. Based on the classification, the typing rule for let-expression is given such that if the let-bound expression is not a value, then generalization is allowed only for applicative type variables; otherwise (i.e., the let-bound expression is a value), generalization is possible with no restriction. On the other hand, we have no distinction of the type variables, and a single typing rule for (let) is used. There is a simple translation from the ML-programs to the λ^v_{exc}-terms, such that the two subclasses of type variables in ML are degenerated into a single class.

$$\lceil \texttt{callcc}(M) \rceil = \{y\} \lceil M \rceil (\lambda x. yx); \qquad \lceil \texttt{throw } M \ N \rceil = \texttt{raise}(\lceil M \rceil \lceil N \rceil),$$
and $\lceil \tau \texttt{ cont} \rceil = \neg \lceil \tau \rceil$.

Then the typing relation is preserved under the translation in the following sense:

Proposition 6 *If $\Gamma \vdash M : \tau$ in ML, then $\lceil \Gamma \rceil \vdash \lceil M \rceil : \lceil \tau \rceil$ in λ^v_{exc}.*

However, there are some distinctions; according to Harper et al. [12], the program

$\qquad \texttt{let } f = \texttt{callcc}(\lambda k.\lambda x.\texttt{throw } k \ (\lambda x'.x)) \texttt{ in } (\lambda x'.\lambda x.x)(f1)(f\texttt{true})$

is not typable in ML, since $\texttt{callcc}(\lambda k.\lambda x.\texttt{throw } k \ (\lambda x'.x))$ with imperative type variables is not a value, and in the case of non-value expressions, polymorphism is allowed only for expressions with applicative type variables. On the other hand, under the translation $\lceil \ \rceil$, we have

$\qquad \texttt{let } f = \{y\}\lambda x.\texttt{raise}(y(\lambda x'.x)) \texttt{ in } (\lambda x'.\lambda x.x)(f1)(f\texttt{true})$ with type $\texttt{bool}$,

and it is reduced to $\texttt{true}$.

With respect to the implemented system (Standard ML of New Jersey, Version 110, December 9, 1997 [CM&CMB]),

$\qquad (\texttt{let } f = \lambda x.\texttt{callcc}(\lambda k.\texttt{throw } k \ x) \texttt{ in } \lambda x.x(f1)(f(\lambda x.x))) \ (\lambda x'.\lambda x.x) \ 0$

with type $\texttt{int}$ is evaluated to 0. However, the type checking of the subterm

$\qquad \texttt{let } f = \lambda x.\texttt{callcc}(\lambda k.\texttt{throw } k \ x) \texttt{ in } \lambda x.x(f1)(f(\lambda x.x))$

gives this warning: type vars not generalized because of value restriction are instantiated to dummy types, i.e., the result alone could not be used polymorphically. On the other hand, under the translation, we have

$\qquad \texttt{let } f = \lambda x.\{y\}\texttt{raise}(yx) \texttt{ in } \lambda x.x(f1)(f(\lambda x.x))$

with type $(\texttt{int} \rightarrow (\alpha \rightarrow \alpha) \rightarrow \beta) \rightarrow \beta$, and it is evaluated to $\lambda x.x1(\lambda x.x)$.
    rule,

The construction of λ^v_{exc} is, in part, under the influence of the work by Parigot [21] and de Groote [6]; The call-by-name version of our system is isomorphic to

Parigot's $\lambda\mu$-calculus with respect to finite types, in the sense that equivalent $\lambda\mu$-terms are translated into equivalent λ_{exc}^v-terms and vice versa.

Ph.de Groote [6] introduced a call-by-value language based on the simply typed lambda calculus for formalizing an exception-handling mechanism. At first appearance, our finite type fragment is a small subsystem of his; however, under some translations, they are logically equivalent with respect to finite types. Moreover, our system can be regarded as a meaningful simplification of his, see Fujita [10]. For comparison with the call-by-value style, λ_c of Felleisen [8,9], see also [10].

Ong&Stewart [20] extensively studied a call-by-value programming language based on a call-by-value variant of Parigot's $\lambda\mu$-calculus. There are some distinctions between Ong&Stewart and our finite type fragment; their reduction rules have type annotations like the Church-style, and, using the annotation, more reduction rules are defined than ours, which can give a stronger normal form. In addition, our notion of values is an extended one, which would be justified by observation based on the CPS-translation.

References

1. Y.Andou: A Normalization-Procedure for the First Order Classical Natural Deduction with Full Logical Symbols, *Tsukuba Journal of Mathematics*, Vol.19, No.1, pp.153-162, 1995.
2. H.P.Barendregt: *The Lambda Calculus, Its Syntax and Semantics* (revised edition), North-Holland, 1984.
3. H.P.Barendregt: *Lambda Calculi with Types*, Handbook of Logic in Computer Science Vol.II, Oxford University Press, pp.1-189, 1992.
4. F.Barbanera, S.Berardi: Extracting Constructive Context from Classical Logic via Control-like Reductions, LNCS 664, pp.45-59, 1993.
5. L.Damas, R.Milner: Principal type-schemes for functional programs, *Proc. 9th Annual ACM Symposium on POPL*, pp.207-212, 1982.
6. P.de Groote: A Simple Calculus of Exception Handling, LNCS 902, pp.201-215, 1995.
7. M.Felleisen, D.P.Friedman: The Seasoned Schemer, The MIT Press, 1996.
8. M.Felleisen, D.P.Friedman, E.Kohlbecker, B.Duba: Reasoning with Continuations, *Proc. Annual IEEE Symposium on LICS*, pp.131-141, 1986.
9. M.Felleisen, R.Hieb: The Revised Report on the Syntactic Theories of Sequential Control and State, *Theor.Comput.Sci.* 103, pp.131-141, 1992.
10. K.Fujita: Calculus of Classical Proofs I, LNCS 1345, pp.321-335, 1997.
11. T.G.Griffin: A Formulae-as-Types Notion of Control, *Proc. 17th Annual ACM Symposium on POPL*, pp.47-58, 1990.
12. R.Harper, B.F.Duba, D.MacQueen: Typing First-Class Continuations in ML, *J.Functional Programming*, 3 (4) pp.465-484, 1993.
13. R.Harper, M.Lillibridge: Polymorphic type assignment and CPS conversion, *LISP and Symbolic Computation* 6, pp.361-380, 1993.
14. W.Howard: *The Formulae-as-Types Notion of Constructions*, Academic Press, pp.479-490, 1980.
15. S.Kobayashi: Monads as modality, *Theor.Comput.Sci.* 175, pp.29-74, 1997.

16. R.Milner: A Theory of Type Polymorphism in Programming, *J. Comput. Syst. Sci.* 17, pp.348-375, 1978.
17. C.R.Murthy: An Evaluation Semantics for Classical Proofs, *Proc. 6th Annual IEEE Symposium on LICS*, pp.96-107, 1991.
18. A.Meyer, M.Wand: Continuation Semantics in Typed Lambda-Calculi, LNCS 193, pp.219-224, 1985.
19. C.-H.L.Ong: A Semantic View of Classical Proofs: Type-Theoretic, Categorical, and Denotational Characterizations, *Linear Logic '96 Tokyo Meeting*, 1996.
20. C.-H.L.Ong, C.A.Stewart: A Curry-Howard Foundation for Functional Computation with Control, *Proc. 24th Annual ACM Symposium of POPL*, Languages, 1997.
21. M.Parigot: $\lambda\mu$-Calculus: An Algorithmic Interpretation of Classical Natural Deduction, LNCS 624, pp.190-201, 1992.
22. G.Plotkin: Call-by-Name, Call-by-Value and the λ-Calculus, *Theor. Comput. Sci.* 1, pp. 125-159, 1975.
23. D.Prawitz: Ideas and Results in Proof Theory, *Proc. 2nd Scandinavian Logic Symposium*, edited by N.E.Fenstad, North-Holland, pp.235-307, 1971.
24. N.J.Rehof, M.H.Sørensen: The λ_Δ-Calculus, LNCS 789, pp.516-542, 1994.
25. M.Tofte: Type Inference for Polymorphic References, Information and Computation 89, pp.1-34, 1990.
26. P.H.Winston, B.K.P.Horn: *LISP* (3rd, Ed), Addison Wesley, 1989.

Inference and Verification in Medical Appropriateness Criteria Using Gröbner Bases *

L. M. Laita[1], E. Roanes-Lozano[2], and V. Maojo[1]

[1] Universidad Politécnica de Madrid, Dept. I.A. (Fac. Informática)
Campus de Montegancedo, Boadilla del Monte, 28660-Madrid, Spain
laita@fi.upm.es; vmaojo@infomed.dia.fi.upm.es
[2] Universidad Complutense de Madrid, Dept. Algebra
Edificio "La Almudena", Paseo Juan XXIII s/n, 28040-Madrid, Spain
eroanes@eucmos.sim.ucm.es

Abstract. In this article techniques borrowed from Computer Algebra (Gröbner Bases) are applied to deal with Medical Appropriateness Criteria including uncertainty. The knowledge was provided in the format of a table. A previous translation of the table into the format of a "Rule Based System" (denoted RBS) based on a three-valued logic is required beforehand to apply these techniques. Once the RBS has been obtained, we apply a Computer Algebra based inference engine, both to detect anomalies and to infer new knowledge. A specific set of criteria for coronary artery surgery (originally presented in the form of a table) is analyzed in detail.

Keywords. Verification. Inference Engines. RBSs in Medicine. Gröbner Bases.

Topics: Integration of Logical Reasoning and Computer Algebra. Symbolic Computation for Expert Systems and Machine Learning.

1 Introduction

"Appropriateness criteria" are ratings of the appropriateness for a given diagnostic or therapeutic procedure. Whereas other policies such as clinical practice guidelines are designed to assist practitioners in patient decision making, those criteria are developed primarily to evaluate retrospectively the appropriateness of clinical decisions ([7], [4]). Nevertheless, these Medical Appropriateness Criteria are usually tested for consistency by using ad hoc techniques, and no mechanical procedures for extracting new information are usually provided.

In this article we present a method of verification and knowledge extraction that, using Computer Algebra and Logic, deals with appropriateness criteria translated into RBSs.

A considerable amount of work has been done in the field of dealing with medical information from logic (for instance [15]), and under the form of RBSs

* Partially supported by projects FIS 95/1952 and DGES PB96-0098-C04 (Spain).

(for instance, but not only, [3]). The difference with the method described in this paper is that it uses an algebraic theory that can cope with medical information (or other) that can be expressed as an RBS based on a multivalued and modal logic. This kind of logic is very convenient both to represent incomplete knowledge and to be treated with a Computer Algebra based inference engine. A short comment about the efficiency of this article's approach is included at the end of the Appendix.

2   Table Description and Translation into IF-THEN Statements

A set of medical data regarding coronary diseases [12] (asymptomatic; effort-test positive or negative; one, two, three blood vessel disease; anterior proximal descendent affected or not; LVEF values) was presented to a panel of ten experts, and they were asked about the appropriateness of taking action accordingly (Revascularization, PTCA, CABG)[1]. Data and actions were resumed in a table under the following format.

LVEF	Revasc. Appropriate	PTCA, CABG Approp.
$L > 50$	1: 1 2 3 4 5 6 7 8^1 $*^9$ $+A$	2: 1 2 3 4 5 6 7 8^1 $*^9$ $-+A$
$50 \geq L > 30$	5: 1 2 3 4 5 6 7 8 $*^{10}$ $+A$	6: 1 2 3 4 5 6 7 8^1 $*^9$ $-+A$
$30 \geq L \geq 20$	9: 1 2 3 4 5 6 7 8^1 $*^9$ $+A$	10: 1 2 3 4 5 6 7^1 8^1 $*^8$ $-+A$

Fig. 1. Asymptomatic.   *A: Effort test: positive.*
1: Left Common trunk disease.   Surgical risk low/moderate.

LVEF	Revasc. Appropriate	PTCA, CABG Approp.
$L > 50$	3: 1 2 3 4 5 6 7 8^4 $*^6$ $+A$	4: 1 2 3 4 5^1 6 7^1 8^1 $*^7$ $-+A$
$50 \geq L > 30$	7: 1 2 3 4 5 6 7 8^3 $*^7$ $+A$	8: 1 2 3 4 5 6^1 7^1 8^2 $*^6$ $-+A$
$30 \geq L \geq 20$	11: 1 2 3 4 5 6 7^1 8^2 $*^7$ $+A$	12: 1 2 3 4 5 6^2 7 8^2 $*^6$ $-+A$

Fig. 2. Asymptomatic.   *A: Effort test: positive.*
1: Left Common trunk disease.   Surgical risk high.

LVEF	Revasc. Appropriate	PTCA, CABG Approp.
$50 \geq L > 30$		20: 1 2 3^3 4^2 5 $*$ 7 8^2 9^3 $++D$
$L > 50$	39: 1^2 2 3^1 4^1 $*^3$ 6 7^3 8 9 $?D$	40: 1^5* 2^1 3^3 4 5^1 6 7 8 9 $?-A$

Fig. 3. Asymptomatic.   *A: Effort test: positive.*
1: Three blood vessel disease.   Surgical risk high.

---

[1] LVEF means "Left Ventricle Ejection Fraction", CABG means "Coronary Artery Bypass Grafting", and PTCA means "Percutaneous Transluminal Coronary Angioplasty".

The table assigns a number (**1**, **2**, etc), to each row of digits, symbols $+, -$, *, and letters A, D, I.

The superscripts, for instance (^{1}) in **1** (appropriateness of Revascularization): 1 2 3 4 5 6 7 8^1 *^{9} $+A$, express the number of experts that have assigned that value (8 in a scale from 1 to 9) to the appropriateness of the action. The symbol "*" stands for the median. The symbol $+$ means "appropriateness of the treatment (Revascularization or PTCA or CABG)". The symbol "$-$" that appears in other rows of the table means "inappropriateness" and the symbol "?" means "undecided appropriateness".

The letter A (after the symbol $+$ or $-$ or ?), means that there is "Agreement" among the panelists about the appropriateness ($+$), inappropriateness ($-$) or undecided appropriateness (?) of Revascularization, PTCA or CABG. A letter D means that there is "Disagreement" and a letter I means "Undecided Agreement". There exists disagreement (D) if of the ten panelists, the opinion of three of them ranks between 1 and 3, and the opinion of another three ranks between 7 and 9 (i.e. big standard deviation). There is agreement (A) when the opinion of no more than two of them is outside the interval of $[1, 3]$, $[4, 6]$, $[7, 9]$ containing the median. Otherwise there exists undecided agreement (I).

Observe that we have transcribed here just 17 of the more than 400 rows in the table (these rows are the ones needed in the examples given in this paper). The rest of the table is similar with the following changes.

Instead of the heading "A: Effort test positive", two other sections of the table begin respectively with: "B: Effort test negative", and "C: Effort test not performed or not determined". In each of sections A,B,C, the heading "1: Left common trunk disease", can also be: "2: Three blood vessel disease", "3: Two blood vessel disease with proximal anterior descendent affected", "4: Two blood vessel disease with proximal anterior descendent not affected", "5: One blood vessel disease with proximal anterior descendent affected", or "6: One blood vessel disease with proximal anterior descendent not affected". Under each of these 6 headings there are 12 items of information.

3 Translation of a Set of Criteria into an RBS

3.1 Introductory Note

An RBS rule is usually represented by an implication between a conjunction of propositional variables $X[i]$ and a disjunction (or a conjunction) of propositional variables, as

$$\circ X[1] \wedge \circ X[2] \wedge ... \wedge \circ X[n] \rightarrow \bullet(\circ X[n+1] \vee .... \vee \circ X[s]) \ .$$

Under our three-valued logic, the symbol "$\circ$" can be replaced by symbol L, $\Diamond$, $\neg$, a combination of them (for instance $\Diamond \neg X$), or no symbol at all. LX means "it is necessary that X holds". X means "X holds". $\Diamond X$ means "it is possible that X holds". $\neg X$ means "not-X".

The symbol "•" refers to the degree of certainty of the whole rule or the certainty of the conclusions and is to be replaced by a combination of the symbols $\neg$, L, $\Diamond$.

3.2 Translation of the Information in the Table

We can assign a propositional variable $X[i]$, to each datum and action.

Asymptomatic: no variable is assigned (all other symptoms suppose that this case is assumed).

Surgical risk.- low/moderate: $\neg X[1]$, high: $X[1]$.

Effort test.- positive: $X[2]$, negative: $\neg X[2]$, not done or not decisive: $\Diamond \neg X[2]$ (this case is translated as "it is possible that the effort test is negative").

Left common trunk disease: $X[3]$. Three blood vessel disease: $X[4]$. Two blood vessel disease: $X[5]$. One blood vessel disease: $X[6]$.

Anterior proximal descendent.- affected: $X[13]$, not affected: $\neg X[13]$.

LVEF (L).- $L > 50\%$: $X[7]$, $50\% \geq L > 30\%$: $X[8]$, $30\% \geq L \geq 20\%$: $X[9]$.

Revascularization: $X[10]$, PTCA: $X[11]$, CABG: $X[12]$.

These variables are combined to form rules, under the following conventions (which we have designed, in order to translate as accurately as possible the information in the tables):

(i) Data is written with no symbol preceding it. The reason is that we can suppose that it has been collected with a reasonable degree of certainty.

(ii) The symbols $+$, ?, and $-$, are respectively translated into L, $\Diamond$, and $L\neg$.

(iii) The symbol "•" (see subsection 3.1 above) will be replaced by L, $\Diamond$, or $\Diamond\neg$. They respectively mean that the experts agree, have undecided agreement, or disagree, on the appropriateness of Revascularization, PTCA and CABG.

Item 1 in the table can be reinterpreted as: "IF a patient is asymptomatic AND his/her surgical risk is low/moderate AND his/her effort test is positive AND he/she has a left common trunk disease AND his/her LVEF is strictly larger than 50% THEN the panel of experts have judged that Revascularization is appropriate". Moreover, "there is agreement (A) about this judgment". It is translated as:

R1: $\neg X[1] \wedge X[2] \wedge X[3] \wedge X[7] \rightarrow L(LX[10])$

The following list is a translation of the 2nd to 12th rules of the part of the table that refers to "effort test positive". The whole table can be translated similarly. The system automatically performs logical simplifications such as $LLX \leftrightarrow LX$.

R2: $\neg X[1] \wedge X[2] \wedge X[3] \wedge X[7] \rightarrow L(L\neg X[11] \wedge LX[12])$

R3: $X[1] \wedge X[2] \wedge X[3] \wedge X[7] \rightarrow L(LX[10])$

R4: $X[1] \wedge X[2] \wedge X[3] \wedge X[7] \rightarrow L(L\neg X[11] \wedge LX[12])$

R5: $\neg X[1] \wedge X[2] \wedge X[3] \wedge X[8] \rightarrow L(LX[10])$

R6: $\neg X[1] \wedge X[2] \wedge X[3] \wedge X[8] \rightarrow L(L\neg X[11] \wedge LX[12])$

R7: $X[1] \wedge X[2] \wedge X[3] \wedge X[8] \rightarrow L(LX[10])$

R8: $X[1] \wedge X[2] \wedge X[3] \wedge X[8] \rightarrow L(L\neg X[11] \wedge LX[12])$

R9: $\neg X[1] \wedge X[2] \wedge X[3] \wedge X[9] \rightarrow L(LX[10])$

R10: $\neg X[1] \wedge X[2] \wedge X[3] \wedge X[9] \rightarrow L(L\neg X[11] \wedge LX[12])$
R11: $X[1] \wedge X[2] \wedge X[3] \wedge X[9] \rightarrow L(LX[10])$
R12: $X[1] \wedge X[2] \wedge X[3] \wedge X[9] \rightarrow L(L\neg X[11] \wedge LX[12])$.

An example of a rule in which there exists disagreement, that is, symbols $\Diamond\neg$ precede its conclusion, is (see the Explanatory Note below, that justifies why writing the symbol "$\vee$" in the conclusion):

R20: $X[1] \wedge X[2] \wedge X[4] \wedge X[8] \rightarrow \Diamond\neg(LX[11] \vee LX[12])$.

Other rules that will be used in this example are:

R39: $X[1] \wedge X[2] \wedge X[5] \wedge \neg X[13] \wedge X[7] \rightarrow \Diamond\neg(\Diamond X[10])$,

which translates in the conclusion the disagreement on the undecided appropriateness of revascularization, and:

R40: $X[1] \wedge X[2] \wedge X[5] \wedge \neg X[13] \wedge X[7] \rightarrow L(\Diamond X[11] \wedge L\neg X[12])$,

which translates the agreement in both the undefined appropriateness of PTCA and the unappropriateness of CABG.

Explanatory note: Let us justify why writing "$\wedge$" in the conclusion of some rules, but "$\vee$" in some other rules.

In principle, only $\wedge$ but no $\vee$ symbols would appear in the consequent of each rule. But, in addition to the information given in the table, it is necessary to take into account information that does not appear explicitly in it, but that belongs to medical practice. For instance, PTCA and CABG cannot be performed simultaneously. This is translated into $L\neg(LX[11] \wedge LX[12])$ (the expression $LX[11] \wedge LX[12]$ is an example of what in RBSs context is called an "integrity constraint"). Then $L\neg(LX[11] \wedge LX[12])$ would lead to a logical conflict with rule 20 if rule 20 had an $\wedge$ in its consequent. So, there must be an initial pre-process to obtain the definitive rules, and some $\wedge$ are substituted by $\vee$ (in the consequent of some rules) in order to avoid logical conflicts.

4   Knowledge Extraction and Consistency in the Set of Criteria

For the sake of simplicity, we will consider only rules 1 to 12 and rules 20, 39 and 40, which are enough to clarify the concepts to be described.

(1) Potential facts and facts

It is known in RBSs literature that a potential fact in a set of rules is any literal, which stands in the antecedent of some rule(s) of the set but not in any consequent of any rule(s) in the set. A literal is a variable preceded or not by the symbol $\neg$ (if we are in bivalued logic) or by $\neg$, $\Diamond$, L, or by some combination of these symbols (in the three valued case).

Then, potential facts in our example are $X[1]$, $\neg X[1]$, $X[2]$, $X[3]$, $X[4]$, $X[5]$, $X[7]$, $X[8]$, $X[9]$, $\neg X[13]$.

In each case, those among the potential facts that are stated will be called "facts".

(2) Integrity constraints

In our example, both PTCA and CABG cannot be carried out simultaneously. To take both actions simultaneously (Integrity Constraint: IC) is translated as $LX[11] \wedge LX[12]$. Therefore, the formula (NIC) that strongly negates it, $L\neg(LX[11] \wedge LX[12])$, needs to be added as new information.

(3) Addition of information directly by the experts

As Revascularization can be carried out by PTCA or CABG (under the integrity constraint in *(2)*), we add as information the formula $X[10] \leftrightarrow X[11] \vee X[12]$. The formula $X[10] \leftrightarrow X[11] \vee X[12]$, will be hereafter referred to as "ADDI1" (additional information 1).

Addition of more information suggested by the anomalies found will be explained in *(4)*.

(4) Anomalies

In our implementation, we will ask the computer the following questions.

(4-1) Is the set of rules 3, 4, 7, 8, 11, 12, 20, 39, 40 together with the facts $X[1]$, $X[2]$, $X[3]$ $X[4]$, $X[5]$, $X[7]$, $X[8]$, $X[9]$, $\neg X[13]$ (plus ADDI1 and NIC) consistent? The answer will be NO (see Appendix).

The same occurs with the set of rules 1, 2, 5, 6, 9, 10, 20, 39, 40, together with the facts $\neg X[1]$, $X[2]$, $X[3]$, $X[4]$, $X[5]$, $X[7]$, $X[8]$, $X[9]$, $\neg X[13]$ (plus ADDI1 and NIC).

Note that we do not ask about consistency of all rules 1 to 12 with their potential facts altogether because $\neg X[1]$ and $X[1]$ can never be given as facts simultaneously, as they give inconsistency immediately.

(4-2) Is the set of rules 3, 4, 7, 8, 11, 12 together with their respective potential facts, NIC and ADDI1, consistent? (ADDI2 is not included because it contains another different variable). The answer will be YES.

The same occurs with the set of rules 1, 2, 5, 6, 9, 10 together with their respective potential facts, NIC and ADDI1.

(4-3) Is the set of rules 8, 20 together with their potential facts consistent? The answer will be NO.

The contradiction between rules 8 and 20 is conditional on the case when simultaneously a left common trunk disease ($X[3]$) and a three vessel disease ($X[4]$) occur. But if $X[3]$ (left common trunk disease) holds, rule 8 says that PTCA is absolutely disregarded (under agreement) and if $X[4]$ (three vessel disease) holds, PTCA is considered appropriate (under disagreement).

We wonder if in a situation where $X[3]$ and $X[4]$ hold together it would be better to assess that if a left common trunk disease implies (under agreement) absolutely disregarding PTCA, then a three vessel disease should imply (also under agreement) at least a possibility of disregarding PTCA. In symbols: $(X[3] \rightarrow L(L\neg X[11])) \rightarrow (X[4] \rightarrow L(\Diamond\neg X[11]))$.

This is the kind of information to be added, if the experts agree. The formula above will be denoted "ADDI2" (additional information 2).

(4-4) Is the set of rules 39, 40 together with the potential facts in these two rules and ADDI1 consistent? The answer will be again NO.

The existence of contradictions in the points *(4-3)* and *(4-4)* suggests changes in the rules, and therefore a way to improve the information in the table. In the case *(4-3)* we could leave rule 8 as it is and exchange rule 20 with:

RN20: $X[1] \land X[2] \land X[4] \land X[8] \to \bullet(\Diamond \neg X[11] \lor LX[12])$,

where $\bullet$ could be L or $\Diamond$ (according to the judgement of the experts). "RN" means "new rule".

In the case *(4-4)* we could exchange rule 39 with:

RN39: $X[1] \land X[2] \land X[5] \land \neg X[13] \land X[7] \to \Diamond(\Diamond X[10])$

and leave rule 40 as it is.

It will have to be checked that no contradiction appears under these changes.

(4-5) Exchange in question *(4-3)* rule 20 with RN20 plus ADDI2, and check the consistency of this new set. The answer will be YES.

(4-6) Exchange in question *(4-4)* rule 39 with RN39 and check the consistency of this new set. The answer will be YES.

(5) Knowledge extraction

Theorem proving is an important field of research in A.I. A particular case is the extraction of knowledge which is implicit in an RBS. The implementation to be described in the Appendix allows asking the computer questions in the form of propositional formulae that are written using the propositional variables which appear in the RBS. Let us illustrate this with two examples÷

i) Is $\Diamond(LX[11] \lor LX[12])$ a consequence of rules 3, 4, 7, 8, 11, 12, their potential facts and ADDI1? The answer will be YES.

ii) Is ADDI2 a consequence of rules 8 and RN20 together with their potential facts? The answer will be YES, which is a nice result because it certifies that exchanging rule 20 with RN20 can be proposed to the panelists.

As a consequence of these and other suggestions, they have decided to make substantial changes in the configuration of the tables. For instance the $+/-$ is not considered any longer. The knowledge included has also been revised.

5   Theory Description: Two Basic Items and the Main Result

In this section we briefly summarize the RBSs theory of verification and of knowledge extraction on which the treatment of this paper is based. The theoretical model is developed in [14] and [11], and therefore only some important basic items and an informal statement of the main result together with its application to consistency checking are stated below.

This theory is related to term-rewriting and theorem proving (see for instance [8], [9], [13]). The main theorem first appeared in [2], was improved in [6] and is proved in a different way in [14] and [11]. The method differs substantially from other known verification methods (a state of the art in Knowledge Systems verification is studied in [10]).

5.1   Tautological Consequences and Contradictory Domains

A propositional formula A_0 is a tautological consequence of the propositional formulae $A_1, A_2, ..., A_m$, denoted $\{A_1, A_2, ..., A_m\} \models A_0$ iff for any truth-valuation v such that if $v(A_1) = v(A_2) = ... = v(A_m) = 2$, then $v(A_0) = 2$ ($0, 1, 2$ are the values respectively assigned to "false", "indeterminate" and "true" in our three valued logic).

In this context, $\{A_1, A_2, ..., A_m\}$ is called "a contradictory domain" iff $\{A_1, A_2, ..., A_m\} \models A$, where A is any formula of the language in which $A_1, A_2, ..., A_m$ are expressed. The name *contradictory domain* comes from the fact that, if all formulae follow from $\{A_1, A_2, ..., A_m\}$, in particular contradictory formulae follow.

5.2   Translation of Logical Formulae into Polynomials

The information contained in the RBS is translated into polynomials. From here onwards the polynomial ring will be $(\mathbb{Z}/(3)\mathbb{Z})[x_1, ..., x_n]/I$ where the x_i are the polynomial variables that correspond to the propositional variables that appear in the RBS and I is the ideal generated by the polynomials of the form $x_i^3 - x_i$. In the three-valued case (Lukasiewicz's Logic with modal connectives), the polynomial translation of the connectives is (see [14] for details):

$$f_\neg(q) = (2 - q) + I$$
$$f_\diamond(q) = 2q^2 + I$$
$$f_L(q) = (q^2 + 2q) + I$$
$$f_\vee(q, r) = (q^2 r^2 + q^2 r + qr^2 + 2qr + q + r) + I$$
$$f_\wedge(q, r) = (2q^2 r^2 + 2q^2 r + 2qr^2 + qr) + I$$
$$f_\rightarrow(q, r) = (2q^2 r^2 + 2q^2 r + 2qr^2 + qr + 2q + 2) + I$$
$$f_\leftrightarrow(q, r) = (q^2 r^2 + q^2 r + qr^2 + 2qr + 2q + 2r + 2) + I$$

5.3   The Main Result

Theorem 1. * *A formula A_0 is a tautological consequence of $\{A_1, A_2, ..., A_m\}$ ($\{A_1, A_2, ..., A_m\} \models A_0$) iff the polynomial translation of the negation of A_0 belongs to the ideal generated by the polynomial translation of the negations of $A_1, A_2, ..., A_m$:*

$$f_\neg(A_0) \in\, <f_\neg(A_1), ..., f_\neg(A_m)> + I \ .$$

In particular the $A_1, ..., A_m$ can be rules, facts, integrity constraints and additional information.

Corollary 1. * *If* 1 *belongs to the ideal above, the ideal is the whole ring. But this is the same as to say that the set of the formulae in the RBS is a contradictory domain. Therefore the condition of inconsistency is: the RBS is inconsistent iff*

$$1 \in < f_\neg(Rules), f_\neg(Facts), f_\neg(NICs), f_\neg(ADDIs) > +I .$$

6   Conclusion

We have suggested the possibility of translating medical knowledge in terms of RBSs based on a multivalued and modal logic. This approach is more dynamic than the one using tables: it allows the addition of new knowledge (with no changes in the inference devices), to improve knowledge, and even to extract knowledge that is not explicit in the RBS. Despite the fact that dealing with medical knowledge under an RBS interpretation is not new, we think this approach using a multivalued and modal logic and Computer Algebra is.

7   Acknowledgments

We would like to thank Dr. P. Lázaro (Institute of Health, Carlos III Hospital) for providing the set of criteria used in this research (that is now in a process of improvement and simplification) and for his most valuable comments.

We would also like to thank Ana María Díaz, for her contribution to the article.

Appendix: CoCoA Implementation of Our Set of Criteria

The language CoCoA[2] (see [5]) is very well suited for our purposes, as it is specialized in computing "Gröbner Bases" (GB) and "Normal Forms" (NF) in polynomial rings over finite fields (see [16], and [1]).

In this subsection we describe step by step in CoCoA the example of section 4.

(1) Declare the ring of polynomials to be $\mathbb{Z}/(3)[x_1, ..., x_{13}]$:
```
A := Z/(3)[x[1..13]];
USE A;
```

(2) Declare the ideal generated by the polynomials $x_i^3 - x_i$, for $i = 1$ to 13.
```
I:=Ideal(x[1]^3-x[1],...., x[13]^3-x[13]);
```
Explanation: CoCoA has not yet implemented quotient rings. Instead of checking an ideal J in A/I, the ideal $J + I$ will be studied in A.

---

[2] A. Capani, G. Niesi, L. Robbiano, CoCoA, a system for doing Computations in Commutative Algebra. Available via anonymous ftp from: cocoa@dima.unige.it

(3) Translate connectives into polynomials (see subsection 5.2)

```
NEG(M) := NF(2-M, I)
POS(M) := NF(2*M^2,I);
NEC(M) := NF(M^2+2*M, I);
OR1(M,N) := NF(2*M^2*N^2+M^2*N +M*N^2+2*M*N+M+N, I);
AND1(M,N) := NF(2*M^2*N^2+2*M^2*N +2*M*N^2+M*N, I);
IFF(M,N) := NF(2*M^2N^2+ 2*M^2*N + 2*M*N^2+ M*N + 2*M + 2, I);
IMP(M,N) := NF(M^2*N^2 + M^2*N + M*N^2+ 2*M*N +2*M +2*N + 2, I);
```

(4) Declare the rules in prefix form. As an illustration, rule 20 is rewritten as:

```
R20 := IMP(AND1(AND1(AND1(X[1],X[2]),X[4]),X[8]),
      OR1(NEC(X[11]),NEC(X[12])));
```

(5) Declare the potential facts

```
F1 := NEG(X[1]);                F2 := X[1];
F3 := X[2];                     F4 := X[3];
F5 := X[4];                     F6 := X[5];
F7 := X[7];                     F8 := X[8];
F9 := X[9];                     F10 := NEG(X[13]);
```

(6) Declare integrity constraints and other additional information (see *4(3)*):

```
NIC := NEC(NEG(AND1(NEC(X[11]), NEC(X[12]))));
```

For example, ADDI1, that is, $X[10] \leftrightarrow X[11] \vee X[12]$, is introduced as:

```
ADDI1 := AND1(IMP(X[10],OR1(X[11],X[12])),IMP(OR1(X[11],X[12]),
      X[10]));
```

and ADDI2, that is, $(X[3] \to L(L\neg X[11])) \to (X[4] \to L(\Diamond\neg X[11]))$, as:

```
ADDI2 := IMP(IMP(X[3],NEC(NEC(NEG(X[11])))),IMP(X[4],
      NEC(POS(NEG(X[11])))));
```

(7) Declare the ideal J generated by the negations of rules 3, 4, 7, 8, 11, 12, 20, 39, 40 and all potential facts (with the exception of F1), integrity constraints and additional information:

```
J := IDEAL(NEG(F2),NEG(F3),...,NEG(F9),NEG(F10),NEG(R3),NEG(R4),
      NEG(R7),NEG(R8),NEG(R11),NEG(R12),NEG(R20),NEG(R39),NEG(R40),
      NEG(NIC),NEG(ADDI1));
```

(8) Check the consistency of the set of rules, facts... of *(7)*:

```
GBasis(I+J);
```

As the answer is [1], there is inconsistency (as advanced in *4(4-1)*).

(9) Let us suppress rules 20, 39, 40 and facts F5, F6, F10 from the generators of *J* and let us denote *H* the new ideal. As

```
GBasis(I+H);
```

returns a set of polynomials, there is consistency (as advanced in *4(4-2)*).

(10) Similarly, the set {R8, R20, F2, F3, F4, F5, F8} and also the set {R39, R40, F2, F3, F4, F6, F7, F10, ADDI1} give inconsistency (as advanced in 4*(4-3)* and *(4-4))*.

(11) The existence of inconsistency in *(9)* suggest (subsection 4*(4-5)* and *(4-6))*, exchanging {R8, R20} and {R39, R40}, respectively with {R8, RN20, ADDI2} and {RN39, R40}. Accordingly we define two new ideals S and T.

```
S := Ideal(NEG(F2),NEG(F3),NEG(F4),NEG(F5),NEG(F8),NEG(F13),
        NEG(R8),NEG(RN20),NEG(ADDI2) );
T := Ideal(NEG(F2),NEG(F3),NEG(F6),NEG(F7),NEG(RN39),NEG(R40),
        NEG(ADDI1) );
GBasis(I+T);
```

Now, in both cases, CoCoA returns a set of polynomials, which means consistency, as advanced in 4*(4-5)* and *(4-6)*.

(12) Knowledge extraction. Let us check, for instance, if the first formula in *(5)* of section 4: $\Diamond(LX[11] \lor LX[12])$, that is:

```
FOR1 := POS(OR1(NEC(X[11]),NEC(X[12])));
```

is a consequence of the information given by the ideal H above:

```
NF(NEG(FOR1), I+H );
```

The answer is 0, and therefore the formula FOR1 follows from the generators of ideal H.

Another example from *(5)* of section 4: Is ADDI2 a consequence of rules 8 and RN20, together with all their potential facts?

```
U := Ideal(NEG(F2),NEG(F3),NEG(F4),NEG(F5),NEG(F8),NEG(R8),
        NEG(RN20));
NF(NEG(ADDI2), I+U );
```

The answer is 0, and therefore ADDI2 follows from rules 8 and RN20.

(13) Detecting the formula that produces the inconsistency. Our implementation also contains a program CONSIST, that can be applied to check where the inconsistency is. It adds one new formula each time (in the order in which they are given) and points out the first element that produces inconsistency.

The whole process of processing all the examples in this section (translating rules, facts, integrity constraints and additional information, checking for consistency, answering if the given formulae are consequences of the corresponding ideal, and the application of the program CONSIST) takes about fifteen seconds in a standard 64 MB RAM Pentium-based PC.

References

1. V. Adams, P. Loustanau, An Introduction to Gröbner Bases. Graduate Studies in Mathematics 3, American Mathematical Society, Providence, RI, (1994).

2. J. A. Alonso and E. Briales, Lógicas Polivalentes y Bases de Gröbner. Procs. of the V Congress on Natural Languages and Formal Languages, Ed. M. Vide, Barcelona, (1989), 307-315.

3. B.G. Buchanan and E.H. Shortliffe, Rule Based Expert Systems: The MYCIN experiments of the Stanford Heuristic Programming Project. Addison-Wesley, Reading, MA. (1984).

4. S. Bernstein, J. Kahan, Personal communication. RAND Corporation (1993).

5. A. Capani and G. Niesi, CoCoA User's Manual (v. 3.0b). Dept. of Mathematics, University of Genova (1996).

6. J. Chazarain, A. Riscos, J.A. Alonso, E. Briales, Multivalued Logic and Gröbner Bases with Applications to Modal Logic. Journal of Symbolic Computation, 11, 181-194 (1991).

7. M. Field, M., K. Lohr, (Eds). Guidelines for Clinical Practice. From Development to Use. National Academy Press, Washington D.C. (1992).

8. J. Hsiang, Refutational Theorem Proving using Term-rewriting Systems, Artificial Intelligence, 25 (1985), 255-300.

9. D. Kapur and P. Narendran, An Equational Approach to Theorem Proving in First-Order Predicate Calculus. 84CRD296 General Electric Corporate Research and Development Report, Schenectady, NY, March 1984, rev. Dec. 1984. Also in , Proceedings of IJCAI-85 (1985), 1446-1156.

10. L.M. Laita, L. de Ledesma, Knowledge-Based Systems Verification, Encyclopedia of Computer Science and Technology, Eds. A Kent, J.G. Williams. Marcel Dekker, New York (1997), 253-280.

11. L.M. Laita, E. Roanes-Lozano, J.A. Alonso, L. de Ledesma, Automated Multi-Valued Logic reasoning in rule based Expert Systems. Preprint, AI Dept., Universidad Politécnica de Madrid, sent for tentative publication in Soft Computing (1998).

12. P. Lázaro, K. Fitch, Criterios de uso apropiado para by-pass coronario. Unpublished Report (1996).

13. G. C. Moisil, The Algebraic Theory of Switching Circuits. Pergamon Press, Oxford (1969).

14. E Roanes-Lozano, L.M. Laita, E. Roanes-Macías, A Polynomial Model for Multivalued Logics with a Touch of Algebraic Geometry and Computer Algebra. Special Issue "Non-Standard Applications of CA", in Mathematics and Computers in Simulation, 45/1-2 (1998), 83-99.

15. J.K. Slaney, Formal Logic and its application in medicine. Pillips CI. Logic in Medicine, British Medical journal (1988).

16. F. Winkler, Introduction to Computer Algebra. Lecture Notes WS 93/94, RISC-Linz (1994).

The Unification Problem for One Relation Thue Systems*

Christopher Lynch

Department of Mathematics and Computer Science Box 5815, Clarkson University,
Potsdam, NY 13699-5815, USA,
clynch@sun.mcs.clarkson.edu,
http://www.clarkson.edu/~clynch

Abstract. We give an algorithm for the unification problem for a generalization of Thue Systems with one relation. The word problem is a
special case. We show that in many cases this is a decision procedure with
at most an exponential time bound. We conjecture that this is always a
decision procedure.

1 Introduction

In this paper we study the unification problem and word problem for Thue
Systems. This basic problem appears under several different names. It is also
known as the unification and word problem for semigroups, terms with monadic
function symbols, and ground terms with one associative operator.

In particular, we are interested in Thue Systems with only one equation,
but we have generalized our results to larger classes. The word problem for one
equation can be stated simply: Given an equation $s = t$, and two words u_0 and
u_n, is there a sequence of words $u_0, \cdots, u_n$ such that each u_{i+1} is the result of
replacing an occurrence of s in u_i by t, or replacing an occurrence of t in u_i by
s?

Despite the very simple formulation of the problem, it is unknown whether
the problem is decidable. It has been shown to be undecidable when there are
three equations instead of one [9], but the case for two equations is also unknown.
The word problem for groups with one defining equation has been known to be
decidable for 65 years [8]. However, despite considerable work in the area [2,4]
(see [6] for a survey), the decidability for one equation Thue systems is unknown.

In this paper, we address (but do not solve) the problem, and we also generalize the problem in some ways. One of our generalizations is to consider the
unification problem, which is a generalization of the word problem. The unification problem is as follows: Given an equation $s = t$ and words u and v, are there
words x and y such that vy can be reached from ux with a sequence of replacements of s by t and t by s. We also generalize from one equation Thue systems
to allow more than one equation but require a certain syntactic structure. Our

* This work was supported by NSF grant number CCR-9712388 and partially done
while visiting INRIA Lorraine and CRIN.

result is a procedure that decides the unification problem when it halts, and also produces the most general unifier. We have not been able to prove that it halts for all instances of our generalization of the one equation unification problem, but we conjecture that it does.

Although we have not proved a decidability result, we believe our work is important. We have provided some theorems showing how to automatically detect that the algorithm is a decision procedure for certain Thue systems. We even give a complexity result, showing that the algorithm is at most exponential for a large class of Thue systems We have implemented an algorithm for one equation Thue systems, based on the one in this paper. On every example we have tried, it always terminates quickly with the answer.

Our main interest in this problem is not just for one equation Thue systems. Our goal is to extend these results to equations over terms. Popular methods for deciding word and unification problems, like the Knuth-Bendix completion method have many examples, even very simple ones, where they do not halt. Our method attempts to avoid those problems. Although our presentation here is only for monadic function symbols, the ideas extend to function symbols of higher arity. The syntactic restrictions on the class used in this paper allow for our algorithm to be deterministic. Relaxing those restrictions is possible if we allow the algorithm to be non-deterministic. Our plan for the near future is to investigate all these extensions. We expect the ideas in this paper will be important for finding decision procedures for interesting classes of equational theories. The main inspiration for our paper is our previous work on SOUR graphs [7]. This paper is actually a simplification of those ideas, although the ideas have evolved quite a lot. We have achieved the two main purposes we sought in the evolution of those ideas: First, they are vastly simplified to allow much easier understanding and implementation. Second, we have shown the use of the method to solve decision problems, which we did not realize before.

The next section of the paper gives some required background. The section after that builds up the necessary machinery for our algorithm. We convert the unification problem into a problem in rewrite systems. The following section develops the rewrite system problem into an algorithm. Interestingly, in this section we show that the unification (and word) problem is equivalent to a problem in termination of rewrite systems. We show how to detect loops in rewrite systems, and conjecture that all nonterminating rewrite sequences are loops, which forms the basis of our algorithm. Interesting this is the same conjecture made for termination of one rule semi-Thue systems [10], another decision problem whose solution is unknown. That gives us the impression that the same techniques used for solving the termination problem will be useful for solving the word (and unification) problem. In the conclusion, we relate our work with other work.

2 Preliminaries

We are given a set A as *alphabet*. In this paper, we use letters a, b, c, d, e, f, g, h as members of the alphabet. A *word* is a sequence of members of the alphabet. We use letters r, s, t, u, v, w to represent words. If w is a word then $|w|$ is the number of symbols in w. If $|w| = 0$, then we write w as ϵ and call it the *empty word*. If u and v are words, then uv represents the concatenation of u and v. Then u is a *prefix* of uv, and v is a *suffix* of uv. Also, v is a subword of uvw. $u \approx v$ is an *equation* if $u \neq \epsilon$ and $v \neq \epsilon$.[1]

A *Thue System* is a set of equations. We assume it is closed under symmetry.[2] Let E be a Thue system. If $s \approx t \in E$ then we write $usv \approx_E utv$. If E is obvious, we may write $usv \approx utv$. We call this an *equational step at position p*, where $p = |u|$. If $p = 0$, then this is called an *equational step at the top*. A *proof of* $u \approx_E v$ is of the form $u_0 \approx_E u_1 \approx_E \cdots \approx_E u_n$ where $n \geq 0$, $u = u_0$, $v = u_n$, and $u_{i-1} \approx_E u_i$ for all $i > 0$. Given a Thue system E and a pair of words u and v, the *uniform word problem for Thue Systems* is the problem of deciding whether $u \approx_E v$. This is also called the *word problem for semigroups*, although in this case the problem is stated semantically [3]. The syntactic version of the word problem for semigroups was shown equivalent to the semantic version by Birkhoff.

Another way to examine the problem is to view the members of A as monadic function symbols. In that case, a set of variables V is added to the language. We refer to members of V with letters x, y, z. Also, a set of constants C is added. A *term* is a variable, or a constant, or a function symbol applied to a term (parentheses are omitted). Equations are of the form $s \approx t$, where s and t are terms. A *substitution* is a mapping from the set of variables to the set of terms, which is the identity almost everywhere. A substitution is extended to its homomorphic extension (i.e., $(ft)\sigma = f(t\sigma)$). A *ground instance* of a term (resp. equation) t is anything of the form $t\sigma$, where $x\sigma$ is ground for all x in t. If t is a term or equation, then $Gr(t)$ is the set of all ground instances of t. If E is a set of equations, then $v \approx_E v$ if and only if every ground instance of $u \approx v$ is in the congruence closure of the set of all ground instances of equations in E. That is the semantic definition. This also could be defined syntactically by saying that $us \approx_E ut$ if $s \approx t$ is an instance of an equation of E. Proofs are defined as for Thue Systems, and the word problem is stated in the same way. Birkhoff showed that the semantic and syntactic definition are equivalent. In this paper the syntactic definition will be more useful.

Given terms u and v and a set of equations E, σ is an *E-unifier* of u and v if $u\sigma \approx_E v\sigma$. The unification problem for monadic function symbols is to find all E-unifiers of u and v. A set of equations E is said to be *unitary* if for every pair of terms u and v, there is one E-unifier σ such that for every E-unifier θ there

---

[1] See [1] for the case where $u = \epsilon$ or $v = \epsilon$.

[2] Therefore one relation Thue Systems are presented with two equations.

[3] $u \approx_E v$ if and only if $u \approx v$ is true in every model of E

is a substitution η such that $x\sigma\eta \approx_E x\theta$ for every variable x in u or v. Then σ is a *most general unifier* of u and v.

It is well-known that the word problem for Thue Systems can be expressed as a word problem for monadic function symbols. Given the word problem for Thue Systems $u \approx_E v$, where $E = \{s_1 \approx t_1, \cdots, s_n \approx t_n\}$, we transform it into a word problem for monadic function symbols by asking if $uc \approx_E vc$, such that $E = \{s_1 x \approx t_1 x, \cdots, s_n x \approx t_n x\}$, where x is a variable and c is a constant.[4]

We need the notion of rewriting in terms of words. Let s and t be words (possibly empty), then $s \rightarrow t$ is a *rewrite rule*. If R is a set of rewrite rules, then we write $usv \rightarrow utv$, and say usv *rewrites to* utv if $s \rightarrow t \in R$. The reflexive transitive closure of $\rightarrow$ is written as $\rightarrow^*$. A word u is in R- *normal form* if there is no v such that u rewrites to v. A set of rewrite rules is *confluent* if $s \rightarrow^* t$ and $s \rightarrow^* u$ implies that there is a v such that $t \rightarrow^* v$ and $u \rightarrow^* v$. A set of rewrite rules R is *weakly terminating* if for every u there is a v in normal form such that $u \rightarrow^* v$. R is *strongly terminating* if there is no infinite rewrite sequence. A confluent and strongly terminating set of rewrite rules has the property that every rewrite sequence from u leads to the same v in normal form. We say that a set of rewrite rules R is *non-overlapping* if there are no rules $s \rightarrow t$ and $u \rightarrow v$ such that a nonempty prefix of s is a suffix of u or s is a subword of u. If a set of rewrite rules R is non-overlapping and weakly terminating then R is confluent and strongly terminating.

3   The Word and Unification Problem

In this section we give a class of Thue systems which is a generalization of one equation Thue systems. Then we give the structure of a proof of a unification (or word) problem in this generalized class. Finally, we show how this proof structure leads us to a problem in rewrite systems.

First we give the generalized class. A key idea is the notion of syntacticness from [5].

Definition 1. *A proof $u_0 \approx u_1 \approx \cdots \approx u_n$ is syntactic if there is at most one i such that $u_{i-1} \approx_E u_i$ is an equational step at the top. A Thue System E is syntactic if whenever there is a proof of $u \approx_E v$, then there is a syntactic proof of $u \approx_E v$.*

There is another restriction we need on the class to allow for our final procedure to be deterministic.

Definition 2. *A Thue System $s_1 \approx t_1, \cdots, s_n \approx t_n$ has a repeated top equation if there is an $i \neq j$ and $a, b \in A$ and words u, v, u', v' such that $s_i = au$, $t_i = bv$ and $s_j = au'$, $t_j = bv'$. A Thue System $s_1 \approx t_1, \cdots, s_n \approx t_n$ has a repeated top symbol if there is an i and j ($i \neq j$) and $a \in A$ and words u, v such that $s_i = au$ and $s_j = av$, or if there is an i and an a and an s_i and t_i and words u and v such that $s_i = au$ and $s_j = av$.*

---

[4] We sometimes confuse the notation of words and terms. When the distinction is important, we clarify it.

Every word problem for Thue systems of one relation can be reduced to a simpler word problem which is either known to be solvable or has a different top symbol on the left and right side [3]. Therefore, for one equation Thue systems, it is only necessary to consider word problems where the one equation is of the form $as \approx bt$, where a and b are different symbols and s and t are words. Such theories are syntactic [2]. Such theories also have no repeating equation. Below, we show that any theory with no repeating top symbol is syntactic and has no repeating top equation.

Theorem 1. *Let E be a Thue system that has no repeating top symbol. Then E has no repeating top equation and E is syntactic*

Proof. The fact that E has no repeating top equation follows by definition. We prove that E is syntactic by contradiction. Consider the set of all shortest proofs between any pair of terms. Consider the shortest proof $u_0 \approx_E u_1 \cdots \approx_E u_n$ of that set with more than one equational step at the top. Then there is a step from some u_i to u_{i+1} at the top using some equation $au \approx bv$, and later there is another step from some u_j to u_{j+1} using $bv \approx au$. Since this is the shortest proof, every proper subproof must be syntactic. But then there can be no intermediate steps involving u and v, so the steps from u_i to u_{i+1} and from u_j to u_{j+1} can be removed from the proof resulting in a shorter proof of $u_1 \approx_E u_n$.

The results in this paper apply to syntactic Thue systems with no repeating top equation. Next we look at the structure of proofs of the unification problem in such theories. First we consider the case of unifying two terms with a different top symbol.

Lemma 1. *Let E be a syntactic Thue System. Let aux and bvy be terms. If σ is an E-unifier of aux and bvy then σ is of the form $[x \mapsto u'z, y \mapsto v'z]$ and there exists*

- *an equation $asx \approx btx \in E$,*
- *words r_1, r_2 such that $u' = r_1 r_2$,*
- *words w_1, w_2 such that $v' = w_1 w_2$,*
- *and words s', t' such that $ur_1 \approx_E ss'$, $tt' \approx_E vw_1$, and $s'r_2 \approx_E t'w_2$.*

Proof. Let $u_0 \approx_E \cdots \approx_E u_n$ be the proof of $aux\sigma \approx_E bvy\sigma$. There must be exactly one equational step at the top. The proof can be divided up into four parts. First $aux\sigma$ must be changed to a new word with as as prefix. The second part is to change as to bt. The third part is all the steps below bt, and the fourth part is to change bt into a word with bv as prefix. The second and third parts can be exchanged, but wlog we assume they happen in the order given.

Suppose that $u_i \approx_E u_{i+1}$ is the first equational step at the top. Then u_i has as as a prefix. This means that $x\sigma$ must have some r_1 as a prefix, such that there is an s' where $ur_1 \approx_E ss'$. Therefore u_{i+1} has bts' as a prefix. That gives us the first part of the proof. The third part is all the steps below bt so there must be r_2, t', w_2 such that $s'r_2 \approx_E t'w_2$. The fourth part changes tt' to something with a v as a prefix, so there must be a w_1 such that $tt' \approx_E vw_1$.

To sum it all up, the proof looks like: $aur_1r_2 \approx_E ass'r_2 \approx_E bts'r_2 \approx_E btt'w_2 \approx_E bvw_1w_2$.

Now we look at the proof structure when unifying two terms with the same top symbol.

Lemma 2. *Let E be a syntactic set of monadic equations containing no equation of the form $as = at$. Then σ is an E-unifier of u and v if and only if σ is an E-unifier of au and av.*

Proof. Suppose there is an equational step at the top of the proof of $au \approx_E bv$. Since there is no equation of the form $as \approx at$, there must be two equational steps at the top of the proof. But that cannot be, because E is syntactic. Therefore, there is no equational step at the top of the proof.

Note that the condition of no equation of the form $as \approx at$ is implied by the condition of no repeated equation, since each Thue System is assumed to be closed under symmetry.

Our next step is to convert each unification problem to a rewrite system over an extended language, where the above two lemmas are applied as rewrite rules. First we define a new alphabet $\bar{A} = \{\bar{a} \mid a \in A\}$. Let $B = A \cup \bar{A}$. We define an inverse function on words in B^* such that

- If $a \in A$ then $a^{-1} = \bar{a}$.
- If $\bar{a} \in \bar{A}$ then $(\bar{a})^{-1} = a$.
- $(b_1 \cdots b_n)^{-1} = b_n^{-1} \cdots b_1^{-1}$ for $n \geq 0$, and $b_i \in B$ for all $1 \leq i \leq n$.

Any word $w \in B^*$ can be represented uniquely in the form $u_1v_1^{-1} \cdots u_nv_n^{-1}$ where $n \geq 0$, each $u_i, v_i \in A^*$, $u_i = \epsilon$ only if $i = 1$ and $v_i \neq \epsilon$, and similarly $v_i = \epsilon$ only if $i = n$ and $u_i \neq \epsilon$. We say that w has n *blocks*.

Given a Thue System $E = \{a_1s_1 \approx b_1t_1, \cdots, a_ns_n \approx b_nt_n\}$ over A, we define a rewrite system R_E over B containing

- $\bar{a}_ib_i \rightarrow s_it_i^{-1}$ for all $1 \leq i \leq n$, and
- $\bar{a}a \rightarrow \epsilon$ for all $a \in A$. These are called *cancellation rules*.

Example 1. Given the Thue System $\{abaa \approx bbba, bbba \approx abaa\}$, the associated rewrite system R_E is

1. $\bar{a}b \rightarrow baa\bar{a}\bar{b}\bar{b}$
2. $\bar{b}a \rightarrow bba\bar{a}\bar{a}\bar{b}$
3. $\bar{a}a \rightarrow \epsilon$
4. $\bar{b}b \rightarrow \epsilon$

Given an E-unification problem $G = ux \approx vy$, we associate a word $w_G = u^{-1}v$. Then we have the following theorem which translates a unification problem into a rewrite problem.

Theorem 2. *Let E be a syntactic Thue System with no repeating equation, and G be a unification problem over A. Then G has a solution if and only if w_G has an R_E-normal form of the form $u'(v')^{-1}$ with u' and v' in A^*.[5] Furthermore $\sigma = [x \mapsto u'z, y \mapsto v'z]$ is the most general unifier of G in E.*

Proof. Given a word w of n blocks, where $w = u_1 v_1^{-1} \cdots u_n v_n^{-1}$, we think of w as representing the unification problem $(x = u_1 z_1) \wedge \bigwedge_{2 \le i \le n}(v_{i-1} z_{i-1} = u_i z_i) \wedge (v_n z_n = y)$ for some $z_1, \cdots, z_n$. Therefore, if $G = ux \approx_E vy$, then $w_G = u^{-1}v$ represents the unification problem $x = z_1 \wedge u z_1 = v z_2 \wedge z_2 = y$, that is $ux \approx_E vy$.

Since each word w represents a unification problem, the solution to the unification problem has a corresponding proof. We will give an induction argument based on the lexicographic combination of the length of that corresponding proof and the number of symbols in w.

Suppose we are given a word $w = u^{-1}v$, representing the unification problem $ux \approx_E vy$. If $ux = au_1 x$ and $vy = av_1 y$ for some u_1 and v_1, then $ux \approx_E vy$ has most general unifier σ if and only if $u_1 x \approx_E v_1 y$ has most general unifier σ. Then the word $w = u^{-1}v = u_1^{-1}\bar{a}av_1 \to u_1^{-1}v_1$, which is a smaller unification problem.

Suppose that $ux = au_1 x$ and $vy = bv_1 y$, and suppose there is an equation $as \approx bt \in E$. Then $\bar{a}b \to st^{-1}$. Furthermore, the unification problem $ux \approx_E vy$ is satisfiable and has most general unifier $\sigma = [x \mapsto u'z, y \mapsto v'z]$ if and only if there are words r_1, r_2 such that $u' = r_1 r_2$, words w_1, w_2 such that $v' = w_1 w_2$, and words s', t' such that $u_1 r_1 \approx_E ss'$, $tt' \approx_E v_1 w_1$, and $s'r_2 \approx_E t'w_2$. So $u_1^{-1}s \to r_1 s'^{-1}$, $t^{-1}v_1 \to t'w_1^{-1}$, and $s'^{-1}t' \to r_2 w_2^{-1}$, Then the word $w = u^{-1}v = u_1^{-1}\bar{a}bv_1 \to u_1^{-1}st^{-1}v_1 \to r_1 s'^{-1}t'w_1^{-1} \to r_1 r_2 w_2^{-1}w_1^{-1}$. This is in normal form, and it represents the unification problem $x = r_1 r_2 z_1 \wedge w_1 w_2 z_1 = y$, which has σ as most general unifier.

Suppose that $ux = au_1 x$ and $vy = bv_1 y$, and there is no equation $as \approx bt \in E$. Then $w = u^{-1}v = u_1^{-1}\bar{a}bv_1$ has no redex at the subword $\bar{a}b$, and therefore has no normal form with zero or one block.

This theorem also shows that any syntactic Thue system with no repeating equation has a unitary E-unification problem, because the rewriting is deterministic and leads to at most one most general unifier.

The following corollary shows how the theorem applies to word problems.

Corollary 1. *Let E be a syntactic Thue System with no repeated equation, and G be a word problem over A. Then G is true in E if and only if the normal form of w_G in R_E is ϵ.*

Proof. The corollary follows from the theorem because the word problem is true if and only if $\sigma = [x \mapsto z, y \mapsto z]$ is a most general unifier.

Example 2. For example, consider the Thue System $\{aa \approx ba, ba \approx aa\}$. Then R_E is

---

[5] Notice that if A has two symbols, then every normal form is of this form.

1. $\bar{a}b \rightarrow a\bar{a}$
2. $\bar{b}a \rightarrow a\bar{a}$
3. $\bar{a}a \rightarrow \epsilon$
4. $\bar{b}b \rightarrow \epsilon$

Let G be the unification problem $abb \approx_E bb$. Then $w_G = \bar{b}\bar{b}abb$. This gives us the following rewrite sequence: $w_G = \bar{b}\bar{b}abb \rightarrow \bar{b}\bar{b}a\bar{a}b \rightarrow \bar{b}\bar{b}aa\bar{a} \rightarrow \bar{b}a\bar{a}a\bar{a} \rightarrow \bar{b}a\bar{a} \rightarrow a\bar{a}\bar{a}$, which is in normal form. That means that the most general unifier of G in E is $\sigma = [x \mapsto az, y \mapsto aaz]$. The word problem for G is not true, because the normal form is not ϵ. However, if we consider the word problem $G' = abba \approx bbaa$, then $w_{G'} = \bar{a}w_Gaa \rightarrow^* \bar{a}a\bar{a}\bar{a}aa \rightarrow^* \epsilon$. Therefore, the word problem G' is true in E.

Example 3. For another example, consider the Thue System $E = \{a \approx bba, bba \approx a\}$. Then R_E is

1. $\bar{a}b \rightarrow \bar{a}\bar{b}$
2. $\bar{b}a \rightarrow ba$
3. $\bar{a}a \rightarrow \epsilon$
4. $\bar{b}b \rightarrow \epsilon$

Consider the unification problem $G = ba \approx_E a$. Then $w_G = \bar{a}\bar{b}a \rightarrow \bar{a}ba \rightarrow \bar{a}ba = w_G$. So this rewrite sequence loops, and there are no other possible rewrite sequences. Therefore $\bar{a}\bar{b}a$ has no normal form in R_E which implies that the unification problem $ba \approx_E a$ (and also the word problem) has no solution in E.

4   Deciding the Unification Problem

In this section we first show why the condition of syntacticness and no repeating equations leads to a deterministic procedure. Then we define an ordering to show termination of rewrite sequences. This ordering is used to define a decision procedure for certain classes of problems. In some cases, we can even bound the complexity by an exponential of the goal size. Finally, we give an algorithm that we conjecture decides the unification problem in all cases.

First we show the interest of the class of problems we are considering.

Theorem 3. *If E is a syntactic Thue System with no repeating top equation, and G is a satisfiable unification problem in E, then R_E is confluent and strongly terminating on w_G.*

Proof. It follows from the fact that R_E is non-overlapping and is weakly terminating on w_G.

This gives us a deterministic procedure to decide the word problem. We can assume that we always apply the rightmost rewrite step. Unfortunately, this deterministic procedure may not always halt. So we need some ways to determine non-termination or rewrite sequences. One way to do this is to find loops. First, let's define a useful ordering to determine terminating rewrite sequences.

We define an ordering on words of three or fewer blocks.

Definition 3. *Let w be an n block word of the form $u_1 v_1^{-1} \cdots u_n v_n^{-1}$, with each $u_i, v_i \in A^*$ and $n \leq 3$. Define $\mu(w)$ to be the ordered pair (i, j) such that if $n \geq 1$ then $i = |v_1|$ else $i = 0$, and if $n = 3$ then $j = |u_3|$ else $j = 0$. Ordered pairs are compared lexicographically, i.e., $(i, j) > (k.l)$ if and only if $i > k$, or $i = k$ and $j > l$. Note that this ordering is well-founded.*

Now we define a set of words that we will later use to show that if we can find the normal forms of these words, then we can find the normal form of any given word, or determine that it does not have one.

Definition 4. *Let A be a set of words and R be a rewrite system on $B = A \cup \bar{A}$. Let C be a set of words in $(\bar{A})^*$, such that every non-empty prefix of C is in C. A word ua is called an extended word of C if $u \in C$ and $a \in A$. Let C' be the set of all R-normal forms w of extended words of C. Then R is C-complete if for all w in C*

1. *w contains one at most one block, and*
2. *if w contains one block (i.e., $w = u_1 v_1$ with $u_1 \in A^*$ and $v_1 \in (\bar{A})^*$) then v_1 is in C if $v_1 \neq \epsilon$.*

Note that condition 2 is trivially true if a is the inverse of the last letter in u.

Definition 5. *Let R be a rewrite system of the form $\{\bar{a}_1 b_1 \rightarrow s_1 t_1^{-1} \cdots \bar{a}_n b_n \rightarrow s_n t_n^{-1}\}$, with each $a_i, b_i \in A$, and $s_i, t_i \in A^*$. C is said to be a completion of R if R is C-complete and $t_i^{-1} \in C$ if $t_i \neq \epsilon$.*

If we rewrite in a certain way, we can force one of these special words to appear in a certain place in the word.

Definition 6. *Let C be a set of words. The word w is C-reducible if and only if w has at most three blocks, and if w has three blocks (i.e., is of the form $u_1 v_1^{-1} u_2 v_2^{-1} u_3 v_3^{-1}$) then $v_2^{-1} \in C$.*

We use the previous four definitions in the following crucial lemma. It shows that any word of a particular form can be reduced to a smaller word of the same form, or we can detect that it will not have an appropriate normal form.

Lemma 3. *Let E be a Thue System and C be a completion of R_E. Let w be a C-reducible word in B^* of three or fewer blocks. Suppose that for all extended words ua of C, it is decidable whether ua has an R_E-normal form. If w is not in normal form, then we can find a smaller C-reducible w' with three or fewer blocks such that $w \rightarrow w'$ or we can detect that w has no normal form with one or fewer blocks.*

Proof. If w has at most one block, then w is in normal form. Suppose w has two blocks, then $w = u_1 v_1^{-1} u_2 v_2^{-1}$ with $u_1, v_1, u_2, v_2 \in A^*$. Suppose $v_1 = av$ and $u_2 = bu$. Then $w = u_1 v^{-1} \bar{a} b u v_2^{-1}$. If $a = b$ then $w \rightarrow u_1 v^{-1} u v_2^{-1} = w'$, which is smaller than w because $\mu(w) = (|v_1|, 0) > (|v|, 0) = \mu(w')$. Also, w' has fewer than three blocks, so it is C-reducible.

Suppose $a \neq b$ and there is a rule in R_E of the form $\bar{a}b \to st^{-1}$. Then $w \to u_1 v_1^{-1} st^{-1} u v_2^{-1} = w'$, which is smaller than w because $\mu(w) = (|v_1|, 0) > (|v|, |u|) = \mu(w')$. [6] w' has three or fewer blocks. Also, note that $t^{-1} \in C$ by definition of C-complete, therefore w' is C-reducible.

If $a \neq b$ and there is no rule in R_E of the form $\bar{a}b \to st^{-1}$, then any normal form of w must have more than one block.

Now suppose w has three blocks, then $w = u_1 v_1^{-1} u_2 v_2^{-1} u_3 v_3^{-1}$ with $u_1, v_1, u_2, v_2, u_3, v_3 \in A^*$. Suppose $v_2 = av$ and $u_3 = bu$. Then $w = u_1 v_1^{-1} u_2 v^{-1} \bar{a}buv_3^{-1}$. If $a = b$ then $w \to u_1 v_1^{-1} u_2 v^{-1} uv_3^{-1} = w'$, which is smaller than w because $\mu(w) = (|v_1|, |u_3|) > (|v_1|, |u|) = \mu(w')$. w' has at most three blocks. $v_2^{-1} \in C$ since w is C-reducible, therefore $v^{-1} \in C$ since C is closed under prefixes, so w' is C-reducible.

Suppose $a \neq b$. Then, $v_2^{-1} \in C$, because w is C-reducible. So we can decide whether $v_2 b$ has an R_E-normal form. If it has no R_E-normal form then neither does w. Otherwise, we can calculate the normal form of $v_2 b$. By definition of C-complete, the normal form $v_2 b$ is of the form $u' v'^{-1}$, with $v'^{-1} \in C$. [7] Then $w \to u_1 v_1^{-1} u_2 u' v'^{-1} u v_3^{-1} = w'$, which is smaller than w because $\mu(w) = (|v_1|, |u_3|) > (|v_1|, |u|) = \mu(w')$. w' has at most three blocks. Also, w' is C-reducible, since $v'^{-1} \in C$.

If $a \neq b$ and there is no rule in R_E of the form $\bar{a}b \to st^{-1}$, then any normal form of w must have more than one block.

The following theorem is the main result used to decide the word and unification problem.

Theorem 4. *Let E be a Thue System and G a goal over A. Let C be a completion of R_E.*

1. *Suppose that for all extended words ua of C it is decidable whether ua has an R_E-normal form. Then the word and unification problem for E is decidable.*

2. *If C is finite, then the word and unification problem is decidable in time at most exponential in the size of the goal.*

Proof. We construct R_E and w_G. Note that w_G has two blocks. Let w be a C-reducible word with three or fewer blocks. We perform induction on $\mu(w)$. The induction hypothesis is that we can find the normal form of all smaller words or prove they do not have one with one or fewer blocks. By the previous lemma, we can either reduce w to a smaller C-reducible w' with three or fewer blocks, or else detect that w has no normal form with one or fewer blocks. In the second

---

[6] In all of these cases, we should consider the case where $u = \epsilon$, but then $\mu(w') < \mu(w)$ because the second number in the ordered pair of $\mu(w')$ is 0.

[7] Here we do not consider the simpler cases where the normal form is ϵ or only contains members of A or A^{-1}.

case, we are done. In the first case, w has the same normal form as w', so we are also done.

This takes care of the first part of the theorem. When C is finite, the above argument still shows that the word problem is decidable, since decision problems on finite sets are always decidable. But we must show that the decision procedure runs in at most exponential time in the size of the goal. For that we must analyze the procedure induced by the previous lemma. If $\mu(w) = (i, j)$, then there are at most j rewrite steps before i gets smaller. But during that time, w can increase by a product of k, where k is the maximum size of u_1 for a normal form $u_1 v_1^{-1}$ of ua with $u \in C$. Therefore, to calculate the normal form of w, we potentially multiply w by k, $|w|$ times, at most. So the word can become as big as $k^{|w|}$ at most. And since each operation is linear in the size of the goal, the running time as also bounded by an exponential.

We give some examples to illustrate.

Example 4. Let $E = \{aba \approx bab, bab \approx aba\}$. Then R_E is

1. $\bar{a}b \rightarrow ba\bar{b}\bar{a}$
2. $\bar{b}a \rightarrow ab\bar{a}\bar{b}$
3. $\bar{a}a \rightarrow \epsilon$
4. $\bar{b}b \rightarrow \epsilon$

Let $C = \{\bar{b}, \bar{a}, \bar{b}\bar{a}, \bar{a}\bar{b}\}$. The normal forms of $\bar{b}a$, $\bar{a}b$, $\bar{b}\bar{a}b$ and $\bar{a}\bar{b}a$ are respectively $ab\bar{a}\bar{b}$, $ba\bar{b}\bar{a}$, $a\bar{b}\bar{a}$, and $b\bar{a}\bar{b}$. Each of these normal forms contains only one block. Since all nonempty prefixes of $\bar{a}\bar{b}$ and $\bar{b}\bar{a}$ are in C, then C is a completion of R_E.

Example 5. Let $E = \{abb \approx baa, baa \approx abb\}$. Then R_E is

1. $\bar{a}b \rightarrow bb\bar{a}\bar{a}$
2. $\bar{b}a \rightarrow aa\bar{b}\bar{b}$
3. $\bar{a}a \rightarrow \epsilon$
4. $\bar{b}b \rightarrow \epsilon$

Let $C = \{\bar{a}, \bar{b}, \bar{a}\bar{a}, \bar{b}\bar{b}\}$. The normal forms of $\bar{a}b$ and $\bar{b}a$ are respectively $bb\bar{a}\bar{a}$ and $aa\bar{b}\bar{b}$. Note that $\bar{a}\bar{a}b \rightarrow \bar{a}bb\bar{a}\bar{a} \rightarrow bb\bar{a}\bar{a}b\bar{a}\bar{a}$ which contains $\bar{a}\bar{a}b$ as subword. Therefore $\bar{a}\bar{a}b$ has no normal form. Similarly, $\bar{b}\bar{b}a$ has no normal form. We only need to consider the normal forms $bb\bar{a}\bar{a}$ and $aa\bar{b}\bar{b}$. Since all the nonempty prefixes of $\bar{a}\bar{a}$ and $\bar{b}\bar{b}$ are in C, then C is a completion of R_E.

Here is an example for which C is infinite.

Example 6. Let $E = \{bab \approx a, a \approx bab\}$. Then R_E is

1. $\bar{b}a \rightarrow ab$
2. $\bar{a}b \rightarrow \bar{b}a$
3. $\bar{a}a \rightarrow \epsilon$
4. $\bar{b}b \rightarrow \epsilon$

Let $C = \{(\bar{b})^n \mid n > 0\} \cup \{(\bar{b})^n \bar{a} \mid n \geq 0\}$. Given n, the normal form of $(\bar{b})^n a$ is ab^n, and the normal form of $(\bar{b})^n \bar{a} b$ is $(\bar{b})^{n+1} \bar{a}$. These can be proved by induction on n. The cases where $n = 0$ are trivial. If $n > 0$, we have $(\bar{b})^n \bar{a} b = \bar{b}(\bar{b})^{n-1} \bar{a} b \rightarrow (\bar{b})^n \bar{a}$. Also, $(\bar{b})^n a = \bar{b}(\bar{b})^{n-1} a \rightarrow \bar{b} a b^{n-1} = ((\bar{b})^{n-1} \bar{a} b)^{-1} \rightarrow ((\bar{b})^n \bar{a})^{-1} = ab^n$.

Here we used the fact that $u^{-1} \rightarrow v^{-1}$ if $u \rightarrow v$.

Now we must address the question of how to determine if a word has a nonterminating rewrite sequence. We say a word w *loops* if there exist words u and v such that $w \rightarrow^+ uwv$. We conjecture that every nonterminating rewrite sequence loops. This is the same as the conjecture for one rule semi-Thue systems in [10].

Conjecture 1. Let E be a syntactic Thue System with no repeated equations, and G be a unification problem. Then w_G has a nonterminating rewrite sequence in R_E if and only if there are some words u, v, w such that $w_G \rightarrow uvw$ and v loops.

It is possible to detect loops, so a proof of the conjecture would imply that the unification and word problem are decidable. We now give an algorithm for deciding the unification problem, whose halting relies on the truth of the conjecture.

For the algorithm, we are given a Thue System E, and a goal G. We construct R_E and w_G. The intention of the algorithm is to reduce the goal to its normal form at the same time we are creating a subset of the extensions of C (the completion of R_E), and keeping track of the normal forms or lack of normal forms of those extensions of C.

The algorithm involves w which is initially set to w_G and any applicable cancellation rules are applied. w is always a reduced version of w_G with at most three blocks. The algorithm also involves a stack T of ordered pairs. Each element of T is an ordered pair (u, v) such that u is of the form $u'^{-1} a$ with $u' \in A^*$ and $a \in A$, and v is a word of at most three blocks. The values of u will be words that we are trying to find the normal form of, and v will be a reduced version of u. There is a set of ordered pairs S involved in the algorithm. An element of S is an ordered pair (u, v) where u is of the form $u'^{-1} a$ with $u' \in A^*$ and $a \in A$, and v is a word of one or fewer blocks which is the normal form of u. S and T are both initially empty.

The algorithm proceeds as follows:

First check if T is empty. If T is empty and w is in normal form, then check if w has one or fewer blocks. If it does, then return w. That is the normal form of w_G. If it does not, then return FALSE, because w_G has no normal form of one or fewer blocks, thus the unification problem is false.

Suppose T is empty and w is not in normal form, we examine the rightmost redex position of w. Either w has two blocks and is of the form $u_1 v_1^{-1} u_2 v_2^{-1}$, or w has three blocks and is of the form $u_1 v_1^{-1} u_2 v_2^{-1} u_3 v_3^{-1}$. If w has two blocks, set $u' = v_1^{-1}$ and set c to be the first letter of u_2. If w has three blocks, set $u' = v_2^{-1}$ and set c to be the first letter of u_3. If $\bar{d}$ is the last character in u' and there is no c such that $\bar{d} c \in R_E$, then return FALSE. Search for an ordered pair

$(u'c, v)$ in S for some v. If it exists, then replace $u'c$ in w by v and perform any cancellation rules that now apply. Note that w still has at most three blocks. If no $(u'c, v)$ exists in S, then push $(u'c, u'c)$ onto T.

If T is not empty, let (u, v) be on top of the stack. Either v has two blocks and is of the form $u_1 v_1^{-1} u_2 v_2^{-1}$, or v has three blocks and is of the form $u_1 v_1^{-1} u_2 v_2^{-1} u_3 v_3^{-1}$. If v has two blocks, set $u' = v_1^{-1}$ and set c to be the first letter of u_2. If v has three blocks, set $u' = v_2^{-1}$ and set c to be the first letter of u_3. If $\bar{d}$ is the last character in u' and there is no c such that $\bar{d}c \in R_E$, then return FALSE. Search for an ordered pair $(u'c, v')$ in S for some v'. If it exists then replace $u'c$ in v by v' and perform applicable cancellations. Note that v still has at most three blocks. If v is now in normal form, then if v has at most one block, then we add (u, v) to S and remove (u, v) from T, else return FALSE. If v contains u as a subword, or if v contains s as a subword with (s, t) in T for some T, we return FALSE. If no $(u'c, v')$ exists in S, then push $(u'c, v)$ onto T.

Keep repeating this process until it halts.

Based on our implementation, this algorithm appears to be very efficient, and we conjecture that it always halts. Note that the algorithm constructs extensions of a completion of R_E. Based on theorem 4, we can see that this algorithm will halt in time at most exponential in the size of the goal if R_E has a finite completion.

There is another interesting generalization of the class of problems. We consider Thue systems to only contain equations of the form $ux \approx vx$. Suppose we allowed other monadic terms. For example $ux \approx vy$. If all our equations are of this type, then lemma 1 is still true, with the removal of the condition that $s'r_2 \approx_E t'w_2$. We could say a simliar thing for equations of the form $ua \approx vb$. This would allow us to modify the definition of R_E so that the right hand side of the rewrite rules have a marker between the two halves, preventing interaction between the two. This allows us to solve the unification problem in polynomial time in terms of the goal if E is a syntactic set of monadic terms, with no repeated equations, and no equations of the form $ux \approx vx$. Space prevents us from giving the details of this argument. But it is interesting to note that the problem becomes easier when the equations are not linear.

5   Conclusion

We have given a method for trying to solve the unification (and word) problem for one equation Thue systems and other monadic equational theories. Our method works on a larger class of problems, which we have defined. We have shown certain cases where we can prove that the method is a decision procedure. We gave an algorithm, which has been implemented, and appears to be efficient. It halts and serves as a decision procedure for every input we have tried. This is opposed to the Knuth-Bendix procedure which often runs forever. The closest work to our approach is given in [4]. This is based on an algorithm in [2] for Thue systems with one equation. The algorithm does not always halt. In [4], a rewrite system is given to help determine when the algorithm of [2] halts. They also needed to prove the termination of the rewrite system. But their method and rewrite system is quite different from ours. For example, our rewrite system halts on different problems than theirs. They also gave an example of a rewrite system with a word that did not terminate but had no loop (called a *simple loop* in their paper). It would be interesting to do a more detailed comparison of our two methods. We think that methods used to decide termination of one rule semi-Thue systems might be helpful for us. Our ultimate goal is to extend our method to all unification problems over terms, and find a large class of problems for which our approach halts. This approach in this paper was designed with that intention.

References

1. S. Adian. Definining relations and algorithmic problems for groups and semigroups. *Trudy Matem. in-ta im. Steklova AN SSSR*, 85, 1996 (Russian).
2. S. Adian. Transformations of words in a semigroup presented by a system of defining relations. *Algebra i logika*, 15(6),611-621, 1976 (Russian).
3. S. Adian, and G. Oganesian. On the word and divisibility problems in semigroups with a single defining relation. *Izv. An. SSSR Ser. Matem.*, 42(2),219-225, 1978 (Russian).
4. J. Bouwsma. Semigroups Presented by a Single Relation. PhD dissertation at Pennsylvania State University, 1993.
5. C. Kirchner. Computing unification algorithms. In *Proceedings of the First Symposium on Logic in Computer Science*, Boston, 200-216, 1990.
6. G. Lallement. The word problem for Thue rewriting systems. In *Spring School in Rewriting*, ed. H. Comon and J. P. Jouannaud, Lecture Notes in Computer Science, 1994.
7. C. Lynch. Goal Directed Completion using SOUR Graphs. In *Proceedings of the Eighth International Conference on Rewriting Techniques and Applications (RTA)*, Sitges, Spain, June 2-4, 1997.
8. W. Magnus. Das Identitätsproblem für Gruppen mit einer definierenden Relation. *Math Ann.*, 106,295-307, 1932.
9. Y. Matisaevich. Simple examples of unsolvable associative calculi. *Dokl. Akad. Nauk. SSSR*, 173,1264-1266, 1967.
10. R. McNaughton. Well-behaved derivations in one-rule Semi-Thue Systems. Tech. Rep 95-15, Dept. of Computer Science, Rensselaer Polytechnic Unstitute, Troy, NY, Nov. 1995.
11. W. Savitch. How to make arbitrary grammars look like context-free grammars. *SIAM Journal on Computing*, 2(3),174-182, September 1973.

Basic Completion with E-cycle Simplification

Christopher Lynch[1]* and Christelle Scharff[2]

[1] Department of Mathematics and Computer Science Box 5815, Clarkson University, Potsdam, NY 13699-5815, USA - Christopher.Lynch@clynch.mcs.clarkson.edu - http://www.clarkson.edu/~clynch
[2] LORIA BP 239, 54506 Vandoeuvre-les-Nancy Cedex, France - Christelle.Scharff@loria.fr - http://www.loria.fr/~scharff

Abstract. We give a new simplification method, called E-cycle Simplification, for Basic Completion inference systems. We prove the completeness of Basic Completion with E-cycle Simplification. We prove that E-cycle Simplification is strictly stronger than the only previously known complete simplification method for Basic Completion, Basic Simplification, in the sense that every derivation involving Basic Simplification is a derivation involving E-cycle Simplification, but not vice versa. E-cycle Simplification is simple to perform, and does not use the reducibility-relative-to condition. We believe this new method captures exactly what is needed for completeness. *ECC* implements our method.

1 Introduction

In automated theorem proving, it is important to know if an inference system is complete, because a complete inference system guarantees that a proof will be found if one exists and if it halts without a proof, then the theorem is false. However, in practice, incomplete inference systems are often used because complete ones are not efficient.

An example of this phenomenon is the case of Basic Completion [BGLS95], [NR92]. This is a restriction on Knuth-Bendix Completion [KB70] such that the most general unifier is saved as a constraint, instead of being applied to the conclusion of an inference [KKR90]. The effect of this restriction is that much of a term is stored in a constraint, and therefore the variable positions appear closer to the root than in the non-basic case, or else variable positions occur where there are no variable positions in the non-basic case. In Knuth-Bendix Completion, there is a restriction that inferences are not allowed at variable positions. This restriction then becomes much more powerful in Basic Completion. In [BGLS95,NR92], it was shown that Basic Completion is complete.

Simplification rules are crucial in any form of completion. However, in [NR92] it was shown that the combination of Basic Completion and Standard Simplification is not complete (see [BGLS95] for more incompleteness examples). In [BGLS95], a new form of simplification, called *Basic Simplification*, is shown

---

* This work was supported by NSF grant number CCR-9712388 and partially done during a visit in the PROTHEO group in Nancy.

to be complete in combination with Basic Completion. Unfortunately, Basic Simplification can only be performed under certain circumstances. So, to retain completeness, a theorem prover must either not simplify under these circumstances, or else apply the constraint of the simplifying equation before simplifying. The first solution is unsatisfactory because it does not allow as much simplification. The second solution is unsatisfactory because it removes the advantages of Basic Completion.

These results lead us to an analysis of simplification strategies for Basic Completion. The goal is to understand when simplification will destroy completeness and when it will not. We provide an abstract setting to develop and prove the completeness of a concrete simplification method for Basic Completion, called *E-cycle Simplification*, which does not use the reducibility-relative-to condition of Basic Simplification. We prove that E-cycle Simplification is complete and strictly stronger than Basic Simplification, in the sense that every derivation involving Basic Simplification is a derivation involving E-cycle Simplification, but not vice versa (see section 5). Also, there are many examples where E-cycle Simplification may be performed but Basic Simplification may not (see section 5 for an example) .

The idea behind E-cycle Simplification is simple. No equation may simplify one of its *ancestors* [1]. In the inference procedure we build a *dependency (directed) graph*. The nodes of the dependency graph are labelled by the equations. When we deduce a new equation, we add a node to the graph labelled by the new equation to show dependencies and ancestors. A Basic Critical Pair inference adds an *Inference* edge to indicate that the conclusion depends on the premises if it has an irreducible constraint. A Simplification adds *Simplification* edges. When the rule deduces a constrained equation, *Constraint* edges are added from the constrained equation to its original ancestors in E. These dependencies are only needed if the constraint of the equation is reducible, to be able to create a reduced version of the constrained equation. Edges are associated with reducibility constraints, which may conflict with each other. We define *E-paths* and *E-cycles* in the dependency graph to be paths and cycles with no conflict in reducibility constraints. E-cycles may only occur when an equation simplifies an ancestor. Whenever a simplification would create an E-cycle in the dependency graph, we disallow the simplification.

Our completeness proof is based on the model construction proof of [BG94], which is also used in the completeness proofs of Basic Completion in [BGLS95], [NR92]. Like those proofs, we build a model of irreducible equations, based on an ordering of the equations. The difference is that we do not use the multiset extension of the term ordering but a different ordering $\succ_g$ directly based on the dependency graph. If there is an edge from a node labelled with equation e_1 to a node labelled with equation e_2, then e_1 is larger than e_2 in our ordering $\succ_g$ and we write $e_1 \succ_g e_2$. The ordering $\succ_g$ is well-founded, because the dependency graph does not have any E-cycles or infinite E-paths.

---

[1] We define the notion of *ancestor* in the paper.

The paper is organized as follows. Section 2 contains some definitions and notions useful for the comprehension of the paper. Section 3 defines dependency graphs, E-cycles, E-cycle Simplification and the construction of the dependency graphs. In section 4, we show that Basic Completion with E-cycle Simplification is complete. Then, in section 5, we show that E-cycle Simplification is strictly more powerful than Basic Simplification.

The full version of this paper that includes complete details and full proofs is available in [LSc97].

2   Preliminaries

We assume the reader is familiar with the notation of equational logic and rewriting. A survey of rewriting is available in [DJ90]. We only define important notions for the comprehension of the paper and new notions and definitions we introduce.

Let $=^?$ be a binary infix predicate. An *equational constraint* φ is a conjunction $s_1 =^? t_1 \wedge \ldots \wedge s_n =^? t_n$ of syntactic equality $s_i =^? t_i$. $\top$ is the true equational constraint and $\bot$ is the false constraint. The symbol $\approx$ is a binary symbol, written in infix notation, representing semantic equality. In this paper $\succeq_t$ will refer to an *ordering* on terms ($\succ_t$ in its strict version), which is a well-founded reduction ordering total on ground terms. $\approx$ is symmetric and, when we write the equality $s \approx t$, we assume that $s \not\prec_t t$. We extend the ordering $\succeq_t$ to ground equations and we call the new ordering $\succeq_e$ ($\succ_e$ in its strict version). Let $s \approx t$ and $u \approx v$ be two ground equations. We define the ordering $\succ_e$ such that $s \approx t \succ_e u \approx v$ if either $s \succ_t u$ or, $s = u$ and $t \succ_t v$ [2]. A pair $s \approx t[\![\varphi]\!]$ composed of an equation $s \approx t$ and an equational constraint φ is called a *constrained equation*. An equation $s\sigma \approx t\sigma$ is a ground *instance* of a constrained equation $t[\![\varphi]\!]$ if σ is a ground substitution solution of φ. We denote by $Gr(e[\![\varphi]\!])$ the set of ground instances of an equation $e[\![\varphi]\!]$. This is extended to a set E by $Gr(E) = \bigcup_{e \in E} Gr(e)$. We call $e\sigma_1[\![\varphi_2]\!]$ a retract form of a constrained equation $e[\![\varphi]\!]$ if $\sigma = mgu(\varphi)$, $\sigma_2 = mgu(\varphi_2)$ and $\forall x \in Dom(\sigma), x\sigma = x\sigma_1\sigma_2$. For example, $g(f(y)) \approx b[\![y =^? a]\!]$ is a retract of $g(x) \approx b[\![x =^? f(a)]\!]$.

Reducibility Constraints: We define a predicate symbol Red, which is applied to a term.

A *reducibility constraint* is:
- $\top$ denoting the empty conjunction and the true reducibility constraint or
- $\bot$ denoting the false reducibility constraint or
- of the form $\varphi_{r_1} \wedge \cdots \wedge \varphi_{r_n}$, where φ_{r_i} is of the form $(\bigvee_j Red(t_j))$ or $\neg Red(t)$ or $\top$ where t, $t_j \in \mathcal{T}$ (where $\mathcal{T}$ is the set of terms built on a particular signature).

The syntax of the Red predicate is extended in [LSc97]. First instances of reducibility constraints can be found in [Pet94] and in [LS95].

A *ground reducibility constraint* is a reducibility constraint such that the parameter of the predicate Red is a ground term. Let φ_r be a ground reducibility constraint, and R be a ground rewrite system. Then φ_r is *satisfiable in R*, if and only if one of the following conditions is true:

---

[2] Recall that $s \not\prec_t t$ and $u \not\prec_t v$

- $\varphi_r = \top$.
- $\varphi_r = Red(t)$ and t is reducible in R.
- $\varphi_r = \neg\varphi_r'$ and φ_r' is not satisfiable in R.
- $\varphi_r = \varphi_r' \wedge \varphi_r''$ and φ_r' and φ_r'' are satisfiable in R.
- $\varphi_r = \varphi_r' \vee \varphi_r''$ and φ_r' is satisfiable in R or φ_r'' is satisfiable in R.

A reducibility constraint φ_r is *satisfiable* iff there exists a rewrite system R and a ground substitution σ such that $\varphi_r\sigma$ is satisfiable in R. Satisfiability is a semantic notion. In our inference procedure, we deal with syntactic objects. For that, we need the notion of *consistency*.

Definition 1. *A reducibility constraint φ_r is inconsistent if and only if, $\varphi_r = \bot$ or there exist $u_1[t_1\sigma_1], \cdots, u_n[t_n\sigma_n]$ such that σ_i are substitutions and $(\bigvee_{i\in\{1,\cdots,n\}} Red(t_i))$ appears in φ_r and $\neg Red(u_i[t_i\sigma_i])$ ($i \in \{1, \cdots, n\}$) appear in φ_r.*

A reducibility constraint is consistent if and only if it is not inconsistent.

Note that it is simple to test if a reducibility constraint is consistent using this definition. There is a close relationship between consistency and satisfiability.

Theorem 2. *Let φ_r be a reducibility constraint. If φ_r is satisfiable, then φ_r is consistent.*

Let φ be an equational constraint. We define $RedCon(\varphi)$ as the reducibility constraint $\bigvee\{Red(x\sigma) \mid x \in Dom(\sigma) \text{ and } \sigma = mgu(\varphi)\}$ and, in particular, $RedCon(\top) = \bot$.

Inference Systems: Our inference system is based on Basic Completion.

The main inference rule of Basic Completion is the *Basic Critical Pair* inference rule:

Basic Critical Pair

$$\frac{u[s'] \approx v[\![\varphi_1]\!] \qquad s \approx t[\![\varphi_2]\!]}{u[t] \approx v[\![s =^? s' \wedge \varphi_1 \wedge \varphi_2]\!]} \quad \text{if :}$$

- s' is not a variable,
- there exists a substitution σ such that $\sigma \in Sol((s =^? s') \wedge \varphi_1 \wedge \varphi_2)$, $s\sigma \not\preceq_t t\sigma$ and $u[s']\sigma \not\preceq_t v\sigma$.

Let Γ be a set of equations. This inference means that the set of equations $\{u[s'] \approx v[\![\varphi_1]\!], s \approx t[\![\varphi_2]\!]\} \cup \Gamma$ is transformed to $\{u[s'] \approx v[\![\varphi_1]\!], s \approx t[\![\varphi_2]\!], u[t\sigma] \approx v[\![s =^? s' \wedge \varphi_1 \wedge \varphi_2]\!]\} \cup \Gamma$ [3].

We now present *Standard Simplification* and *Basic Simplification* deletion rules.

---

[3] In rules, we assume the two premises have disjoint sets of variables. If the two equations share some variables, we first rename one premise so that they no longer share any variables, before performing the rule. We denote by *"into" equation*, the equation $u[s'] \approx v[\![\varphi_1]\!]$, by *"from" equation*, the equation $s \approx t[\![\varphi_2]\!]$ and by *conclusion equation*, the deduced equation.

The *Standard Simplification* deletion rule is the following:

Standard Simplification

$$\frac{u[s'] \approx v[\![\varphi_1]\!] \quad s \approx t[\![\varphi_2]\!]}{\mathbf{u[t\sigma_2\mu] \approx v[\![\varphi_1 \wedge \varphi_2\mu]\!]}} \quad \text{if} :$$

- s' is not a variable,
- there exists a substitution μ, $\sigma_1 = mgu(\varphi_1)$ and $\sigma_2 = mgu(\varphi_2)$ such that $s'\sigma_1 = s\sigma_2\mu$, $s\sigma_2\mu \succ_t t\sigma_2\mu$, and $v\sigma_1 \succ_t t\sigma_2\mu$ if $u = s'$.

Let Γ be a set of equations. In this rule, the set of equations $\{u[s'] \approx v[\![\varphi_1]\!], s \approx t[\![\varphi_2]\!]\} \cup \Gamma$ is transformed to $\{s \approx t[\![\varphi_2]\!], \mathbf{u[t\sigma_2\mu] \approx v[\![\varphi_1 \wedge \varphi_2\mu]\!]}\} \cup \Gamma$.[4]

Basic Simplification is based on the notion of reduced-relative-to and is described in [BGLS95].

Basic Simplification

$$\frac{u[s'] \approx v[\![\varphi_1]\!] \quad s \approx t[\![\varphi_2]\!]}{\mathbf{u[t\mu] \approx v[\![\varphi_1 \wedge \varphi_2\mu]\!]}} \quad \text{if} :$$

- s' is not a variable,
- there exists a match μ and $\sigma_1 = mgu(\varphi_1)$ such that $s' = s\mu$ and $(u \approx v)\sigma_1 \succ_e (s \approx t)\sigma_2\mu$, and
- $(s \approx t[\![\varphi_2]\!])\mu$ is substitution reduced relative to $u[s'] \approx v[\![\varphi_1]\!]$.

There are two optional but useful rules. If the conditions for the application of Basic Simplification are not true, it is possible to apply the *Retraction* rule which consists of retracting the "from" equation of the inference to make the application of Basic Simplification possible. *Basic Blocking* is a deletion rule based on the reduced-relative-to condition that deletes an equation with a reducible constraint.

We call *BCPBS* the Basic Completion inference system consisting of Basic Critical Pair and Basic Simplification plus Basic Blocking and Retraction. In this paper we give a new Basic Completion inference system **BCPES** which uses the Basic Critical Pair rule and a restricted version of the **Standard Simplification** rule, that we call *E-cycle Simplification*. In the full version of the paper [LSc97], *BCPES* is extended by *E-cycle Retraction* and *E-cycle Blocking* rules.

3   E-cycle Simplification

In this section, we describe the framework used in the paper. We first describe the *dependency graph* of a set of unconstrained equations E to complete and then we give the definition of an E-cycle. We describe the way the dependency graph is constructed using the rules of *BCPES* using *Graph Transitions*.

---

[4] This formulation resolves the ambiguity of the first notation. The ambiguity can be resolved by remembering that inference rules add equations, while simplification deletion rules add and delete an equation.

3.1 The Dependency Graph and E-cycles

The dependency graph is a directed graph. The vertices of the dependency graph are labelled by equations. We associate a set of vertices $C\_ancestor(v)$ to each vertex. There are three kinds of edges in the dependency graph: C edges, I edges and S edges. C stands for Constraint, I stands for Inference and S stands for Simplification. Each edge has a reducibility constraint associated with it determined by the type of the edge (C, I and S) and the constraints of equations labelling the vertices at the extremities of the edge.

Let e_d be an edge from a vertex v_1 labelled by an equation $e_1[\![\varphi_1]\!]$ to a vertex v_2 labelled by an equation $e_2[\![\varphi_2]\!]$ in the dependency graph. If e_d is a C edge, then the constraint associated with e_d is $RedCon(\varphi_1)$. e_d is denoted by (v_1, v_2, C). If e_d is an I edge, the constraint associated with e_d is $\neg RedCon(\varphi_2)$. e_d is denoted by (v_1, v_2, I). If e_d is an S edge, then the constraint associated with e_d is $\top$. e_d is denoted by (v_1, v_2, S).

An E-path is a path of C, I and S edges in the dependency graph such that the conjunction of the reducibility constraints associated to the edges is consistent. An E-cycle is an E-path which begins and ends at the same vertex and which contains at least a C and an S edge. The problem of finding an E-path and so an E-cycle in the dependency graph is NP-complete [HM98].

3.2 Construction of the Dependency Graph and E-cycle Simplification

At the beginning of the Basic Completion process, the initial set E is represented by the initial dependency graph G_{init} that is defined as follows. Each equation of the set of equations E to complete is a label of a vertex of the initial dependency graph $G_{init} = (V_{init}, ED_{init})$ and $ED_{init} = \emptyset$. $C\_ancestor(v) = \{v\}$ for all $v \in V_{init}$. When an inference of $BCPES$ is performed, the dependency graph is updated. A new vertex labelled by the conclusion of the rule is added and edges are added.

We now present the E-cycle Simplification rule and explain how Basic Critical Pair inferences and E-cycle Simplification update the dependency graph using *Graph Transitions*.

The **BCPES** inference system is composed of the Basic Critical Pair Inference rule and of the following E-cycle Simplification deletion rule.

E-cycle Simplification

$$\frac{u[s']\approx v[\![\varphi_1]\!] \qquad s\approx t[\![\varphi_2]\!]}{u[t\sigma_2\mu]\approx v[\![\varphi_1\wedge\varphi_2\mu]\!]} \quad \text{if :}$$

– $u[s']\approx v[\![\varphi_1]\!]$ can be standard simplified by $s\approx t[\![\varphi_2]\!]$ and
– the addition of S edges from the "into" premise to the "from" premise and from the "into" premise to the conclusion equation does not create an E-cycle in the dependency graph.

Definition 3. *A Graph Transition is denoted by* $(E_i, G_i) \rightarrow (E_{i+1}, G_{i+1})$, *where* E_i *and* E_{i+1} *are sets of equations such that* E_{i+1} *is obtained from* E_i *by per-*

*forming a Basic Critical Pair Inference or a deletion rule [5] and $G_i = (V_i, ED_i)$
and $G_{i+1} = (V_{i+1}, ED_{i+1})$ are dependency graphs such that G_{i+1} is obtained
from G_i by:*

- *A Basic Critical Pair Inference.*
  *We have the following Graph Transition $(\{e_0, e_1\} \cup \Gamma, G_i) \rightarrow (\{e_0, e_1, e_2\} \cup \Gamma, G_{i+1})$ where e_0 is the "into" equation, e_1 is the "from" equation and e_2
  is the conclusion equation of the Basic Critical Pair inference.*
  Let e_0 be the label of v_0 and e_1 be the label of v_1.
  - $V_{i+1} = V_i \cup \{v_2\}$ *such that* $label(v_2) = e_2$
  - $ED_{i+1} = ED_i \cup E_C \cup E_I$ *where:*
  If e_2 is an unconstrained equation then $E_C = \emptyset$, otherwise $E_C = \bigcup_{v \in C\_ancestor(v_0)} (v_2, v, C) \cup \bigcup_{v \in C\_ancestor(v_1)} (v_2, v, C)$.
  $C\_ancestor(v_2) = C\_ancestor(v_0) \cup C\_ancestor(v_1)$.
  $E_I = \{(v_0, v_2, I)\}$
- *A deletion rule.*
  We have the following Graph Transition $(\{e_0, e_1, \cdots, e_n\} \cup \Gamma, G_i) \rightarrow (\{e_1, e_2, \cdots, e_n\} \cup \Gamma, G_{i+1})$ where e_0 is removed because of $e_1, \cdots, e_n$.
  Let e_i be the label of v_i for $i \in \{0, \cdots, n\}$.
  - $V_{i+1} = V_i$
  - $ED_{i+1} = ED_i \cup E_S$ *where:*
  $E_S = \bigcup_{i \in \{1, \cdots, n\}} (v_0, v_i, S)$.

We now summarize the above definition. A C edge is created from a constrained equation to its initial *ancestors*, initial unconstrained equations of E.
An I edge is added from the vertex labelled by the "into" premise of an inference to the vertex labelled by the conclusion of the inference. This indicates that the "into" premise depends on the conclusion. We can notice that E-paths and also E-cycles do not contain an I edge followed by a C edge. This is due to the reducibility constraints associated to I and C edges. An S edge is from the simplified equation to the simplifier, and also from the simplified equation to the conclusion of the simplification. This indicates that the simplified equation depends on the other two. The dependency graph will not contain an E-cycle, because only a S edge could create an E-cycle (see theorem 7) and E-cycle Simplification forbids creation of E-cycles.

Definition 4. *Given a sequence of equations $E_0, E_1, \cdots$, the limit E_∞ is $\bigcup_i \bigcap_{j \geq i} E_j$. Given a sequence of graphs $G_0, G_1, \cdots$ where $G_i = (V_i, ED_i)$ for all i, the limit G_∞ is (V_∞, ED_∞), where $V_\infty = \bigcup_i \bigcap_{j \geq i} V_j$, $label(v) = \bigcup_i \bigcap_{j \geq i} label(v_j)$ for all $v \in V_\infty$, and $ED_\infty = \bigcup_i \bigcap_{j \geq i} ED_j$.*

Definition 5. *A Graph Transition Derivation from E is a possibly infinite derivation $(E_0 = E, G_0 = G_{init}) \rightarrow (E_1, G_1) \rightarrow \cdots$, where for all i, $(E_i, G_i) \rightarrow (E_{i+1}, G_{i+1})$ is a Graph Transition. The Transition Limit is denoted by $T_\infty = (E_\infty, G_\infty)$.*

---

[5] A Simplification rule consists of a Critical Pair inference that adds an equation plus a deletion rule.

The two following theorems are consequences of the way the dependency graph is constructed. The first theorem proves, in particular, that an E-cycle does not contain only C edges. The second theorem proves that it is only a deletion rule, so the addition of an S edge that could create an E-cycle. It also proves that an E-cycle contains at least an S edge.

Theorem 6. *An E- cycle does not contain only* C *edges.*

Theorem 7. *Let* $(E_0 = E, G_0 = G_{init}) \rightarrow (E_1, G_1) \rightarrow \cdots \rightarrow (E_{n-1}, G_{n-1}) \rightarrow (E_n, G_n) \cdots$ *be a Graph Transition Derivation. If* G_{n-1} *does not contain an E-cycle and* G_n *contains an E-cycle, then* E_n *was obtained from* E_{n-1} *by a deletion rule.*

To illustrate *BCPES*, we now develop the counter-example of Nieuwenhuis and Rubio [NR92], that proves that Basic Completion with Standard Simplification is incomplete. We adopt the same execution plan.

Example 8. Let $E = \{a \approx b\ (1), f(g(x)) \approx g(x)\ (2), f(g(a)) \approx b\ (3)\}$. We assume a lexicographic path ordering based on the precedence $f \succ_{prec} g \succ_{prec} a \succ_{prec} b$.

The dependency graph for the two inferences processed here is in figure 1. The full development of this example can be found in the full version of the paper [LSc97]. The saturated set, we obtain, is $E_\infty = \{a \approx b\ (1), f(g(x)) \approx g(x)\ (2), f(b) \approx b\ (5), f(g(b)) \approx b\ (7), g(x) \approx b[\![x =^? b]\!]\ (8)\}$.

1.
$$\frac{f(g(x)) \approx g(x)\ (2) \qquad f(g(a)) \approx b\ (3)}{g(x) \approx b[\![x =^? a]\!]\ (4)}$$

We add C edges from equation (4) to the initial equations (2) and (3). The reducibility constraint associated to these edges is $Red(a)$.

We add an I edge from equation (2) to equation (4). The reducibility constraint associated to this edge is $\neg Red(a)$.

2.
$$\frac{f(g(a)) \approx b\ (3) \qquad g(x) \approx b[\![x =^? a]\!]\ (4)}{f(b) \approx b\ (5)}$$

We add no C edge because equation (5) is an unconstrained equation. However, the set of initial equations equation (5) depends on is recorded. Equation (5) depends on the initial equations (2) and (3).

We add an I edge from equation (3) to equation (5). The reducibility constraint associated to this edge is $\top$.

Equation (3) can be standard simplified by equation (4). However, there is no E-cycle Simplification. Indeed, if we add S edges, an E-cycle is created. The S edge from (3) to (4) (whose associated reducibility constraint is $\top$) and the C edge from (4) to (3) (whose associated reducibility constraint is $Red(a)$) describe an E-cycle.

If we delete equation (3) as in Standard Simplification, then we cannot construct a confluent system (equation $g(b) \approx b$ has no rewrite proof), therefore the inference system would not be complete. The presence of the E-cycle prevents us from deleting equation (3). Thus we have used the dependency graph to detect incompleteness. The reducibility-relative-to condition of Basic Simplification also detects this. However, that condition also prevents some simplifications that would not cause loss of completeness, which E-cycle Simplification allows.

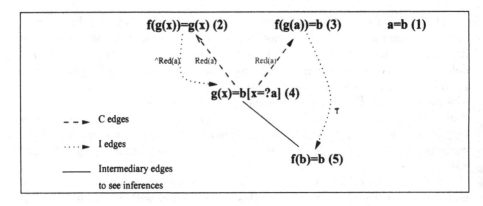

Fig. 1. Dependency graph of Basic Completion with E-cycle Simplification of $E = \{a \approx b, f(g(x)) \approx g(x), f(g(a)) \approx b\}$: the first two inferences.

4   *BCPES* Is Complete

In this section, we give the completeness result of *BCPES*. In the completeness proof, we need to construct a ground dependency graph which is an instance of the dependency graph we created in the previous section. Our proof is based on the model construction proof of [BG94]. The ground dependency graph is used to built a model of irreducible equations and a well-founded ordering $\succ_g$ of the equations.

4.1   The Ground Dependency Graphs

In the ground dependency graph, the labels of vertices are ground equations. Edges are added only if the reducibility constraints associated to them are satisfiable in a particular set of ground equations.

We first define GG_{init} for a (non-ground) set of equations E to complete. GG_{init} is the initial ground dependency graph (V_{init}, ED_{init}), where V_{init} is the set of vertices such that each $e \in Gr(E)$ labels one vertex of V_{init} and $ED_{init} = \emptyset$. As in the non-ground case, we set $C\_ancestor(v) = \{v\}$ for every $v \in V_{init}$. In a ground dependency graph, C edges are added from ground instances of constrained equations to ground instances of unconstrained initial equations. An I edge is added as previously from the "into" premise to the conclusion of an inference. Furthermore, we add an I edge from the "into" premise to the "from" premise of an inference. We also add I edges from the "into" equation to the "from" and the conclusion equation of inferences at the ground level that simulates "inferences" at a variable position not in the constraint at the non-ground level [6]. In a ground dependency graph, we no longer speak about E-cycle but only about cycle.

---

[6] At the non-ground level, no inference or simplification is performed at a variable position. However, we refer to it as an inference. At the ground level, the inference or the simplification must be performed. This remark applies to the rest of section 4.

The consequences of a rule on the ground dependency graph are formalized using *Ground Graph Transitions modulo an equational theory E'* distinguishing the applied rule at the ground level. E' is a set of ground equations with respect to which the reducibility constraints are tested. In the completeness proof, this set is instantiated by $Gr(E_\infty)$. Ground Graph Transitions modulo an equational theory are described in detail in [LSc97]. We need to lift from ground level to non-ground level. It is why we speak about Ground Graph Transition Derivation associated to (non-ground) Graph Transition Derivation.

4.2   Completeness Proofs

In this section, we first give completeness results concerning the Ground Graph Transition Derivation and then completeness results concerning *BCPES*. But first, we give some lemmas describing properties of ground dependency graphs. These properties follow from the construction of ground dependency graphs.

The following theorem provides the result that a cycle in a ground dependency graph contains at least a C and an S edge. The proof is done by contradiction.

Theorem 9. *Let* $(EG_0 = Gr(E), GG_0 = GG_{init}) \rightarrow (EG_1, GG_1) \rightarrow \cdots \rightarrow (EG_n, GG_n) \cdots$ *be a Ground Graph Transition Derivation modulo* E'. *If* GG_n *contains a cycle, then this cycle contains at least a C and an S edge.*

The following lemma proves that if an I edge goes from a vertex v_1 to a vertex v_2 in GG_∞, then $label(v_1)$ is reducible in $Gr(E_\infty)$.

Lemma 10. *Let* $(EG_0 = Gr(E), GG_0) \rightarrow (EG_1, GG_1) \rightarrow \cdots \rightarrow (EG_n, GG_n) \cdots$ *be a Ground Graph Transition Derivation modulo* $Gr(E_\infty)$. *If there is an edge from a vertex* v_1 *to a vertex* v_2 *in* GG_∞, *then* $label(v_1)$ *is reducible in* $Gr(E_\infty)$.

The following lemma proves that a cycle in a ground dependency graph does not contain an I edge from an "into" equation to a "from" equation of an inference.

Lemma 11. *Let* $(EG_0 = Gr(E), GG_0 = GG_{init}) \rightarrow (EG_1, GG_1) \rightarrow \cdots \rightarrow (EG_n, GG_n) \cdots$ *be a Ground Graph Transition Derivation modulo* $Gr(E_\infty)$. *If* GG_i *is a dependency graph containing a cycle* C, *then* C *does not contain an I edge from an "into" equation to a "from" equation of an inference.*

The following lemma proves that if there is a cycle or an infinite path at the ground level then there is an E-cycle at the non-ground level. For the proof of this lemma, we basically need to show that the extra edges we added to the graph in the ground case do not create any cycles that do not already exist at the non-ground level. In particular, lemma 11 and theorem 9 are used.

Lemma 12. *Let* $(E_0 = E, G_0 = G_{init}) \rightarrow (E_1, G_1) \rightarrow \cdots \rightarrow (E_n, G_n) \cdots$ *be a Graph Transition Derivation and* $(EG_0 = Gr(E), GG_0 = GG_{init}) \rightarrow (EG_1, GG_1) \rightarrow \cdots \rightarrow (EG_n, GG_n) \cdots$ *its associated Ground Graph Transition Derivation modulo* $Gr(E_\infty)$, *then for all* i,
  – *if* GG_i *contains a cycle then* G_i *contains an E-cycle.*
  – *if* GG_i *contains an infinite path, then* G_i *contains an E-cycle.*

The first completeness theorem provides a completeness result for Ground Graph Transition derivations. The proof of this theorem is based on the construction of a model of $Gr(E_\infty)$ which is a convergent rewrite system. For doing that the ordering $\succ_g$ is constructed directly from the ground dependency graph GG_∞.

Definition 13. *Let $\succ_g$ be the ordering such that $e \succ_g e'$ if and only if there are two vertices v and v' in GG_∞ such that $label(v) = e$, $label(v') = e'$, and there is a path in GG_∞ from v to v'.*

The ordering may not be total, but it is defined on the equations we use in the completeness proof. GG_∞ contains no infinite path or cycle if we do Basic Completion with E-cycle Simplification (see lemma 12) and so $\succ_g$ is well-founded.

We define redundancy in terms of this ordering.

Definition 14. *A ground equation e is g-redundant in a set of ground equations E if there are equations $e_1, \cdots, e_n \in E$ such that $e_1, \cdots, e_n \models e$ and $e_i \prec_g e$ for all i.*

E-cycle simplification is an example of g-redundancy as expressed in the following lemma.

Lemma 15. *Let E be a set of unconstrained equations. Let $(E_0 = E, G_0 = G_{init}) \to (E_1, G_1) \to \cdots \to (E_n, G_n) \cdots$ be a Graph Transition and $(EG_0 = Gr(E), GG_0 = GG_{init}) \to (EG_1, GG_1) \to \cdots \to (EG_n, GG_n) \cdots$ its associated Ground Graph Transition Derivation modulo $Gr(E_\infty)$ where, GG_∞ does not contain a cycle and an infinite path. Let e be an equation that is E-cycle simplified in some E_i. Then every ground instance e' of e is g-redundant in $Gr(E_\infty)$.*

Theorem 16. *Let E be a set of unconstrained equations. Let $(EG_0 = Gr(E), GG_0 = GG_{init}) \to (EG_1, GG_1) \to \cdots \to (EG_n, GG_n) \cdots$ be a Ground Graph Transition Derivation modulo $Gr(E_\infty)$ where, GG_∞ does not contain a cycle or an infinite path. Then this Ground Graph Transition Derivation is complete in the sense that $Gr(E_\infty)$ is convergent.*

The following main theorem proves the completeness of *BCPES*. The proof is based on the correspondence between the procedural construction of the non-ground dependency graph presented in section 3.2 and the abstract construction of the ground dependency graph presented in section 4. Lemma 12 is mainly used for the proof.

Theorem 17. *Basic Completion with E-cycle Simplification is complete.*

5 Comparison with Basic Simplification

In this section, we compare E-cycle Simplification with Basic Simplification. We prove that if we simplify because of a Basic Simplification then there is no E-cycle in the dependency graph we construct and so there is an E-cycle Simplification. So Basic Simplification is a subset of E-cycle Simplification. The proof of this theorem is based on a series of lemmas that show what patterns of edges can be added to the graph. So we show that the reducibility-relative-to condition never allows an edge to be added that would create an E-cycle. Furthermore, we give an example where we can simplify with E-cycle Simplification and where Basic Simplification does not permit us to simplify.

Lemma 18. *Let* $(E_0 = E, G_0 = G_{init}) \rightarrow (E_1, G_1) \rightarrow \cdots \rightarrow (E_n, G_n) \cdots$ *be a Graph Transition Derivation such that* $E_0 \rightarrow E_1 \rightarrow \cdots \rightarrow E_n$ *is a derivation of BCPBS. Then for all* i, G_i *contains no E-path consisting of an I edge followed by S edges and then by a C edge.*

Lemma 19. *Let* $(E_0 = E, G_0 = G_{init}) \rightarrow (E_1, G_1) \rightarrow \cdots \rightarrow (E_n, G_n) \cdots$ *be a Graph Transition Derivation such that* $E_0 \rightarrow E_1 \rightarrow \cdots \rightarrow E_n$ *is a derivation of BCPBS. Then if there is an* i *such that* G_i *contains an E-cycle, then this E-cycle does not contain any I edge.*

Theorem 20. *Let* $(E_0 = E, G_0 = G_{init}) \rightarrow (E_1, G_1) \rightarrow \cdots \rightarrow (E_n, G_n) \cdots$ *be a Graph Transition Derivation such that* $E_0 \rightarrow E_1 \rightarrow \cdots \rightarrow E_n$ *is a derivation of BCPBS. Then there is no* i *such that* G_i *contains an E-cycle.*

As a direct corollary, we get that if $E_0, E_1, \cdots, E_n$ is a derivation of *BCPBS*, then it is also a derivation of *BCPES*. Inversely, we provide an example that shows that a derivation of *BCPES* is not a derivation of *BCPBS*.

Example 21. Let $E = \{g(x) \approx f(x)\ (1), g(a) \approx b\ (2), h(f(a)) \approx b\ (3)\}$. We assume a lexicographic path ordering based on the precedence $h \succ_{prec} g \succ_{prec} f \succ_{prec} a \succ_{prec} b$. Let us assume the following execution plan using *BCPES*.

1. $$\frac{g(x) \approx f(x)\ (1) \qquad g(a) \approx b\ (2)}{f(x) \approx b[\![x =^? a]\!]\ (4)}$$
    We add C edges from equation (4) to initial equations (1) and (2). The reducibility constraint associated to these edges is $Red(a)$.
    We add an I edge from equation (1) to equation (4). The reducibility constraint associated to this edge is $\neg Red(a)$.

2. $$\frac{h(f(a)) \approx b\ (3) \qquad f(x) \approx b[\![x =^? a]\!]\ (4)}{h(b) \approx b\ (5)}$$
    The equation $h(f(a)) \approx b\ (3)$ is E-cycle simplified by equation $f(x) \approx b[\![x =^? a]\!]\ (4)$. Indeed, no E-cycle is created when we add S edges from equation (3) to equations (4) and (5).
    With Basic Simplification, the equation $h(f(a)) \approx b$ cannot be deleted because $f(x) \approx b[\![x =^? a]\!]$ is not reduced relative to $h(f(a)) \approx b$.

6   Conclusion

We have presented a new method of Simplification in the Basic Completion of a set of equations E, called E-cycle Simplification. Our approach is easy to understand because it is based on a graph. Indeed, E-cycle Simplification is based on the creation of a dependency graph during the completion process showing the dependencies between equations. It permits us to control completeness of

Completion such that, whenever E-cycle Simplification allows a simplification, completeness is preserved. We compare our method with Basic Simplification and prove that Basic Simplification is a strict subset of E-cycle Simplification.

Our method is shown complete using an abstract proof technique based on model construction. We think that this abstract framework is promising in the sense that this method of proof can lead us to an analysis of different simplification strategies from the point of view of completeness in constrained completion procedures. We conjecture that all complete Simplification methods for Basic Completion can be fit into our framework. We plan to use this method for AC Basic Completion and in particular, for simplification in AC Basic Completion.

We have implemented our method of Basic Completion with E-cycle Simplification. The system is called *ECC* (E-cycle Completion). It is written in *ELAN* [KKV95], which is a language based on rewriting and adapted for prototyping. The system is fully operational. Some implementation and experimental details with the two different methods of simplification are available at http://www.loria. fr/~scharff.

References

[BG94] L. Bachmair and H. Ganzinger. Rewrite-based equational theorem proving with selection and simplification. *Journal of Logic and Computation*, 4(3):217-247, 1994.

[BGLS95] L. Bachmair and H. Ganzinger and C. Lynch and W. Snyder. Basic Paramodulation. *Information and Computation*, 121(2):172-192, 1995.

[DJ90] N. Dershowitz and J.-P. Jouannaud. *Handbook of Theoretical Computer Science*, volume B, chapter 6: Rewrite Systems, pages 244-320. Elsevier Science Publishers B. V. (North-Holland), 1990. Also as: Research report 478, LRI.

[HM98] M. Hermann. Constrained Reachability is NP-complete. http://www.loria.fr/ hermann/publications.html#notes.

[KB70] D.E. Knuth and P.B. Bendix. Simple word problems in universal algebras. *Computational Problems in Abstract Algebra*, pages 263-297. Pergamon Press, Oxford, 1970.

[KKR90] C. Kirchner and H. Kirchner and M. Rusinowitch. Deduction with symbolic constraints. *Revue d'Intelligence Artificielle*, 4(3):9-52, 1990. Special issue on Automatic Deduction.

[KKV95] C. Kirchner and H. Kirchner and M. Vittek. ELAN V 1.17 User Manual Inria Lorraine & Crin, Nancy (France), first edition, november 1995.

[LS95] C. Lynch and W. Snyder. Redundancy criteria for constrained completion. *Theoritical Compluter Science*, volume 142, pages 141-177, 1995.

[LSc97] C. Lynch and C. Scharff. Basic Completion with E-cycle Simplification. 1997, http://www.loria.fr/~scharff.

[NR92] R. Nieuwenhuis and A. Rubio. Basic superposition is complete. In B. Krieg-Brückner, editor, *Proceedings of ESOP'92*, volume 582 of *Lecture Notes in Computer Science*, pages 371-389. Springer-Verlag, 1992.

[Pet94] G.E. Peterson. Constrained Term-Rewriting Induction with Applications. *Methods of Logic in Computer Science*, 1(4):379-412, 1994.

SoleX: A Domain-Independent Scheme for Constraint Solver Extension

Eric Monfroy[1] and Christophe Ringeissen[2]

[1] CWI
P.O. Box 94079, NL-1090 GB Amsterdam, the Netherlands
Eric.Monfroy@cwi.nl, http://www.cwi.nl~eric
[2] LORIA-INRIA
615 rue du Jardin Botanique, BP 101, F-54602 Villers-lès-Nancy Cedex, France
Christophe.Ringeissen@loria.fr, http://www.loria.fr~ringeiss

Abstract. In declarative programming languages based on the constraint programming paradigm, computations can be viewed as deductions and are enhanced with the use of constraint solvers. However, admissible constraints are restricted to formulae handled by solvers and thus, declarativity may be jeopardized. We present a domain-independent scheme for extending constraint solvers with new function symbols. This mechanism, called **SoleX**, consists of a collaboration of elementary solvers. They add and deduce information related to constraints involving new functions, complete the computation domain and purify constraints. Some extensions of computation domains have already been studied to demonstrate the broad scope of **SoleX** potential applications.

1 Introduction

In the last decade constraint programming (CP) [8] emerged as a new programming paradigm. The basic notion of this framework is the separation between (1) a programming language to specify requirements (the *constraints*) on objects (the *computation domain*) and (2) a mechanism (the *solver*) for solving constraints. CP has to face the dilemma "declarativity vs. efficiency". Thus, the solvers cannot always handle all the constraints the user manipulate in the programming languages. A solver is said to be *complete* if it is able to solve any constraint defined by the language. However, solvers of CP systems are not always complete: for example CLP($\mathcal{R}$) [9] does not solve non-linear constraints, i.e., they are suspended till they become linear. Although this kind of technique is sufficient for some applications, it is not satisfactory in the general case.

Designing a solver that handles all the constraints the programming language provides is a hard task. Possibly, there may be no solver for this computation domain. Thus, we are concerned with a general framework and mechanisms for extending/completing efficient solvers so they can handle new function symbols. In [6] a decision procedure on $\mathcal{R}$ is extended to a decision procedure on $\mathcal{R} + \mathcal{M}$ [1].

---

[1] CLP($\mathcal{R} + \mathcal{M}$) is obtained by extending the domain of CLP($\mathcal{R}$) with some special nonarithmetic function symbols.

Jacques Calmet and Jan Plaza (Eds.): AISC'98, LNAI 1476, pp. 222–233, 1998.
© Springer-Verlag Berlin Heidelberg 1998

In this paper, we extend the method of [6] with other syntactical manipulations and semantic transformations as well. Moreover, our framework is independent from the computation domain and the programming language.

The inverse robot kinematics problem [3] illustrates our motivations. We want to determine, for a given robot, a position and an orientation of the end-effector, the distances at the prismatic joints and the angles at the revolute joints (see Figure 1). The problem for a robot having two degrees of freedom can be described by a system of equations (see Section 5) that also involves trigonometric functions. However, neither trigonometric solvers nor trigonometric simplifications automatically return the solution we expect, i.e., a symbolic expression describing the relation between parameters and variables. Thus, we would like to extend a solver for non-linear polynomial constraints with the trigonometric functions *sine* and *cosine*. Let us give a second example. One may have to consider unification problems together with constraints on depths of ground terms, i.e., constraints such as $depth(X) = 4 - depth(g(Z)) \wedge g(g(Z)) = g(g(g(b))) \wedge g(Y) = g(g(X))$ We have here two disjoint sorts: terms (solved by unification) and integers. Since no solver over the integers can handle the function *depth*, we want to extend a Diophantine solver (or a finite domain like solver) with the *depth* function.

Nowadays, some methods may be investigated for extending solvers. Solver combination methods [1,11,15,7] aim at designing a general solver (corresponding to a new mixed domain which is a conservative extension of the original ones) based on the cooperation of elementary solvers. Since we want to stay on the same interpretation domain, such frameworks are not well suited. Independently of these theoretical results, more practical issues have been explored for the cooperation of several solvers on a single domain [14,4,2], or on several domains [12]. However, such systems cannot directly handle extra function symbols.

The need of integrating deduction techniques into computer algebra is now well-established [16], and standard computer algebra systems (such as Mathematica [17]) already provides some equation simplification tools. Although they are powerful, no method/technique are available for designing a solver extension or to insure its soundness. Similar comments can be done about CHRs [5] and ELAN [10] [2]. Some works were also conducted in the area of constraint transformation [12], but these techniques act only as a pre-processing.

To overcome the problems of solver extension and to generalize/formalize some of the previous works, we designed **SoleX**, a mechanism for extending[3] constraint solvers. **SoleX** enables one increasing the declarativity of CP systems without jeopardizing completeness of solvers, nor designing new solvers from scratch. The aim of **SoleX** is to enrich the solvers so they can treat new function symbols called *alien* symbols. Their semantics can be of different kinds. First, they can be syntactic sugar to replace the extensional definition of a function (e.g., $3.x^2 + 2.x + 1$ may be named $p(x)$). Second, they can be standard functions that are not handled by the solver. For example, usual solvers for arithmetic constraints cannot manipulate the functions *sine* and *cosine*. Thus, it is important

---

[2] However, these systems are really well suited for implementing our framework.

[3] This mechanism can also be viewed as a way to complete constraint solvers.

to be able to extend methods such as Gaussian elimination or Gröbner bases for solving constraints with occurrences of these trigonometric functions. Unlike to the previous cases, the last class of alien symbols corresponds to functions with no defined meaning on the domain. For example, a function can be characterized by experimental measures that can be expressed as constraints. The solved form may define the extensional definition of the function or of a class of functions.

SoleX is the ordered application of four phases (collections of rules or *component solvers*) to process alien function symbols and deduce related information. The Reduction phase reduces the search space by adding semantic and syntactical information carried by the functions. The Expansion phase completes the constraints with always valid (w.r.t. the extended domain) constraints, i.e., characteristics of the functions (e.g., an absolute value is always greater than or equal to zero). Then, the constraint store is purified by abstracting remaining function symbols that cannot be processed by the solver. After application of the built-in solver (Solving phase), the Contraction phase replaces abstraction variables with their related alien terms (this is the "opposite" of abstraction) and removes "redundancies" added by the expansion phase. Several applications of **SoleX** may be necessary to reach a fixed point and to solve the constraints.

The paper is organized as follows. Section 2 formalizes our framework. Section 3 describes the (rule-based) elementary solvers. We then examine (Section 4) the problem of controlling solvers. Section 5 describes some applications of **SoleX** over different domains. Finally, comparisons, conclusions and future works are discussed in Section 6. A longer version of this paper [13] includes the transformation rules that formalize the solver extensions, and some proofs as well.

2   Basic Concepts

Let us first introduce some standard notations about terms and substitutions of variables by terms. Given a first-order signature Σ and a denumerable set $\mathcal{V}$ of variables, $T(\Sigma, \mathcal{V})$ denotes the set of $\mathcal{F}_\Sigma$-terms with variables in $\mathcal{V}$. Terms (resp. variables) are denoted by $t_1, \ldots, t_n$ (resp. $x_1, \ldots, x_n$). A *ground* term is a term without variables. The terms $t_{|\omega}$, $t[s]_\omega$ and $t[\omega \hookleftarrow s]$ denote respectively the subterm of t at the position ω, the term t with the subterm s at the position ω and the replacement in t of $t_{|\omega}$ by s. The symbol of t occurring at the position ω (resp. the top symbol of t) are written $t(\omega)$ (resp. $t(\epsilon)$). The term $t[s]$ denotes a term t with some subterm s. The term $t[s \hookleftarrow u]$ denotes the term where s is replaced by u in *all* occurrences of s in t. $\mathcal{V}(t)$ denotes the set of variables occurring in the term t. A *substitution* $\{x_1 \mapsto t_1, \ldots, x_n \mapsto t_n\}$ is an assignment from $\mathcal{V}$ to $T(\Sigma, \mathcal{V})$. We use letters $\sigma, \mu, \gamma, \phi, \ldots$ to denote substitutions. The application of a substitution σ to a term t is written in postfix notation $t\sigma$. We now define the objects handled by **SoleX**: solvers, and constraint systems.

Definition 1 (Constraint system).
A constraint system is a 4-tuple $(\Sigma, \mathcal{D}, \mathcal{V}, \mathcal{L})$ where:

- Σ *is a first-order signature given by a set of function symbols $\mathcal{F}_\Sigma$, and a set of predicate symbols $\mathcal{P}_\Sigma$,*

- $\mathcal{D}$ is a Σ-structure (its domain is denoted by $|\mathcal{D}|$),
- $\mathcal{V}$ is an infinite denumerable set of variables,
- $\mathcal{L}$ is a set of constraints: it is a non-empty set of $(\Sigma, \mathcal{V})$-atomic formulas closed under conjunction and disjunction. The unsatisfiable constraint is denoted by $\perp$ and the truth constraint is denoted by $\top$. An assignment is a mapping $\alpha : \mathcal{V} \to |\mathcal{D}|$. The set of all assignments is denoted by $ASS_{\mathcal{D}}^{\mathcal{V}}$. An assignment α extends uniquely to a homomorphism $\underline{\alpha} : T(\Sigma, \mathcal{V}) \to \mathcal{D}$. The set of solutions of a constraint $c \in \mathcal{L}$ is the set $Sol_{\mathcal{D}}(c)$ of assignments $\alpha \in ASS_{\mathcal{D}}^{\mathcal{V}}$ such that $\underline{\alpha}(c)$ holds. A constraint c is valid in $\mathcal{D}$ (denoted by $\mathcal{D} \models c$) if $Sol_{\mathcal{D}}(c) = ASS_{\mathcal{D}}^{\mathcal{V}}$.

The enrichment of a constraint system CS consists of some additional functions defined on the original domain. The interpretation of symbols defined in CS is unchanged.

Definition 2 (Constraint system enrichment).
Let $CS^+ = (\Sigma^+, \mathcal{D}^+, \mathcal{V}^+, \mathcal{L}^+)$ and $CS = (\Sigma, \mathcal{D}, \mathcal{V}, \mathcal{L})$ be two constraint systems. Then, CS^+ is an enrichment of CS if:

- $\mathcal{F}_{\Sigma} \subseteq \mathcal{F}_{\Sigma^+}$ and $\mathcal{P}_{\Sigma} = \mathcal{P}_{\Sigma^+}$
- $|\mathcal{D}| = |\mathcal{D}^+|$ and $\forall r \in \Sigma$, $r_{\mathcal{D}^+} = r_{\mathcal{D}}$ [4]
- $\mathcal{V} = \mathcal{V}^+$, $\mathcal{L} \subseteq \mathcal{L}^+$

A set of constraints $\{c_1, \ldots, c_n\}$ where $c_i \in \mathcal{L}^+$ for $i = 1, \ldots, n$ (a *constraint store*) is represented by a conjunction of constraints $c_1 \wedge \cdots \wedge c_n$ [5]. This conjunction can be split into an impure component in $\mathcal{L}^+$ and a pure component in $\mathcal{L}$. Hence, we represent a constraint in $\mathcal{L}^+$ by a pair (C, P) where $C \in \mathcal{L}^+$, $P \in \mathcal{L}$, and (C, P) means the conjunction $C \wedge P$. If C is in $\mathcal{L}$, then (C, P) is said *pure*.

Definition 3 (Aliens, pure and impure constraints). A pure *constraint (resp. term)* is a constraint (resp. term) in $\mathcal{L}$. An Alien *subterm in a term t* is a term with a top-symbol in $\Sigma^+ \backslash \Sigma$ such that its super-terms (whenever they exist) have top-symbols in Σ. The set of aliens in C denoted by $Alien(C)$ is the set of alien subterms of terms occurring as arguments of atomic constraints in C. A constraint C is impure if $Alien(C)$ is non-empty.

Intuitively, a component solver is an algorithm which transforms a constraint C into a new constraint C' "simpler" than C, but equivalent to C in the structure $\mathcal{D}$ (a solver preserves the solutions). Moreover, the repeated application of a solver always reaches a fixed-point which is a constraint in *solved form*.

Definition 4 (Component Solver). A component solver *(or solver in short)* for a constraint system $(\Sigma, \mathcal{D}, \mathcal{V}, \mathcal{L})$ is a computable function $S : \mathcal{L} \to \mathcal{L}$ s.t.:

1. $\forall C \in \mathcal{L}$, $Sol_{\mathcal{D}}(S(C)) \subseteq Sol_{\mathcal{D}}(C)$ *(correctness)*
2. $\forall C \in \mathcal{L}$, $Sol_{\mathcal{D}}(C) \subseteq Sol_{\mathcal{D}}(S(C))$ *(completeness)*
3. $\forall C \in \mathcal{L}$, $\exists n \in \mathbb{N}$, $S^{n+1}(C) = S^n(C)$

---

[4] $r_{\mathcal{D}}$ (resp. $r_{\mathcal{D}^+}$) represents the interpretation of r on the Σ-structure $\mathcal{D}$ (resp. $\mathcal{D}^+$).
[5] The c_i's are elements of $\mathcal{L}^+$ and not necessarily atomic constraints.

A constraint C is in solved form w.r.t. S if $S(C) = C$. We denote by S^+, S^n for some $n \geq 0$. Similarly S^* denotes the repeated application of S till reaching the solved form.

In the following, we are interested in the design of a rule-based solver for an enrichment CS^+ of CS based on a solver S known for CS together with a given set of domain-independent transformation rules (solver extensions).

Example 1. The following rule defines a solver for CS^+ if S is a solver for CS.
Solve
$$\frac{(C, P)}{(C, S(P))}$$

We will develop a rule-based solver for CS^+ using the rule Solve. In addition to Solve, some other solvers (solver extensions, see Section 3) are applied on C.

3   Solver Extensions

The solver extensions have been grouped together w.r.t. the kind of action they have on the constraint store. On one hand, semantic rules (Section 3.2) make use of the properties of the domain or of the properties of the alien functions. On the other hand syntactical rules (Section 3.1) are based on syntactical transformations like *Abstraction* or its opposite *Alien Replacement*.

3.1   Syntactical Solver Extensions

The following transformation rules are sound in any constraint system enrichment.

Variable Abstraction The rule Abstraction transforms impure constraints into pure ones by adding new variables to name aliens. These variables X_u replace terms u not in CS and the related equations $X_u = u$ are added to the constraint store. The reader should note that equations $X_u = u$ are no more transformed and remain in the constraint store. In the following, we use a bijective mapping which associates to each non-variable term u a unique variable X_u. Hence, two occurrences of the same term will be automatically replaced by the same variable.

Example 2. Consider the constraint $C = (\sin^2(x + y) + \cos^2(x + y)) = 1 - \sin^3(2x)*(\sin(x+y)+\cos(x+y))$. Abstraction* transforms C into $C' = (X^2+Y^2 = 1 - Z^3 * (X + Y)) \wedge ((X = \sin(x + y)) \wedge (Y = \cos(x + y)) \wedge (Z = \sin(2x)))$.

Alien Replacement In the previous paragraph, we have seen how to purify the constraint store with Abstraction. Then, we can apply Solve (see Example 1). This leads to a conjunction of impure solved forms (involving new variables called *abstraction variables*) together with a simpler pure constraint. In order to be able to apply semantic solver extensions (described later on), it is necessary to re-build an impure constraint without abstraction variables. Hence, AlienRep consists in replacing abstraction variables by their related alien subterms. Obviously, the rule is the converse of **Abstraction** and so (**AlienRep*** $\circ$ **Abstraction***) is the identity solver, where $\circ$ represents the usual composition of functions.

Inter-Reduction The idea of the transformation rule **InterRed** is to consider equations occurring in the constraint store as (ground) rewrite rules and then to use rewriting to simplify the store. For this purpose, we use a total ordering $\prec$ on $T(\Sigma^+, \mathcal{V}^+)$ such that non-ground terms are greater than ground terms and non-ground (resp. ground) impure terms are greater than non-ground (resp. ground) pure terms. For comparing non-ground (resp. ground) terms, we can use a *lpo*-ordering on $T(\Sigma^+ \cup \mathcal{V}^+)$ based on a precedence such that additional function symbols in $\Sigma^+ \backslash \Sigma$ are greater than function symbols in Σ and function symbols in Σ^+ are greater than variables in $\mathcal{V}^+$. This ordering is total, closed under contexts and satisfies the subterm property.

Example 3. Consider the constraint $C = (y \le \sin 2x + \sin z) \wedge (\sin z = 1) \wedge (\sin 2x = 0)$. **InterRed** transforms C into $(y \le 0 + 1) \wedge (\sin z = 1) \wedge (\sin 2x = 0)$ provided that $\sin z \succ 1$ and $\sin 2x \succ 0$.

Moving constraints As said before, a constraint store is represented by a couple (C, P). When a constraint c in C becomes pure (e.g., after **Abstraction**), it can be carried to P, the pure part of the store. This is realized with the rule **ToPure**. In a similar way, constraints in P that become impure (e.g., after **AlienRep**) are moved to C with the rule **FromPure**.

3.2   Semantic Solver Extensions

In this section, some meta-transformation rules (mainly based on rewriting) are proposed. These transformations must be regarded as solvers and aim to integrate properties relevant to functions and predicates in CS^+. For instance, a solver S on CS is usually able to normalize any term in CS (which is more or less its internal representation) and it would be interesting to enhance this normalization on terms in CS^+.

Normalization We assume that the solver S is equipped with a normalizing mapping NF, that is an idempotent computable mapping $NF : T(\Sigma, \mathcal{V}) \rightarrow T(\Sigma, \mathcal{V})$ such that $\forall t \in T(\Sigma, \mathcal{V})$, $\mathcal{D} \models t = NF(t)$. Moreover, the computation of $NF(t)$ does not depend on the names of variables in $\mathcal{V}(t)$ but just depends on the total ordering of variables occurring in t. This ordering is given by the restriction of $\prec$ to variables.

Definition 5. *The mapping* $NF^+ : T(\Sigma^+, \mathcal{V}^+) \rightarrow T(\Sigma^+, \mathcal{V}^+)$ *is defined by:*

- $NF^+(f(t_1, \ldots, t_m)) = f(NF^+(t_1), \ldots, NF^+(t_m))$ *if* $f \in \Sigma^+ \backslash \Sigma$.
- *If* t *is pure, then* $NF^+(t) = NF(t)$
- *If* t *is an impure term with a top symbol in* Σ, *then* $NF^+(t)$ *is recursively obtained as follows:*

1. *Compute* $b_i = NF^+(a_i)$ *for every alien* $a_i \in Alien(t)$.
2. *Compute the term* t^π *obtained by replacing in* t *aliens* $a_1, \ldots a_n$ *by new variables* $x_1, \ldots, x_n$ *such that* $x_k \prec x_l \Leftrightarrow b_k \prec b_l$ *and* $x_k = x_l \Leftrightarrow b_k = b_l$ *for* $1 \leq k, l \leq n$.
3. $NF^+(t)$ *is* $NF(t^\pi)\{x_k \mapsto b_k\}_{k=1,\ldots,n}$.

The transformation rule **Normalize** consists in the replacement of any term t by its normal form $NF^+(t)$ following an innermost strategy.

Example 4. Let us consider the function symbols $+, *, -$, the predicate symbol $\leq$ in Σ (interpreted as usual over reals), the new function symbol f in Σ^+, and the normalizing mapping NF such that: $NF(x - x) = 0$, $NF(x + 0) = x$ and $NF(1 * x) = x$. Then $NF^+(f(1 * x) - f(x + 0)) = 0$ and applying **Normalize** on the constraint $C = (f(1 * x) - f(x + 0) \leq x - y)$ leads to $0 \leq x - y$. The built-in solver can now treat this constraint and so we get the solution $y \leq x$.

More generally, rewriting is a very natural concept for replacing a term by another one which is supposed to be simpler but equivalent in the constraint system. Termination is required in order to get an extended solver. Intuitively, the database of rewrite rules (*TGR* for terms and *CGR* for constraints) must simplify the impure subpart of the constraint. We define in the following how to apply such rules coming from a database of properties. Guarded rules are considered and applied only if the current constraint store entails the related guard (or constraint). Matching of left-hand sides of rules is performed syntactically but one should note that the equality in the built-in constraint system has been incorporated thanks to the **Normalize** solver.

Term Dependent Guarded Reduction The term rewrite system TGR is a finite set of guarded rules $(l \rightarrow r \| g)$ where g is a constraint in CS^+ and l, r are terms such that $\mathcal{D}^+ \models g \Rightarrow l = r$. An instance of the term l, say $l\sigma$, occurring in the constraint store C can be replaced by $r\sigma$ when $g\sigma$ is entailed by C. This transformation rule is called **TermRed**.

Example 5. Consider the guarded rules $(|x| \rightarrow x \| x \geq 0)$ and $(|x| \rightarrow -x \| x < 0)$. The constraint $C = (|y - 2| = x + |x| + 1) \wedge (y \geq 3) \wedge (x * y < 0)$ can be reduced to $y = 3 \wedge x < 0$ thanks to **TermRed** and **Solve** (see Example 1).

Constraint Dependent Guarded Reduction The constraint rewrite system CGR consists of a finite set of guarded rules $(L \rightarrow R \| G)$ where G is a constraint in CS^+ and L, R are conjunctions of atomic constraints in CS^+ such that $\mathcal{D}^+ \models G \Rightarrow (L \Leftrightarrow R)$. A rewrite relation is defined as previously, except that matching is now performed modulo the associativity-commutativity of $\wedge$ and $\vee$. The corresponding transformation rule is called **ConsRed**.

Example 6. Consider the guarded rule $(\sqrt{x} = y \rightarrow x = y^2 \| x \geq 0)$. The constraint $(x > 2) \wedge ((x - 1) * (y - 3) > 0) \wedge (\sqrt{y - 2} = y - 4)$ can be reduced to $(x > 2) \wedge ((x - 1) * (y - 3) > 0) \wedge (y - 2) = (y - 4)^2$ since $y - 2 \geq 0$. Finally we get the solutions for y by calling the built-in solver.

Formally, checking the implication (entailment) requires a validity checker for the enriched constraint system. If such a decision algorithm is not provided, then the semantic entailment can be approximated by a syntactic constraint inclusion test.

Domain Dependent Completion/Deletion In order to Complete/Delete the information encoded in the constraint store, we consider a database of valid facts, i.e. a finite set DDR of valid conjunctions of constraints in CS^+. This leads to a pair of quite opposite transformation rules, namely DomComp and DomDel. DomComp completes the constraint store C by an instance of a constraint $C' \in DDR$ provided this instance is not yet entailed by C. Conversely DomDel deletes an instance of $C' \in DDR$ occurring in the constraint store. For trigonometric functions, examples of valid constraints in DDR are: $-1 \leq \sin x \leq 1$, $-1 \leq \cos x \leq 1$, $\cos^2(x) + \sin^2(x) = 1$.

Example 7. Consider the constraint $C = (1 - \sin 2x = y)$ and the valid constraint $(-1 \leq \sin X \leq 1) \in DDR$. The constraint C is transformed into $C \wedge (-1 \leq \sin 2x \leq 1)$ thanks to DomComp. We do not have to add the constraint $(-1 \leq \sin 2x \leq 1)$ to $C' = (y * (y - 1) = 0) \wedge C$ since $(y * (y - 1) = 0) \wedge (1 - V = y)$ already implies $(-1 \leq V \leq 1)$, where V stands for $\sin 2x$.

4   SoleX: The Solver Collaboration

The solver for CS^+ is described as a set of transformation rules (presented in Section 3) together with control. The basic operation we used for combining solvers is the composition (of functions). The extended solving process **SoleX** is (Contraction ∘ Solve ∘ Expansion ∘ Reduction) where the four phases are as follows:

- Reduction phase (Reduction = (ConsRed$^+$∘TermRed$^+$∘InterRed$^+$∘Normalize$^+$)): the constraint store is transformed using semantic and syntactical solver extensions introduced in the two previous sections.
- Expansion phase (Expansion = (ToPure* ∘ Abstraction* ∘ DomComp*)) : the constraint store is completed by valid constraints which may be helpful in the next phase and may be purified thanks to Abstraction.
- Solve phase (Solve): the built-in solver is applied on the pure part of the constraint store.
- Contraction phase (Contraction = (DomDel* ∘ AlienRep* ∘ FromPure*)): the impure equations introduced in the second phase are merged with the new pure part of the constraint store. The remaining valid constraints added in the same phase are removed.

It is important to notice that a transformation rule is not necessarily a solver since its repeated application may not terminate. For the same termination problem, a composition of two solvers yields a new function which is not necessarily a solver (Definition 4). For proving the termination of a composition of solvers, we may need to embed all orderings related to elementary solvers into a Noetherian

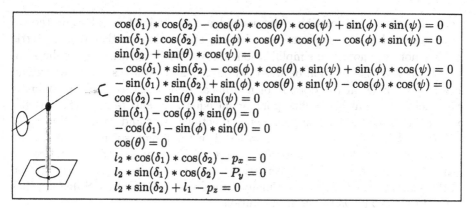

$$\cos(\delta_1) * \cos(\delta_2) - \cos(\phi) * \cos(\theta) * \cos(\psi) + \sin(\phi) * \sin(\psi) = 0$$
$$\sin(\delta_1) * \cos(\delta_2) - \sin(\phi) * \cos(\theta) * \cos(\psi) - \cos(\phi) * \sin(\psi) = 0$$
$$\sin(\delta_2) + \sin(\theta) * \cos(\psi) = 0$$
$$- \cos(\delta_1) * \sin(\delta_2) - \cos(\phi) * \cos(\theta) * \sin(\psi) + \sin(\phi) * \cos(\psi) = 0$$
$$- \sin(\delta_1) * \sin(\delta_2) + \sin(\phi) * \cos(\theta) * \sin(\psi) - \cos(\phi) * \cos(\psi) = 0$$
$$\cos(\delta_2) - \sin(\theta) * \sin(\psi) = 0$$
$$\sin(\delta_1) - \cos(\phi) * \sin(\theta) = 0$$
$$- \cos(\delta_1) - \sin(\phi) * \sin(\theta) = 0$$
$$\cos(\theta) = 0$$
$$l_2 * \cos(\delta_1) * \cos(\delta_2) - p_x = 0$$
$$l_2 * \sin(\delta_1) * \cos(\delta_2) - P_y = 0$$
$$l_2 * \sin(\delta_2) + l_1 - p_z = 0$$

Fig. 1. Robot-arm with two degrees of freedom

ordering $<$. As solver extensions are parameterized by $NF, \prec$, rewrite rules in TGR and CGR, or by valid constraints in DDR, we have to give more precise sufficient conditions in order to insure the termination of the **SoleX** process.

Theorem 1. *Let* $\geq$ *be a quasi-ordering on* $\mathcal{L}^+\backslash\mathcal{L}$ *such that* $>$ *is Noetherian.* **SoleX** *is a solver for* CS^+ *if the following conditions are satisfied:*

1. *TermRed and ConsRed are solvers,*
2. *For any* $E \in \{$Normalize, InterRed, TermRed, ConsRed$\}$, *we have*
$$E(C, P) = (C', P) \text{ and } C' \in \mathcal{L}^+\backslash\mathcal{L} \Rightarrow C \geq C'$$
3. *ExtSolve = Contraction ∘ Solve ∘ Expansion is a solver such that*
$$\text{ExtSolve}(C, P) = (C', P') \Rightarrow C > C'$$

The proof is quite obvious since **SoleX** is ExtSolve ∘ Reduction and we assume a complexity measure that does not increase by Reduction but strictly decreases by ExtSolve. **SoleX** becomes a solver in the particular case where DDR, TGR, CGR are empty, and NF is the identity mapping. More generally, one can find more details about a possible ordering in [13], as well as some sufficient conditions to insure that ExtSolve (respectively TermRed, and ConsRed) is a solver.

5   Applications

Inverse Robot Kinematics We can now solve the problem [3] briefly described in Section 1. This problem for a robot (see Figure 1) having two revolute joints (degrees of freedom) can be described by the system of equations presented in Figure 1 where l_1, l_2, p_x, p_z are parameters and $P_y, \delta_1, \delta_2, \phi, \theta, \psi$ are variables. l_1, l_2 are the lengths of the robot arms, (p_x, P_y, p_z) is the position of the end-effector, ϕ, θ, ψ are the Euler angles of the orientation of the end-effector, and δ_1, δ_2 are the rotation angles of the revolute joints. The expected solution is a symbolic expression describing the dependence of the joint variables on the geometrical and position parameters. For this application, neither trigonometric solvers nor trigonometric simplifications automatically return a symbolic solution expressing the relation between parameters and variables. Thus, we extend a solver working on the domain of non-linear polynomial constraints (namely Gröbner bases which simplify polynomial equations and return relations between the variables) with trigonometric functions (*sine* and *cosine*). Hence, let DDR be $\{\sin^2(X) + \cos^2(X) - 1 = 0\}$. The DomComp solver completes the system by adding for each angle $(\delta_1, \delta_2, \phi, \theta, \psi)$ the property of sine and cosine $(\sin^2(X) + \cos^2(X) = 1)$. The Abstraction solver replaces every remaining sine and cosine with new variables. Finally after Solve and AlienRep, **SoleX** reaches a fixed-point which is the desired solution:

$$
\begin{aligned}
&\sin(\psi) + k_1 * \cos(\phi) * \sin(\delta_1) * \cos(\delta_1) = 0 && \wedge\; P_y + k_6 * \sin(\delta_1) * \cos(\delta_1) = 0 \\
&\wedge \sin(\theta) + k_2 * \cos(\phi) * \sin(\delta_1) = 0 && \wedge\; \sin(\delta_2) + k_7 = 0 \\
&\wedge \sin(\phi) + k_3 * \cos(\phi) * \sin(\delta_1) * \cos(\delta_1) = 0 && \wedge\; \sin^2(\delta_1) + k_8 = 0 \\
&\wedge \cos(\psi) + k_4 * \cos(\phi) * \sin(\delta_1) = 0 && \wedge\; \cos(\delta_2) + k_9 * \cos(\delta_1) = 0 \\
&\wedge \cos(\theta) = 0 && \wedge\; \cos^2(\delta_1) + k_{10} = 0 \\
&\wedge \cos^2(\phi) + k_5 = 0
\end{aligned}
$$

where $k_1, \ldots, k_{10}$ are constants, depending on the parameters l_1, l_2, p_x, p_z.

Constraint solving over integers and terms This example illustrates how to extend a constraint solver working on conjunctions of two-sorted constraints: the integers and the terms. Constraints over integers are equations and inequations between linear polynomials with integer coefficients. Constraints over terms are equations between terms. Formally, the signature is as follows:

$$
\Sigma = \begin{cases} \leq: \mathbb{Z} \times \mathbb{Z}; \ +, - : \mathbb{Z} \times \mathbb{Z} \to \mathbb{Z}; \ 0, 1 : \mathbb{Z} \\ f : \mathbb{T} \times \mathbb{T} \to \mathbb{T}; \ g : \mathbb{T} \to \mathbb{T}; \ a : \mathbb{T} \end{cases}
$$

$\mathbb{T}$ (resp. $\mathbb{Z}$) denotes ground terms (resp. integers) and the function symbols and the predicate symbol $\leq$ are interpreted as usual. We consider two new function symbols $depth : \mathbb{T} \to \mathbb{Z}$ and $max : \mathbb{Z} \times \mathbb{Z} \to \mathbb{Z}$, interpreted respectively as the depth of a term and the maximum of two integers. Since we want to extend the constraint solvers associated to integers and terms, we choose the sets of Term-dependent (Constraint-dependent) Guarded Reduction and DDR as follows:

$$
TGR = \begin{cases} (1) \ depth(f(X,Y)) \to 1 + max(depth(X), depth(Y)) \\ (2) \ depth(g(X)) \to 1 + depth(X) \\ (3) \ depth(a) \to 1 \\ (4) \ (max(x,y) \to y\|x \leq y) \\ (5) \ (max(x,y) \to x\|y \leq x) \end{cases}
$$

$CGR = \{(1) \ depth(X) = 1 \to X = a, (2) \ depth(X) < 1 \to \bot\}, DDR = \{1 \leq depth(X)\}$ where a rule $l \to r$ is an abbreviation for $(l \to r\|\top)$. Consider the repeated application of **SoleX** to

$$
\begin{cases} (I) \quad z' - z && = depth(g(Y)) - depth(Y) \\ (II) \quad max(z', z) - z = u \\ (III)\, 0 && \leq 1 - v + u \\ (IV) \ depth(g(X)) && \leq v \\ (V) \quad f(W, W') && = f(g(W'), f(X, X)) \end{cases}
$$

First, TermRed applies rule (2) of TGR on equation (I). After Normalize, (I) becomes $z' = z + 1$. TermRed applies rule (4) of TGR on (II), and after Normalize (II) becomes $u = 1$. Then, after ToPure, Solve applies the solver for integers and (III) is transformed into $v \leq 2$. TermRed can now apply rule (2) of TGR on (IV): then application of Normalize, DomComp and ConsRed (rule (1)) leads to $X = a$. Finally, after Normalize and ToPure, Solve (solver for terms, i.e., unification) transforms (V) into $W = g(f(a, a)) \wedge W' = f(a, a)$. The solved form is:
$z' = z + 1 \wedge u = 1 \wedge v \leq 2 \wedge X = a \wedge W = g(f(a, a)) \wedge W' = f(a, a)$
Here, the complexity measure for proving the termination of **SoleX** is based on a combination of elementary measures corresponding to the number of *depth* and *max* occurrences and to the multiset of sizes of *depth* arguments.

6   Conclusion

SoleX enables one extending solvers to handle alien function symbols, i.e., a built-in solver (seen as a black-box) is completed with a glass-box mechanism. The extension is composed of syntactical solvers that process the constraints independently from the computation domain, and semantic solvers that enrich constraints with information on the domain or on the function interpretation.

In [6], Heintze & al. propose an extension of the solver of CLP($\mathcal{R}$) for constraints over $\mathcal{R} + \mathcal{M}$. This extension is based on two methods: simplifications that are similar to our notion of normal form extension, and substitutions that can be seen as our rule InterRed. However, our framework for extension is more complete since we also propose some other syntactic rules, as well as semantic rules. Moreover, we can extend every kinds of domains whereas in [6] novel constraint solvers, simplification algorithms or computation domains are always related to $\mathcal{R}$. In an other hand, Heintze & al. not only extend the solver, but also the programming language. Thus, their work also enables applications such as debuggers or prototyping of novel CP systems.

A first implementation of **SoleX** has been realized into **CoSAc** [14]: two solvers of **CoSAc** (one based on Gröbner bases computation and the other one based on Gaussian elimination) are extended with new function symbols (such as *sin*, *cos* and $\sqrt{}$ with their usual interpretation). Hence, significant problems like the *Inverse Robot Kinematics* problem [3] (which is originally expressed with trigonometric functions) and the *Robot in a Corridor* problem [14] (which uses square roots of polynomials) are solved automatically.

In this paper, **SoleX** has been presented as a set of transformation rules plus a built-in solver. Therefore, we could imagine to prototype this rule-based extended solver with a rule-based programming language. In this context, ELAN [10] is a very good candidate since it provides facilities to express strategies for applying rules and to call external solvers. However, we believe that a more efficient implementation should be based on a collaboration of several component solvers running concurrently. This explains why a more complete implementation is currently under way with **BALI** [12] which provides a logical framework for managing constraints and a language for designing and executing solver collaborations.

BALI and **SoleX** have similarities that have to be studied to completely merge the two concepts and realize a framework including both solver collaboration and solver extension. Furthermore, extensions of solvers with new sorts and new constraints will enable to design solvers on totally different domains. Thus, extending a solver on a "simple" domain could lead to realize solvers on complex domains thanks to solver extension of **SoleX** and solver collaboration of **BALI**.

References

1. F. Baader and K. Schulz. On the combination of symbolic constraints, solution domains, and constraint solvers. In *Proc. of CP'95*, volume 976 of *LNCS*, 1995.
2. F. Benhamou and L. Granvilliers. Combining local consistency, symbolic rewriting, and interval methods. In J. Pfalzgraf, editor, *Proc. AISMC-3*, volume 1138 of *LNCS*, Steyr, Austria, Sep. 1996. Springer-Verlag.
3. B. Buchberger. Applications of Gröbner Bases in Non-Linear Computational Geometry. In D. Kapur and J. Mundy, editors, *Geometric Reasoning*, pages 413–446. MIT Press, 1989.
4. O. Caprotti. Extending risc-clp(cf) to handle symbolic functions. In A. Miola, editor, *Proc. of DISCO'93*, volume 722 of *LNCS*. Springer-Verlag, Sep. 1993.
5. T. Frühwirth. Constraint handling rules. In A. Podelski, editor, *Constraint Programming: Basics and Trends*, volume 910 of *LNCS*. Springer-Verlag, 1995.
6. N. Heintze, S. Michaylov, P. J. Stuckey, and R. H. C. Yap. Meta-Programming in CLP($\mathcal{R}$). *JLP*, pages 221–259, 1997.
7. K. Homann and J. Calmet. Combining Theorem Proving and Symbolic Mathematical Computing. In J.A. Campbell J. Calmet, editor, *Proc. of AISMC-2*, volume 814 of *LNCS*, pages 18–29. Springer-Verlag, 1995.
8. J. Jaffar and M. Maher. Constraint Logic Programming: a Survey. *JLP*, 19,20:503–581, 1994.
9. J. Jaffar, S. Michaylov, P. Stuckey, and R. Yap. The CLP($\mathcal{R}$) Language and System. *ACM Transactions on Programming Languages and Systems*, 14(3):339–395, 1992.
10. C. Kirchner, H. Kirchner, and M. Vittek. Designing constraint logic programming languages using computational systems. In P. Van Hentenryck and V. Saraswat, editors, *Principles and Practice of Constraint Programming. The Newport Papers.*, pages 131–158. MIT press, 1995.
11. H. Kirchner and C. Ringeissen. Combining symbolic constraint solvers on algebraic domains. *JSC*, 18(2):113–155, 1994.
12. E. Monfroy. *Collaboration de solveurs pour la programmation logique à contraintes*. Phd thesis, Université Henri Poincaré - Nancy 1, Nov. 1996. Also available in english.
13. E. Monfroy and C. Ringeissen. SoleX: a Domain-Independent Scheme for Constraint Solver Extension (Extended Version). Research report, INRIA, Jun. 1998. Also available at url http://www.inria.fr.
14. E. Monfroy, M. Rusinowitch, and R. Schott. Implementing Non-Linear Constraints with Cooperative Solvers. In *Proc. of ACM SAC'96*, pages 63–72, Feb. 1996.
15. C. G. Nelson and D. C. Oppen. Simplifications by cooperating decision procedures. *ACM Transactions on Programming Languages and Systems*, 1(2), 1979.
16. The Calculemus Project. *Calculemus Workshop: Systems for Integrated Computation and Deduction*, Edinburgh, Scotland, Sep. 1997.
17. S. Wolfram. *The Mathematica Book, 3rd ed.* Cambridge University Press, 1996.

Optimising Propositional Modal Satisfiability for Description Logic Subsumption

Ian Horrocks[1] and Peter F. Patel-Schneider[2]

[1] University of Manchester, Manchester, UK (horrocks@cs.man.ac.uk)
[2] Bell Labs Research, Murray Hill, NJ, U.S.A. (pfps@research.bell-labs.com)

Abstract. Effective optimisation techniques can make a dramatic difference in the performance of knowledge representation systems based on expressive description logics. Because of the correspondence between description logics and propositional modal logic many of these techniques carry over into propositional modal logic satisfiability checking. Currently-implemented representation systems that employ these techniques, such as FaCT and DLP, make effective satisfiable checkers for various propositional modal logics.

1 Introduction

Description logics are a logical formalism for the representation of knowledge about individuals and descriptions of individuals. Description logics represent and reason with descriptions similar to "all people whose friends are both doctors and lawyers" or "all people whose children are doctors or lawyers or who have a child who has a spouse". The computations performed by systems that implement description logics are based around determining whether one description is more general than (subsumes) another. There have been various schemes for computing this subsumption relationship, depending on the expressive power of the description logic and the degree of completeness of the system. As description logic systems perform numerous subsumption checks in the course of their operations, they need to have a highly-optimised subsumption checker.

Recent work [16] has shown that determining subsumption in expressive description logics is equivalent to determining satisfiability of formulæ in propositional modal or dynamic logics. Thus one part of a system that implements a description logic is equivalent to a satisfiability checker for a propositional modal or dynamic logic. Several description logic systems have been built for such description logics, and thus include what is essentially a satisfiability checker, including KRIS [2] and CRACK [5]. These two systems have incorporated a number of optimisations to achieve better performance of their subsumption checkers.

Description logic systems are also optimised in other ways. In particular, their operations are optimised to avoid the potentially-costly subsumption checks whenever possible. There are also other optimisations to subsumption possible in description logic systems, having to do with the nature of the representation of knowledge in a description logic, but these have little or nothing to do with optimising propositional modal satisfiability.

Jacques Calmet and Jan Plaza (Eds.): AISC'98, LNAI 1476, pp. 234–246, 1998.

We have built two systems that explore the optimisations required to build an expressive description logic system, namely FaCT [11], a full description logic system, and DLP [14], an experimental system providing only a limited description logic interface. FaCT is available at http://www.cs.man.ac.uk/~horrocks; DLP is available at http://www-db.research.bell-labs.com/user/pfps.

We have incorporated a range of known, adapted and novel optimisation techniques into the subsumption checkers for these two systems. The optimisation techniques include: lexical normalisation, semantic branching search, boolean constraint propagation, dependency directed backtracking, heuristic guided search and caching.

These optimisations techniques make a drastic difference to the performance of the overall system. As evidence, KRIS is not able to load a modified version of the GALEN knowledge base because it gets stuck trying to determine one of the thousands of subsumptions required to load the knowledge base. FaCT and DLP, which have higher levels of optimisation, are able to easily load this knowledge base, classifying over two thousand definitions in about two hundred seconds.

We have also performed experiments with both FaCT and DLP on several test suites of propositional modal formulae. The optimisations built into the two systems qualitatively change their behaviour on the test suites, indicating that the optimisations have considerable utility simply taken as optimisations for reasoning in propositional modal logics.

2 Background

FaCT and DLP are designed to build and maintain taxonomies of named concepts. Given a collection of definitions of named concepts and statements about these concepts, they determine the subsumption partial order for the named concepts. To do this they have to determine subsumption relationships between descriptions in a description logic.

The description logic that DLP implements is called $\mathcal{ALC}_{R^+}$. FaCT implements a considerably more-expressive logic, but most of the satisfiability optimisations in FaCT are demonstrable in $\mathcal{ALC}_{R^+}$. $\mathcal{ALC}_{R^+}$ is built up from atomic concepts and two kinds of atomic roles, non-transitive roles and transitive roles. Concepts in $\mathcal{ALC}_{R^+}$ are formed using the grammar $A \mid \top \mid \bot \mid \neg C \mid C \sqcap D \mid C \sqcup D \mid \exists R.C \mid \forall R.C \mid \exists T.C \mid \forall T.C,$[1] where A is an atomic concept, C and D are concept expressions, R is a non-transitive role, and T is a transitive role.

The semantics of $\mathcal{ALC}_{R^+}$ is a standard extensional semantics, using an interpretation $\mathcal{I}$ that is a pair $(\Delta^{\mathcal{I}}, \cdot^{\mathcal{I}})$ consisting of a domain and a mapping from concepts to subsets of the domain and from roles to binary relations on the domain (transitive relations for transitive roles, of course). The semantics for concept expressions are given in Table 1. One concept then subsumes another if

---

[1] Throughout the paper, we will be using the syntax of description logics. To translate into the syntax of modal propositional logics, replace $\forall R$ with $\Box_R$ and $\exists R$ with $\Diamond_R$ and perform several other obvious replacements.

Syntax	Semantics
A	$A^{\mathcal{I}} \subseteq \Delta^{\mathcal{I}}$
$\top$	$\Delta^{\mathcal{I}}$
$\bot$	$\emptyset$
$\neg C$	$\Delta^{\mathcal{I}} - C^{\mathcal{I}}$
$C \sqcap D$	$C^{\mathcal{I}} \cap D^{\mathcal{I}}$
$C \sqcup D$	$C^{\mathcal{I}} \cup D^{\mathcal{I}}$
$\exists R.C$	$\{d \in \Delta^{\mathcal{I}} \mid R^{\mathcal{I}}(d) \cap C^{\mathcal{I}} \neq \emptyset\}$
$\forall R.C$	$\{d \in \Delta^{\mathcal{I}} \mid R^{\mathcal{I}}(d) \subseteq C^{\mathcal{I}}\}$
$\exists T.C$	$\{d \in \Delta^{\mathcal{I}} \mid T^{\mathcal{I}}(d) \cap C^{\mathcal{I}} \neq \emptyset\}$
$\forall T.C$	$\{d \in \Delta^{\mathcal{I}} \mid T^{\mathcal{I}}(d) \subseteq C^{\mathcal{I}}\}$

Table 1. Semantics of $\mathcal{ALC}_{R+}$ concept expressions

and only if the extension of the first concept includes the extension of the second in all interpretations.

The semantics of $\mathcal{ALC}_{R+}$ is a simple transformation of the possible world semantics for propositional modal logics. In this transformation elements of the domain correspond to possible worlds, atomic concepts correspond to propositional variables, and roles correspond to modalities. This transformation shows that fragments of $\mathcal{ALC}_{R+}$ correspond to $\mathbf{K}_{(m)}$ and $\mathbf{K4}_{(m)}$. Transitive roles in $\mathcal{ALC}_{R+}$ are used for $\mathbf{K4}_{(m)}$ and non-transitive roles are used for $\mathbf{K}_{(m)}$. $\mathcal{ALC}_{R+}$ can also express formulae in $\mathbf{KT}_{(m)}$ and $\mathbf{S4}_{(m)}$ via the usual encoding that maps $\forall R.C$ into $C \sqcap \forall R.C$, etc.

Determining subsumption in $\mathcal{ALC}_{R+}$ is PSPACE-complete [15], as is the related problem of determining whether a concept in $\mathcal{ALC}_{R+}$ is satisfiable. However, it is possible to build practical description logic systems based on expressive description logics [2,5,11] that have this sort of computationally intractable subsumption. Systems that are based on description logics like $\mathcal{ALC}_{R+}$ generally determine whether a subsumption holds by transforming the subsumption question into a satisfiability question and then attempting to construct a model for this concept, just as a tableaux satisfiability checker for a propositional logic attempts to construct a model for a formula. During this process, various nodes are created, where each node represents an individual (possible world), and tells whether the individual belongs to various concepts (gives values to formulae at this world). This set of concepts is said to form the label of the node—we will use $\mathcal{L}(x)$ to denote the label of a node x. The nodes are connected by modal relationships in a tree fashion, starting at a root node. If a node is related to another node via role R, the second node is called an R-successor of the first.

The basic algorithm starts out with a single node representing an individual (possible world) that must be in the extension of the concept being tested for satisfiability (must have a formula evaluate to true at it). This concept (formula) is expanded to produce simpler concepts that must have the individual in their extension (simpler formulae that evaluate at the world). Disjunctive concepts (formulae) give rise to choice points in the algorithm (branches in the tableau). Existential role concepts, $\exists R.C$, (existential modal formulae) cause the creation of new successor nodes representing other individuals (possible worlds).

Universal role concepts (universal modal formulae) augment the concepts that these individual must belong to (formulae that are true at these possible worlds). In order to guarantee termination, transitive roles (transitive modalities) require filtration or *blocking*: a check to ensure that no other node has the same set of concepts (formulae)—if so, the two nodes can be collapsed into a cycle. If the algorithm constructs a collection of nodes where there are no compound concepts (formulae) that have not been expanded and where there are no obvious contradictions, called *clashes*, at any of the nodes, then the collection of nodes corresponds to a model for the initial concept (formula). If the algorithm fails to construct such a collection then the initial concept (formula) has no model—it is said to be *unsatisfiable*.

The details of the algorithm, including precise termination conditions, are fairly standard, and can be found in [15].

3   Optimisation Techniques

The basic algorithm given above is too slow to form the basis of a useful description logic system. We have therefore investigated and employed a range of known, adapted and novel optimisations that improve the performance of the satisfiability testing algorithm, including lexical normalisation, semantic branching search, boolean constraint propagation, dependency directed backtracking, heuristic guided search, and caching.

Theoretical descriptions of tableaux algorithms generally assume that the concept expression to be tested is in negation normal form, with negations applying only to atomic concepts. This simplifies the (description of the) algorithm but it means that a clash will only be detected when an atomic concept and its negation occur in the same node label. For example, when testing the satisfiability of the concept expression $\exists R.(C \sqcap D) \sqcap \forall R.\neg C$, where C is an atomic concept, a clash would be detected when the algorithm creates an R-successor y because $\{C, \neg C\} \subseteq \mathcal{L}(y)$. However, if C is a concept expression, then the clash would not be detected immediately because $\neg C$ would have been transformed into negation normal form. If C is a large or complex expression this could lead to costly wasted expansion.

This problem is addressed by transforming concept expressions into a lexically normalised form, and by identifying lexically equivalent expressions. All concepts can then be treated equally, whether or not they are atomic, with a clash being detected whenever a concept expression and its negation occur in the same node label.[2] In lexically normalised form, concept expressions consist only of (possibly negated) atomic concepts, conjunction concepts and universal role

---

[2] KRIS addresses the same problem, in a less complete manner, by lazily expanding named concepts, and retaining their names in node labels [1].

concepts: expressions of the form $\exists R.C$ are transformed into $\neg(\forall R.\neg C)$ and expressions of the form $(C_1 \sqcup, \ldots, \sqcup C_n)$ are transformed into $\neg(\neg C_1 \sqcap, \ldots, \sqcap \neg C_n)$, where the $C_1, \ldots, C_n$ are sorted and duplicates are eliminated. The normalisation process can also include simplifications such as $\forall R.\top \longrightarrow \top$, $(\bot \sqcap \ldots) \longrightarrow \bot$ and $(C \sqcap \neg C \sqcap \ldots) \longrightarrow \bot$; in extreme cases the need for a tableau expansion can be completely eliminated by simplifying expressions to $\top$ or $\bot$. Efficiency can be further enhanced by tagging each lexically distinct expression with a unique code so that equivalent expressions can be identified simply by comparing tags.[3]

Tableau expansion of concepts in this form is no more complex than if they are in negation normal form: $\neg(\forall R.C)$ can be dealt with in the same way as $\exists R.\neg C$ and $\neg(C_1 \sqcap, \ldots, \sqcap C_n)$ can be dealt with in the same way as $(\neg C_1 \sqcup, \ldots, \sqcup \neg C_n)$. The expression $\exists R.(C \sqcap D) \sqcap \forall R.\neg C$ would be transformed into $\neg(\forall R.\neg(C \sqcap D)) \sqcap \forall R.\neg C$, and the $\neg(\forall R.\neg(C \sqcap D))$ term would lead directly to the creation of an R-successor whose label contained both C and $\neg C$. As the two occurrences of C will be lexically normalised and tagged as the same concept, a clash will immediately be detected, regardless of the structure of C.

Standard tableaux algorithms are inherently inefficient because they use a search technique based on syntactic branching. When expanding the label of a node x, syntactic branching works by choosing an unexpanded disjunction in $\mathcal{L}(x)$ and searching the different models obtained by adding each of the disjuncts. As the alternative branches of the search tree are not disjoint, there is nothing to prevent the recurrence of an unsatisfiable disjunct in different branches [9]. The resulting wasted expansion could be costly if discovering the unsatisfiability requires the solution of a complex sub-problem. For example, tableau expansion of a node x, where $\{(C \sqcup D_1), (C \sqcup D_2)\} \subseteq \mathcal{L}(x)$ and C is an unsatisfiable concept expression, could lead to the search pattern shown below, where the unsatisfiability of C must be demonstrated twice.

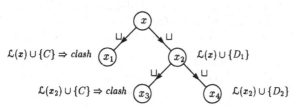

This problem is dealt with by using a semantic branching technique adapted from the Davis-Putnam-Logemann-Loveland procedure (DPLL) commonly used to solve propositional satisfiability (SAT) problems [6,8]. Instead of choosing an unexpanded disjunction in $\mathcal{L}(x)$, a single disjunct D is chosen from one of the unexpanded disjunctions in $\mathcal{L}(x)$. The two possible sub-trees obtained by adding either D or $\neg D$ to $\mathcal{L}(x)$ are then searched. Because the two sub-trees are strictly disjoint, there is no possibility of wasted search as in syntactic branching.

An additional advantage of using a DPLL based search technique is that a great deal is known about the implementation and optimisation of this algorithm.

---

[3] A similar technique is used in KSAT, but without the benefit of tagging [9].

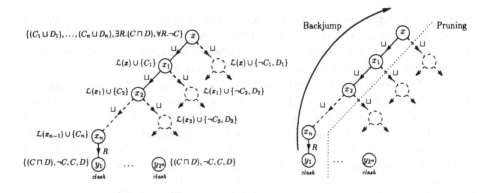

Fig. 1. Thrashing in backtracking search/Backjumping

In particular, both *boolean constraint propagation* and *heuristic guided search* can be used to try to minimise the size of the search tree.

Boolean constraint propagation (BCP) is a technique used to maximise deterministic expansion, and thus pruning of the search tree via clash detection, before performing non-deterministic expansion (branching) [8]. Before semantic branching is applied to the label of a node x, BCP deterministically expands disjunctions in $\mathcal{L}(x)$ which present only one expansion possibility and detects a clash when a disjunction in $\mathcal{L}(x)$ has no expansion possibilities. The number of expansion possibilities presented by a disjunction $(C_1 \sqcup \ldots \sqcup C_n) \in \mathcal{L}(x)$ is equal to the number of disjuncts C_i such that $\neg C_i \notin \mathcal{L}(x)$. In effect, BCP is using the inference rule $\frac{\neg C, C \sqcup D}{D}$ to simplify the expression represented by $\mathcal{L}(x)$.

For example, given a node x such that $\{(C \sqcup (D_1 \sqcap D_2)), (\neg D_1 \sqcup \neg D_2), \neg C\} \subseteq \mathcal{L}(x)$, BCP deterministically expands the disjunction $(C \sqcup (D_1 \sqcap D_2))$ because $\neg C \in \mathcal{L}(x)$. The expansion of $(D_1 \sqcap D_2)$ adds both D_1 and D_2 to $\mathcal{L}(x)$, allowing BCP to identify $(\neg D_1 \sqcup \neg D_2)$ as a clash without any branching having occurred.

Inherent unsatisfiability concealed in sub-problems can lead to large amounts of unproductive backtracking search known as thrashing. The problem is exacerbated when blocking is used to guarantee termination, because blocking may require that sub-problems only be explored after all other forms of expansion have been performed. For example, expanding a node x, where $\mathcal{L}(x) = \{(C_1 \sqcup D_1), \ldots, (C_n \sqcup D_n), \exists R.(C \sqcap D), \forall R.\neg C\}$, would lead to the fruitless exploration of 2^n possible R-successors of x before the inherent unsatisfiability is discovered. The search tree created by the tableau expansion algorithm is illustrated in Fig. 1.

This problem is addressed by adapting a form of dependency directed backtracking called *backjumping*, which has been used in solving constraint satisfiability problems [3] (a similar technique was also used in the HARP theorem prover [13]). Backjumping works by labeling concept expressions with a dependency set indicating the branching points on which they depend. A concept expression $C \in \mathcal{L}(x)$ depends on a branching point when C was added to $\mathcal{L}(x)$

at the branching point or when $C \in \mathcal{L}(x)$ depends an another concept expression $D \in \mathcal{L}(y)$, and $D \in \mathcal{L}(y)$ depends on the branching point. A concept expression $C \in \mathcal{L}(x)$ depends on a concept expression $D \in \mathcal{L}(y)$ when C was added to $\mathcal{L}(x)$ by a deterministic expansion which used $D \in \mathcal{L}(y)$, e.g., if $A \in \mathcal{L}(x)$ was derived from the expansion of $(A \sqcap B) \in \mathcal{L}(x)$, then $A \in \mathcal{L}(x)$ depends on $(A \sqcap B) \in \mathcal{L}(x)$.

When a clash is discovered, the dependency sets of the clashing concepts can be used to identify the most recent branching point where exploring the other branch might alleviate the cause of the clash. The algorithm can then jump back over intervening branching points *without* exploring alternative branches.

For example, when expanding the node x from the previous example, the search algorithm will perform a sequence of n branches, eventually leading to the node x_n with $\{\exists R.(C \sqcap D), \forall R.\neg C\} \subset \mathcal{L}(x_n)$. When $\exists R.(C \sqcap D) \in \mathcal{L}(x_n)$ is expanded the algorithm will generate an R-successor y_1 with $\mathcal{L}(y_1) = \{(C \sqcap D), \neg C\}$. The concept expression $(C \sqcap D)$ will then be expanded and a clash will be detected because $\{C, \neg C\} \subset \mathcal{L}(y_1)$. As neither C nor $\neg C$ in $\mathcal{L}(y_1)$ will have the branching points leading from x to x_n in their dependency sets, the algorithm can either return *unsatisfiable* immediately (if both the dependency sets were empty) or jump directly back to the most recent branching point on which one of C or $\neg C$ did depend. Figure 1 illustrates how the search tree below x is pruned by backjumping, with the number of R-successors explored being reduced by $2^n - 1$.

Heuristic techniques can be used to guide the search in a way which tries to minimise the size of the search tree. A method which is widely used in DPLL SAT algorithms is to branch on the disjunct which has the Maximum number of Occurrences in disjunctions of Minimum Size [8]. By choosing a disjunct which occurs frequently in small disjunctions, this heuristic tries to maximise the effect of BCP. For example, if the label of a node x contains the unexpanded disjunctions $\{C \sqcup D_1, \ldots, C \sqcup D_n\} \subseteq \mathcal{L}(x)$, then branching on C leads to their deterministic expansion in a single step: when C is added to $\mathcal{L}(x)$, all of the disjunctions are fully expanded and when $\neg C$ is added to $\mathcal{L}(x)$, BCP will expand all of the disjunctions. Branching first on any of $D_1, \ldots, D_n$, on the other hand, would only cause a single disjunction to be expanded.

Unfortunately this heuristic interacts adversely with the backjumping optimisation by overriding any "oldest first" order for choosing disjuncts: older disjuncts are those which resulted from earlier branching points and will thus lead to more effective pruning if a clash is discovered [11]. Moreover, the heuristic itself is of little value because it relies for its effectiveness on finding the same disjuncts recurring in multiple unexpanded disjunctions: this is likely in SAT problems, where the disjuncts are propositional variables, and where the number of different variables is usually small compared to the number of disjunctive clauses (otherwise problems would, in general, be trivially satisfiable); it is unlikely in concept satisfiability problems, where the disjuncts are concept expressions, and where the number of different concept expressions is usually large compared to the number of disjunctive clauses. As a result, the heuristic will often discover that all disjuncts have similar or equal priorities, and the guidance it provides is not particularly useful.

An alternative strategy is to employ a heuristic which tries to maximise the effectiveness of backjumping by using dependency sets to guide the expansion. Whenever a choice is presented, the heuristic chooses the concept whose dependency set includes the earliest branching points. This technique can be used both when selecting disjuncts and when ordering R-successors. The use of heuristics is an area of continuing research, but preliminary results suggest that the dependency heuristic is a promising technique.

During a satisfiability check there may be many successor nodes created. These nodes tend to look considerably alike, particularly as the R-successors for a node x each have the same concept expressions for the universal role concepts in $\mathcal{L}(x)$. Considerable time can thus be spent re-performing the computations on nodes that end up having the same label. As the satisfiability algorithm only cares whether a node is satisfiable or not, this time is wasted.

If successors are only created when other possibilities at a node are exhausted, then the entire set of concept expressions that come into a node label can be generated at one time. The satisfiability status of the node is then completely determined by this set of concept expressions. Then, if there exists another node with the same set of initial formulae the two nodes will have the same satisfiability status [7]. Thus work need be done only on one of the two nodes, potentially saving a considerable amount of processing, as not only is the work at one of the nodes saved, but also the work at any of the successors of this node.

The downside of caching is that the dependency information required for backjumping cannot be effectively calculated for the nodes that are not expanded. This happens because the dependency set of any clash detected depend on the dependency sets of the incoming concept expressions, which will differ between the two nodes. Backjumping can still be performed, however, by combining the dependency sets of all incoming concept expressions and using that as the dependency set for the unsatisfiable node.

Another problem with caching is that it requires that nodes, or at least sets of formulae, be retained until the end of a satisfiability test, changing the storage requirements of the algorithm from polynomial to exponential in the worst case.

4   Testing

All the above optimisations are implemented in FaCT and DLP, and we have tested their efficacy on several test suites. (FaCT and DLP differ on their implementation details, how well they implement some of the above optimisations, and the exact heuristic optimisation they do.) All times reported are for runs on machines with approximately the speed of a SPARC Ultra 1.

We would prefer to test on actual description logic knowledge bases, as that is what FaCT and DLP are designed for. However, there are very few description logic knowledge bases that use the more-powerful constructs provided by FaCT and DLP. One test that we have been able to do is to take the GALEN knowledge

		FaCT		DLP		DLP*		KSAT		Kris	
		p	n	p	n	p	n	p	n	p	n
K	branch	6	4	18	12	10	11	8	8	3	3
	d4	>20	8	>20	>20	8	6	8	5	8	6
	dum	>20	>20	>20	>20	10	12	11	>20	15	>20
	grz	>20	>20	>20	>20	>20	>20	17	>20	13	>20
	lin	>20	>20	>20	>20	>20	>20	>20	3	6	9
	path	7	6	>20	>20	7	11	4	8	3	11
	ph	6	7	7	8	6	8	5	5	4	5
	poly	>20	>20	>20	>20	>20	>20	13	12	11	>20
	t4p	>20	>20	>20	>20	6	4	10	18	7	5
KT	45	>20	>20	>20	>20	9	>20	5	5	4	3
	branch	6	4	18	12	16	11	8	7	3	3
	dum	11	>20	>20	>20	9	>20	7	12	3	14
	grz	>20	>20	>20	>20	>20	>20	9	>20	0	5
	md	4	5	3	>20	3	>20	2	4	3	4
	path	5	3	8	8	2	>20	2	5	1	13
	ph	6	7	7	18	5	19	4	5	3	3
	poly	>20	7	>20	8	>20	2	1	2	2	2
	t4p	4	2	>20	>20	1	1	1	1	1	7
S4	45	>20	>20	>20	>20	>20	>20				
	branch	4	4	>20	12	16	12				
	grz	2	>20	>20	>20	0	>20				
	ipc	5	4	10	>20	3	10				
	md	8	4	3	>20	3	>20				
	path	2	1	6	>20	2	>20				
	ph	5	4	4	5	5	15				
	s5	>20	2	19	>20	1	>20				
	t4p	5	3	>20	>20	0	>20				

Table 2. Results for Tableaux'98 Benchmarks

base and construct versions of it that are acceptable to FaCT, DLP and KRIS, by, among other things, making all roles non-transitive and eliminating inclusion axioms. To illustrate the importance of backjumping, caching, and the heuristics, times are also given for DLP with these optimisations disabled—we will refer to this system as DLP*. FaCT and DLP processed the knowledge base in 210 seconds, classifying over two thousand concept definitions requiring tens of thousands of satisfiability tests. Both DLP* and KRIS were unable to complete the processing of the knowledge base in four hours.

Our other testing has been against test suites for propositional modal logics, using the propositional modal logic interface for FaCT and DLP. We have tested against the test suite for the Tableaux'98 propositional modal logic comparison [10] and against a collection of random formulae initially generated by Hustadt and Schmidt [12].

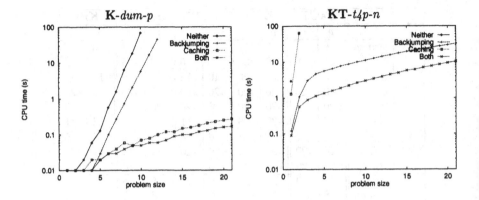

Fig. 2. Solution times for constructed satisfiability problems

The Tableaux'98 test suite consists of several classes of formulae (e.g. *branch*), in both provable (*p*) and non-provable (*n*) forms, for each of **K**, **KT**, and **S4**. For each type of formula, 21 examples of supposedly exponentially increasing difficulty are provided, and the result of a test is the number of the largest formula which the system was able to solve within 100 seconds of CPU time. The results of these tests with FaCT, DLP, DLP*, KSAT [4] and KRIS are summarised in Table 2. In the table, >20 indicates that the hardest problem was solved in less than 100 seconds. (Neither KSAT nor KRIS can reason with transitive roles, so they cannot be used to perform **S4** satisfiability tests.)

In these tests FaCT and DLP outperformed the other systems in this test, with DLP being a clear winner, because of its more-complete caching. Even DLP* performed better than other systems due to the optimizations retained in it. DLP also outperformed the other systems that took part in the the Tableaux'98 comparison [4].

Further analysis of the difference between DLP and DLP*, not presented here because of space limitations, shows that caching is more important than backjumping in these tests, which is more important than the heuristics. In fact the heuristics significantly degraded performance in some cases.

The optimisations in FaCT and DLP often resulted not simply in improved absolute performance but in a different qualitative behaviour. This is illustrated by Fig. 2 which shows the actual solution times for two types of formulae for DLP with backjumping and caching turned off and on. In one of these examples the qualitative improvement is due to caching (a common occurrence); in the other it is due to backjumping (a less-common occurrence).

Our second propositional modal logic test suite uses a method for testing SAT decision procedures that has been adapted for use with propositional modal **K** by Giunchiglia and Sebastiani [9], and further refined by Hustadt and Schmidt [12].

---

[4] The tests here used the original Lisp implementation of KSAT; a much faster C implementation is now available.

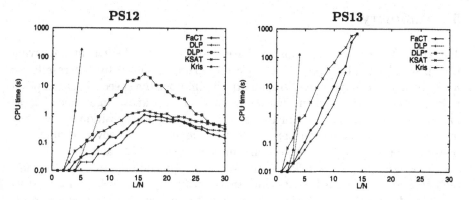

Fig. 3. Median solution times for randomly generated satisfiability problems

The method uses a random generator to produce formulae, with the characteristics of the formulae being controlled by a number of parameters. Each formula is a conjunction of L K-clauses, where a K-clause is a disjunction of K elements, each element being negated with a probability of 0.5. An element is either a modal atom of the form $\forall R.C$, where C is itself a K-clause, or at the maximum modal depth D, a propositional variable chosen from the N propositional variables which appear in the formula. Hustadt and Schmidt used two sets of formulae, denoted **PS12** and **PS13**, choosing $N = 4$ and $N = 6$ respectively, with $K = 3$ and $D = 1$ in both cases. The test sets are created by varying L from N to $30N$, giving formulae with a probability of satisfiability varying from ≈ 1 to ≈ 0, and generating 100 formulae for each integer value of L/N.

The median time required to test the satisfiability of the **PS12** and **PS13** formulae, with a limit of 1,000s per formula, using FaCT, DLP, DLP*, KSAT and KRIS are shown in Fig. 3. It can be seen that in these tests the performance differences between FaCT, DLP and KSAT are much less marked than was the case in the Tableaux'98 tests. This is because the purely propositional problems at depth 1 can always be solved deterministically, and so performance is dependent on the efficiency of propositional reasoning at depth 0. The optimisations which allowed FaCT and DLP to outperform KSAT, notably caching, are of little use with these formulae as there are no hard modal sub-problems.

Although the Tableaux'98 and random test suites show how our optimisations perform on propositional modal logics, neither is very good for our purposes. In particular, for the collection of random formulae most of the computational difficulties have to do with the initial non-modal component. In realistic KBs we expect to encounter problems where the hardness comes from the number of successors that have to be considered and their interaction with the non-modal component. The Tableaux'98 formulae have this form, but there are too few hard collections there to validate our optimisations, and the regular structure of the formulae tends to exaggerate the utility of the caching optimisation, particularly for satisfiable (non-provable) formulae.

5   Summary

The collection of optimizations we have described are effective in improving the speed of modal propositional logic reasoners, as shown by the results we have given above. They can also dramatically improve the speed of subsumption reasoning on description logic knowledge bases. To our knowledge some of these improvements have not been investigated in the modal propositional reasoning literature. The combination appears to be unique and, moreover, results in a powerful reasoner for the propositional modal logics **K**, **KT**, and **S4**.

Unfortunately, the benefits of the various optimizations are not yet completely clear. Caching is best in some areas, backjumping in others. In order to better understand these effects, we continue to analyze and improve the optimisations we have incorporated into our provers. We also plan to create a test suite that emphasizes the modal nature of our description logic. Further, we are embarking on a project to create a description logic system for a description logic that corresponds to a propositional dynamic logic. This project will give us further opportunities to investigate optimisation of satisfiability reasoners.

References

1. F. Baader, E. Franconi, B. Hollunder, B. Nebel, and H.-J. Profitlich. An empirical analysis of optimization techniques for terminological representation systems or: Making KRIS get a move on. In B. Nebel, C. Rich, and W. Swartout, editors, *Principles of Knowledge Representation and Reasoning: Proceedings of the Third International Conference (KR'92)*, pages 270–281. Morgan-Kaufmann Publishers, San Francisco, CA, 1992. Also available as DFKI RR-93-03.

2. F. Baader and B. Hollunder. KRIS: Knowledge representation and inference system. *SIGART Bulletin*, 2(3):8–14, 1991.

3. A. B. Baker. *Intelligent Backtracking on Constraint Satisfaction Problems: Experimental and Theoretical Results*. PhD thesis, University of Oregon, 1995.

4. P. Balsiger and A. Heuerding. Comparison of theorem provers for modal logics — introduction and summary. In H. de Swart, editor, *Automated Reasoning with Analytic Tableaux and Related Methods: International Conference Tableaux'98*, number 1397 in Lecture Notes in Artificial Intelligence, pages 25–26. Springer-Verlag, May 1998.

5. P. Bresciani, E. Franconi, and S. Tessaris. Implementing and testing expressive description logics: a preliminary report. In Gerard Ellis, Robert A. Levinson, Andrew Fall, and Veronica Dahl, editors, *Knowledge Retrieval, Use and Storage for Efficiency: Proceedings of the First International KRUSE Symposium*, pages 28–39, 1995.

6. M. Davis, G. Logemann, and D. Loveland. A machine program for theorem proving. *Communications of the ACM*, 5:394–397, 1962.

7. F. Donini, G. De Giacomo, and F. Massacci. EXPTIME tableaux for $\mathcal{ALC}$. In L. Padgham, E. Franconi, M. Gehrke, D. L. McGuinness, and P. F. Patel-Schneider, editors, *Collected Papers from the International Description Logics Workshop (DL'96)*, number WS-96-05 in AAAI Technical Report, pages 107–110. AAAI Press, Menlo Park, California, 1996.

8. J. W. Freeman. Hard random 3-SAT problems and the Davis-Putnam procedure. *Artificial Intelligence*, 81:183–198, 1996.
9. F. Giunchiglia and R. Sebastiani. A SAT-based decision procedure for $\mathcal{ALC}$. In L. C. Aiello, J. Doyle, and S. Shapiro, editors, *Principles of Knowledge Representation and Reasoning: Proceedings of the Fifth International Conference (KR'96)*, pages 304–314. Morgan Kaufmann Publishers, San Francisco, CA, November 1996.
10. A. Heuerding and S. Schwendimann. A benchmark method for the propositional modal logics k, kt, s4. Technical report IAM-96-015, University of Bern, Switzerland, October 1996.
11. I. Horrocks. *Optimising Tableaux Decision Procedures for Description Logics*. PhD thesis, University of Manchester, 1997.
12. U. Hustadt and R. A. Schmidt. On evaluating decision procedures for modal logic. Technical Report MPI-I-97-2-003, Max-Planck-Institut Für Informatik, Im Stadtwald, D 66123 Saarbrücken, Germany, February 1997.
13. F. Oppacher and E. Suen. HARP: A tableau-based theorem prover. *Journal of Automated Reasoning*, 4:69–100, 1988.
14. P. F. Patel-Schneider. System description: DLP. Bell Labs Research, Murray Hill, NJ, December 1997.
15. U. Sattler. A concept language extended with different kinds of transitive roles. In G. Görz and S. Hölldobler, editors, *20. Deutsche Jahrestagung für Künstliche Intelligenz*, number 1137 in Lecture Notes in Artificial Intelligence, pages 333–345. Springer Verlag, 1996.
16. K. Schild. A correspondence theory for terminological logics: Preliminary report. In *Proceedings of the 12th International Joint Conference on Artificial Intelligence (IJCAI-91)*, pages 466–471, 1991.

Instantiation of Existentially Quantified Variables in Inductive Specification Proofs

Brigitte Pientka* and Christoph Kreitz

Department of Computer Science,Cornell University
Ithaca, NY 14853-7501, U.S.A.
{pientka,kreitz}@cs.cornell.edu

Abstract. We present an automatic approach for instantiating existentially quantified variables in inductive specifications proofs. Our approach uses first-order meta-variables in place of existentially quantified variables and combines logical proof search with rippling techniques. We avoid the non-termination problems which usually occur in the presence of existentially quantified variables. Moreover, we are able to synthesize conditional substitutions for the meta-variables. We illustrate our approach by discussing the specification of the integer square root.

1 Introduction

Constructive type theory [12] offers the unique advantage of total correctness of synthesized programs. In this setting a specification is of the form

$$\forall input. \exists output.\ spec(input, output)$$

where *input* is a vector of arguments, *output* is a result and *spec* is a proposition describing the required relation between them. A program meeting this specification can be extracted from its proof via the proofs-as-programs principle [3]. This style is widely advocated [13] and supported in a number of implementations such as NuPRL [8]. The application of such systems however is limited by its low degree of automation. In order to overcome this drawback, we suggest incorporating techniques from inductive theorem proving.

The first difficult step within a proof is the choice of the appropriate induction scheme. Different induction schemes result in algorithms which differ in their complexity. In this paper we focus on the second crucial step during the induction step, the instantiation of existentially quantified variables. The witness for an existentially quantified variable corresponds to the recursive calls in the program. Sometimes a case split is necessary before decomposing the existential quantifier. The existentially quantified variables are then instantiated according to the cases.

A standard technique to deal with existentially quantified variables is to use meta-variables in place of the existential witness and allow the application of

---

* The research reported is supported by the Gottlieb Daimler and Karl Benz Foundation with a fellowship to the first author.

Jacques Calmet and Jan Plaza (Eds.): AISC'98, LNAI 1476, pp. 247–258, 1998.
© Springer-Verlag Berlin Heidelberg 1998

logical rules that refine the goal. To complete the proof, a unification procedure provides the instantiation of the meta-variables. In inductive specification
proofs, standard unification techniques are not sufficient; we need to rewrite the
expression from the induction conclusion towards the application of the corresponding expression in the induction hypothesis. Both expressions have to be
equal after some rewriting steps. The crucial question is how can we find a chain
of rewriting steps, such that both expressions can be made equal by rewriting in
the presence of meta-variables.

In inductive theorem proving, an annotated rewriting technique, called rippling [7,6], has been used successfully in order to control the rewriting process.
However, only little focus has been devoted to the automatic instantiation of
existentially quantified variables. In this paper we suggest combining the logic
provided by constructive type theory with inductive theorem proving techniques
such as rippling, in order to compute valid instantiations for the existentially
quantified variables. We use first-order meta-variables during proof search within
the sequent calculus. During rippling the meta-variables are treated in the same
way as potential "sink variables". We develop a *reverse rippling match* that
matches the induction conclusion with the induction hypothesis. If this match
is successful, it returns an instantiation for the meta-variables and a rippling
sequence, that rewrites the instantiated induction conclusion to the induction
hypothesis. With this approach we avoid non-termination problems that usually
occur in the presence of existentially quantified variables. Moreover, we check
the consistency of the remaining subgoals under the synthesized substitution.
During this consistency check, we are able to synthesize constraints that form
a case split during the proof. We demonstrate the strength of our approach by
discussing the proof of the integer square root specification.

In Section 2, we give a brief introduction to rippling and discuss the rippling
approaches for dealing with meta-variables. In Section 3 we describe the general
idea of our approach and in 4 we consider the proof of the integer square root. We
show step-by-step how we can derive *conditional substitutions* for the existentially quantified variables. In Section 5 we present a more technical description
of our method using ML-notation. In Section 6 we describe the formalization
in NuPRL. In Section 7 an extension to our technique is presented. We discuss
related work concerning program synthesis in Section 8 and finally in Section 9
we outline future work and draw some conclusions.

2   A Brief Introduction to Rippling

Rippling is an annotated rewriting technique that has been successfully applied in inductive theorem proving. Differences between the induction hypothesis
(*given*) and the induction conclusion (*goal*) are marked by meta-level annotations, called *wave annotations*. Expressions that appear both in the goal and in
the given are called *skeleton*. Expressions that appear in the goal, but not in
the given are called *wave-fronts*. The induction (or recursive) variable that is
surrounded by a wave-front is called *wave-hole*. *Sinks* are parts of the goal which

correspond to universally quantified variables in the given and are marked by $\lfloor sink \rfloor$. We call the annotated rewrite rules *wave-rules*. To illustrate, consider the following wave-rule which is derived from the recursive definitions of $+$.

$$\boxed{\mathsf{s}(\underline{U})}^{\uparrow} + V \overset{R}{\longmapsto} \boxed{\mathsf{s}(\underline{U+V})}^{\uparrow} \tag{1}$$

In this wave-rule $\mathsf{s}(\ldots)$ denotes the wave-front that is marked by a box. The underlined parts $\underline{U}$ resp. $\underline{U+V}$ mark the wave-holes. Intuitively, the position and orientation of the wave-fronts define the direction in which the wave-front has to move within the term tree. An up-arrow $\uparrow$ indicates that the wave-front has to move from a position within the term tree towards the root of the term tree (*rippling-out*). A down-arrow $\downarrow$ moves the wave-front inwards or sideways towards the sink in the term tree, i.e. the sink is filled with the wave-front (*rippling-in*). If rippling succeeds in moving the annotations either to the root of the term tree or to a sink, then rippling terminates successfully and the induction hypothesis matches the induction conclusion. Rippling terminates unsuccessfully if the rewriting process is blocked, i.e. no wave-rule is applicable anymore and the induction hypothesis (given) does not match the induction conclusion (goal).

In Basin & Walsh [2], a calculus for rippling is presented and well-founded measure, called *wave measure* is defined, under which rippling terminates if no meta-variables occur in the goal. The wave measure associates weights to the wave-fronts to measure the width and the size of the wave-front. The *width* of a wavefront is defined by the number of nested function symbols between the root of the wave-front and the wave-hole. The *size* of a wave-front is the number of function symbols and constants in the wave-front. Rewriting is restricted such that each application of a wave-rule is skeleton preserving and measure decreasing according to the defined wave measure.

For instantiating existentially quantified variables via rippling, mainly two approaches have been suggested in the literature. In Bundy *et al.* [6] special existential wave-rules are suggested. Existential wave-rules can be derived from non-existential wave-rules. For example, the existential wave-rule corresponding to wave-rule (1) takes the following form:

$$\exists \boxed{U} : \mathbb{N}. \boxed{U} + V \overset{R}{\longmapsto} \exists \boxed{U'} : \mathbb{N}. \boxed{s(U'+V)}^{\uparrow} \tag{2}$$

Unfortunately, the search problem is exacerbated in the presence of existential quantifiers. Another disadvantage of this approach is that the process of existential rippling does not explicitly record the relationship between the non-existential wave-rule (1) and its existential analogue (2), more precisely the relation between U and U'. This does not matter if we are just interested in provability. In program synthesis, however, the identity of the existential witness plays a vital role of defining the program to be synthesized.

Other approaches [1,10,16] use meta-annotations; the existentially quantified variable x is replaced by $\boxed{F(\underline{X})}$ where capital letters indicate meta-variables. Middle-out reasoning [9] is used in order to instantiate the function F and its argument X. This problem requires a computationally expensive higher-order unification. The presence of higher-order variables also leads to non-termination of rippling, as the width and the size of the wave-front cannot be determined.

Our approach overcomes these drawbacks by combining rippling with first-order theorem proving techniques and using an extended matching procedure for finding the witness for the existentially quantified variables.

3 Automatic Instantiation of Meta-variables

Our research interest is to automate key steps such as the instantiation of the existentially quantified variable in sequent proofs. By the proofs–as–programs paradigm we are then able to extract a program from the proof of a specification. In Figure 1 an overview of our approach is presented. In order to deal with

---

1. **Refinement of the step case formula** by applying sequent rules and using meta-variables in place of existential witness
2. **On atomic subgoals:** Matching of the induction hypothesis with induction conclusion is extended by **reverse rippling match** in order to compute valid substitutions for the meta-variables and a rippling sequence
3. **Consistency check:** test, if all subgoals are true under the found substitution

---

Fig. 1. Automatic instantiation of meta-variables – 3 steps

existentially quantified variables, we suggest first decomposing the existentially quantified formula and use first-order meta-variables in place of the existentially quantified variables. Secondly, an extended matching procedure tries to find a rippling sequence and an instantiation for the meta-variables such that the induction hypothesis term and the corresponding induction conclusion term are equal. Rippling is used to manipulate the atomic subgoals. It rewrites part of the

$$IC \xmapsto{R} C_0 \xmapsto{R} \ldots \xmapsto{R} C_i \xmapsto{R} \ldots \xmapsto{R} C_n \xmapsto{R} IH$$

$$\underbrace{\qquad\qquad}_{\text{rippling}} \qquad \underbrace{\qquad\qquad}_{\text{reverse rippling}}$$

induction conclusion to some formula C_i. Either C_i matches directly with the corresponding induction hypothesis term IH or the rippling sequence $C_i \xmapsto{R} \ldots \xmapsto{R} C_n \xmapsto{R} IH$ is computed by backwards reasoning from the induction hypothesis towards C_i. This process is called *reverse rippling*.

For rippling and reverse rippling we use the *rippling–distance* strategy [11,4]. Each wave-front is mapped to a selected (goal)-sink. The distance between a wave-front and its assigned (goal) sink in the term tree is called *distance measure*. Each application of a wave-rule must reduce this distance until the sink is filled with the wave-front and the distance measure is zero. The main advantage of this approach is the uniform efficient treatment of rippling. Hence, we need not treat the various rippling strategies differently, and it is redundant to mark the wave-fronts with up-arrows ↑ and down-arrows ↓. For uniformly integrating rippling-out the definition of sink has been generalized to arbitrary term positions. By putting a sink around the whole term rippling-out can be simulated by rippling–distance. This approach can be extended easily to incorporate meta-variables. During rippling, meta-variables are treated as potential sink variables. By adopting rippling for meta-variables and using reverse rippling, higher-order variables can be avoided and rippling terminates. For reverse rippling we use the rippling–distance strategy backwards and synthesize a rippling sequence $C_i \xmapsto{R} \ldots \xmapsto{R} C_n \xmapsto{R} IH$ together with a substitution for the meta-variables.

Finally, a consistency checker tests if the remaining subgoals can be proven under the synthesized substitution. If necessary, conditions can be synthesized which constrain the substitution. These *conditional substitutions* form a case split in the proof.

Our approach combines rippling techniques with logical proof search. If it succeeds, the rippling proof is translated back into sequent style.

4   Proving the Specification of the Integer Square Root

In this section we illustrate the automatic instantiation of existentially quantified variables by discussing the proof of the following integer square root specification:

$$\forall x : N.\ \exists y : N.\ y^2 \leq x \wedge x < (y+1)^2 \tag{3}$$

The top–down sequent proof starts by induction on x. We concentrate on the step case of the induction:

$$x : N,\ \exists y : N.\ y^2 \leq x \wedge x < (y+1)^2 \quad \vdash \quad \exists y : N.\ y^2 \leq s(x) \wedge s(x) < (y+1)^2$$

This sequent is proved by a procedure which searches for a rippling proof and an instantiation for the existentially quantified variable y. It proceeds as described in figure 1. In the first step, logical inference rules are used in order to decompose the induction hypothesis and the conclusion. The existentially quantified variable in the conclusion is replaced by a meta-variable Y. This gives us two subgoals, (4) and (5):

$$x : N,\ y : N,\ \underline{y^2 \leq x},\ x < (y+1)^2 \quad \vdash \quad Y^2 \leq \boxed{s(\underline{x})} \tag{4}$$

and

$$x : N,\ y : N,\ y^2 \leq x,\ \underline{x < (y+1)^2} \quad \vdash \quad \boxed{s(\underline{x})} < (Y+1)^2 \tag{5}$$

During the second step, the goal is annotated in such a way, that its skeleton matches the corresponding part in the induction hypothesis. Corresponding parts are underlined in this example. Note, that the meta-variable Y matches the variable y in the given. The following wave-rules are derived from the definitions of functions used in the specification:

$$\boxed{\underline{U}+W)} < \boxed{\underline{V}+W)} \overset{R}{\longmapsto} U < V \tag{6}$$

$$\boxed{s(\underline{U})} + V \overset{R}{\longmapsto} \boxed{s(\underline{U}+V)} \tag{7}$$

$$(\boxed{s(\underline{A})})^2 \overset{R}{\longmapsto} \boxed{\underline{A}^2 + 2A + 1} \tag{8}$$

Note, while in wave-rule (7) and (8) wave-fronts occur on both sides of the wave-rule, in wave-rule (6) the wave-fronts are dropped on the right hand side of the wave-rule. Wave-rules like (6) are usually used to complete proofs[1]. As no wave-rule is applicable, rippling leaves the subgoal unchanged.

The reverse rippling match reasons backwards from the induction hypothesis towards the (rippled) conclusion and extracts a rippling sequence and an instantiation for the meta-variable Y. In the induction hypothesis y is marked as a sink variable. During the first reverse wave-rule application wave-fronts are created by wave-rules of the same type as wave-rule (6). The inserted wave-front is refined step–by–step. This wave-front has to move towards the sink variable y by reverse rippling and results in the instantiation of the meta-variable Y. We start with the second subgoal (5) and try to match $\boxed{s(\underline{x})} < (Y+1)^2$ (induction conclusion IC – goal) and $x < (y+1)^2$ (induction hypothesis IH – given). The induction hypothesis IH represents the final formula in the rippling sequence. In order to determine which formula preceded the induction hypothesis, the wave-rule set is inspected. The given must match the right hand side of a wave-rule. The left hand side of this rule constitutes the predecessor to IH, if further rippling towards the sink variable is possible. By wave-rule (6) it is suggested that the formula before reaching the induction hypothesis $x < (y+1)^2$ is $\boxed{\underline{x}+W} < \boxed{(y+1)^2 + W)}$. This formula can be rippled by wave-rule (8) and W can be instantiated with $2(y+1)+1$. The inserted wave-front moves closer to the sink variable y. Rippling towards the sink variable y is straightforward and the generated rippling sequence is presented in Figure 2. By wave-rule (7) the wave-front s is moved to a position where it surrounds y; therefore our rippling sequence terminates successfully. As no wave-rule is available to justify the final step, we need to prove that the induction conclusion is implied by the last step[2]. Typically these implications can be proven by decision procedure using standard arithmetic. In this case the proof is trivial. Therefore s(y) is a valid substitution for Y.

This substitution and its corresponding rippling sequence (cp. Figure 2) constitute a successful match if the remaining subgoals are true under the found substitution (step 3 in Figure 1). We use a heuristic to check the subgoals and synthesize case splits if necessary. In order to restrict search space, we require

---

[1] We consider here proofs by strong fertilization, which aim for total match between the induction conclusion term and induction hypothesis term.

[2] The rippling rule $C_i \overset{R}{\longmapsto} C_{i+1}$ corresponds to the logical implication $C_{i+1} \Rightarrow C_i$.

$$\boxed{\mathsf{s}(\underline{x})} < (\boxed{\mathsf{s}(\underline{y})} + 1)^2$$

$$\stackrel{R}{\rightleftharpoons} \boxed{x + 2(y+1) + 1} < (\left\lfloor \boxed{\mathsf{s}(\underline{y})} \right\rfloor + 1)^2 \qquad \Big)\; \text{by decision proc.}$$

$$\stackrel{R}{\longmapsto} \boxed{x + 2(y+1) + 1} < \boxed{\mathsf{s}((\lfloor y \rfloor + 1))}^2 \qquad \Big)\; \text{by wr (7)}$$

$$\stackrel{R}{\longmapsto} \boxed{x + 2(y+1) + 1} < \boxed{(\lfloor y \rfloor + 1)^2 + 2(y+1) + 1)} \qquad \Big)\; \text{by wr (8)}$$

$$\stackrel{R}{\longmapsto} x < (\lfloor y \rfloor + 1)^2 \qquad \Big)\; \text{by wr (6)}$$

Fig. 2. Rippling sequence generated by extended matching

that the subgoals can be proven by standard arithmetic, rippling and equational reasoning. If one of the remaining subgoals is not provable by these techniques, we use this subgoal as a constraint of the substitutions. In order to be consistent, we then prove all the remaining subgoals under the negated constraint. In this example, (4) is the only remaining subgoal. We use the subgoal $(\mathsf{s}(y))^2 \leq \mathsf{s}(x)$ as a constraint, and therefore try to prove both cases (4) and (5) for the case $\neg(\mathsf{s}(y))^2 \leq \mathsf{s}(x)$. We use unfolding of the successor and $\neg(\ldots \leq \ldots)$ function s and normal matching, to derive a substitution. We start again by inspecting subgoal (5). Matching between $\mathsf{s}(x) < (Y + 1)^2$ and $\mathsf{s}(x) < (y + 1)^2$ returns the substitution $[y/Y]$. The subgoal (4) is trivially true under this substitution.

By combining logical proof search with rippling and extended matching, we are able to generate automatically a proof for the step case and to synthesize a set of conditional substitutions:$[(\mathsf{s}(y))^2 \leq \mathsf{s}(x), \mathsf{s}(y)/Y], [\neg((\mathsf{s}(y))^2 \leq \mathsf{s}(x)), y/Y]$.

5   An Algorithm for Extended Matching

In this section we present a technical description of the steps performed during the automatic instantiation of existentially quantified variables (see Figure 1). We use ML-notation to describe the algorithm. To automate step 1 standard theorem proving methods can be used. We concentrate on the extended matching procedure (step 2 and 3) which is the core of the automatic instantiation of existentially quantified variables.

Before calling the algorithm extended_matching, the wave-rule set wrs contains potential wave-rules. The wave-rules are annotated dynamically during the rippling and reverse rippling process. The function extended_matching returns (conditional) substitutions and proofs, if

1. The list of subgoals, sgoal_list, contains at least one element sgoal. In this case calling rippling_sequence (conclusion sgoal) (hypothesis sgoal) wrs finds a rippling sequence rip_seq and a substitution subst (Figure 3); (conclusion sgoal) gives us the rippled conclusion and (hypothesis sgoal) returns the corresponding hypothesis.

2. the remaining subgoals, sgoal_list\{sgoal}, are consistent with this substitution (check_subgoals sgoal_list sgoal subst).

The function rippling_sequence (Figure 3) computes a rippling sequence rip_conc $\overset{R}{\longmapsto} \ldots \overset{R}{\longmapsto} \ldots C_i \overset{R}{\longmapsto} \ldots \overset{R}{\longmapsto}$ ind_hyp and a substitution for the meta-variable. First the variable in the induction hypothesis that corresponds to the

```
let rippling_sequence conc ind_hyp wrs =
let ann_conc = annotate conc ind_hyp in
let rip_conc = ripple ann_conc wrs in
let predecessors = poss_predecessors ind_hyp rip_conc wrs in
letrec reverse_rippling_in path predecessors =
        if predecessors = [ ] & filled_sink (hd path) & (hd path) → rip_conc
            then path
            else select p ∈ predecessors
                    let new_predecessors = poss_predecessors p wrs in
                    reverse_rippling_in p::path new_predecessors in
reverse_rippling_in [ ] predecessors
```

Fig. 3. Algorithm for synthesizing rippling sequence $C_n \overset{R}{\longmapsto} \ldots IH$

meta-variable in the induction conclusion is marked with a sink. The function poss_predecessors computes possible predecessors C_{i-1} for a given formula C_i in the rippling sequence by inspecting the wave-rule set wrs. It simulates the backwards application of one rippling rule. The function reverse_rippling_in triggers the reverse rippling process. It proceeds recursively by depth first search if there are several predecessors to a formula. It reasons backwards from the induction hypothesis ind_hyp towards the rippled conclusion rip_conc. In each recursion, the next possible predecessors are computed. It is successful if the sink variable is surrounded by a wave-front, i.e. the sink is filled and this formula C_{i+1} implies C_i. Otherwise backtracking is initiated.

The function check_subgoals checks if the remaining subgoals are provable under the substitution derived by rippling_sequence. If all the remaining subgoals can be proven under the substitution, then the match is successful. If there are subgoals that are not provable by a simple proof procedure simplify[3], then we use one of the remaining unprovable subgoal as a constraint c and prove
1. each g ∈ sgoal_list \{sgoal}. g is provable by simplify under the constraint c
2. each g ∈ sgoal_list. g is provable by simplify and matching under constraint ¬c.
With the presented algorithm we are able to instantiate existentially quantified variables automatically and solve the step case of an inductive proof automatically. Moreover, by a simple heuristic, which is integrated in check_subgoals we are able to synthesize conditional substitutions. These conditions form a case split in the proof. For a more detailed version we refer to [15].

---

[3] simplify is a combination of NUPRL's tactics Unfold, SupInf and Auto. It is a decision procedure that uses standard arithmetic.

6   Integrating into NuPRL

In this section we discuss the integration of our proof method into NuPRL, an interactive, tactic based theorem prover. The described proof procedure is implemented and embedded within the tactic TReverseRipple. If the proof procedure finds a substitution for the meta-variables and all the subgoals can be solved, this proof is translated back into sequent style. Due to the nature of reverse rippling, the eigenvariablen condition is observed, and does not cause any problems for the back translation.

The sequent-style proof in NuPRL for the integer square root specification is presented in Figure 4. In this proof the user specified the induction scheme.

```
⊩ ∀x:N.∃y:N. y²≤x ∧ x<(y+1)²          | \
BY allR                                | ⊩ s(x) < (s(y)+1)²
|                                      | 1 BY Cut ⌜x + 2*(y+1) + 1 < (y+1)² + 2*(y+1) + 1⌝
1. x: N                                | |\
⊩ ∃y:N. y²≤x ∧ x<(y+1)²                | | ⊩ x + 2*(y+1) + 1 < (y+1)² + 2*(y+1) + 1
BY TNatInd ⌜x⌝                         | 1 2 BY Substitution THEN Lemma wave-rule 2
|\                                     | \
| ⊩ ∃y:N. y²≤0 ∧ 0<(y+1)²              | 6. x + 2*(y+1) + 1 < (y+1)² + 2*(y+1) + 1
1 BY exR ⌜0⌝                           | 1 BY Cut ⌜x + 2*(y+1) + 1 < (s(y + 1))²⌝
| |                                    | |\
| ⊩ 0*0 ≤ 0 ∧ 0 < (0+1)*(0+1)          | | ⊩ x + 2*(y+1) + 1 < (s(y + 1))²
1 BY Auto                              | 1 2 BY Substitution THEN Lemma wave-rule 4
\                                      | \
2. ∃y:N. y²≤x ∧ x<(y+1)²               | 7. x + 2*(y+1) + 1 < s((y + 1))²
⊩ ∃y:N. y²≤s(x) ∧ s(x)<(y+1)²          | 1 BY Cut ⌜x + 2*(y+1) + 1 < (s(y)+1)²⌝
BY exL 2 THEN andL 3                   | |\
|                                      | | ⊩ x + 2*(y+1) + 1 < (s(y)+1)²
2. y: N                                | 1 2 BY Substitution THEN Lemma wave-rule 3
3. y² ≤ x                              | \
4. x < (y+1)²                          | 8. x + 2*(y+1) + 1 < (s(y)+1)²
BY Decide ⌜(y+1)² ≤ s(x)⌝ THENW Auto   | 1 BY Unfold ‘s‘ 0 THEN SupInf THEN Auto
|\                                     | 5. ¬( (y+1)² ≤ s(x) )
| 5. (y+1)² ≤ s(x)                     | BY exR ⌜y⌝
1 BY exR ⌜s(y)⌝                        | |
| |                                    | ⊩ y² ≤ s(x) ∧ s(x) < (y+1)²
| ⊩ (s(y))²≤s(x) ∧ s(x)<(s(y)+1)²      | BY andR
1 BY andR                             | |\
| |\                                   | | ⊩ y² ≤ s(x)
| | ⊩ (s(y))²≤s(x)                     | 1 BY Unfold ‘s‘ 0 THEN Auto
1 2 BY Auto                            | \
| \                                    | ⊩ s(x) < (y+1)²
                                       | BY Auto
```

Fig. 4. Proof of the integer square root in NuPRL

First, the universal quantifier on the right hand side is decomposed by tactic allR. The tactic TNatInd 'x' then splits the conjecture into base and step case. Our automation efforts concentrate on the step case of the induction. The step cases of an induction proof are challenging for mainly two reasons: 1) It is harder than in the base case to find the witness for the existentially quantified variable(s). 2) Sometimes, a case split is required and the existentially quantified variable is instantiated according to the different cases. These case splits are not immediately obvious, and often require user insight.

The tactic TReverseRippling synthesizes a conditional substitution set for the existentially quantified variable and translates this information into a sequent proof. The proof displays the subtactics which were applied by TReverseRipple. A case split is performed by tactic Decide based on the conditional substitution set before instantiating the existential quantifier on the right hand side. The tactic exL decomposes the existential quantifier on the left hand side. Applications of tactic andL resp. andR eliminate the conjunction on the left resp. right hand side. The generated rippling sequence is translated back into sequent proof by cut, substitution and lemma applications as described in [11].

7   Extensions to Reverse Rippling

Examples, that can be solved by our method include the specification of quotient remainder, append, half, or last. These examples span the range of specifications usually considered (see [5,10,16]) and do not require any case splits. We also can prove the specification for $\log_2$ which results in a similar proof to the integer square root example. Moreover, we used the extended matching procedure to instantiate universally quantified variables in the hypothesis list. With this extension we are also able to prove the specification of the integer square root and $\log_2$ by using non-standard induction schemes allowing us to synthesize while loops for these two specifications.

To illustrate the flexibility and strength of our technique, we prove the integer square root specification by a different induction scheme[4]. In the step case (induction proceeds over k) we yield the following conjecture:

$$k : N, \forall x, y : N.\ x - y < \mathsf{p}(k) \wedge y^2 \leq x \wedge 0 \leq y \rightarrow \exists n : N.\ y \leq n \wedge n^2 \leq x \wedge x < (n+1)^2$$
$$\vdash \forall x, y : N.\ x - y < k \wedge y^2 \leq x \wedge 0 \leq y \rightarrow \exists n : N.\ y \leq n \wedge n^2 \leq x \wedge x < (n+1)^2$$

Rippling would annotate the term $x - y < k$ to give $\lfloor x \rfloor - \lfloor y \rfloor < \boxed{\mathsf{s}(\mathsf{p}(\underline{k}))}$ as it operates on the induction conclusion. However, no rippling proof for the left hand side of the sequent can be found. Our approach first decomposes the right and left hand side and then uses extended matching to find a match between $x - y < k$ in the hypothesis list and $X - Y < \mathsf{p}(k)$ on the conclusion side. By reverse rippling starting from $x-y < k$ we try to generate a rippling sequence and an instantiation for X and Y. The following additional wave-rules are derived from monotonicity laws and the definition of $-$ is provided:

$$V > 0 \wedge U > 0 \rightarrow \boxed{\mathsf{p}(\underline{U})} < \boxed{\mathsf{p}(\underline{V})} \xmapsto{R} U < V \tag{9}$$

$$\boxed{\mathsf{p}(\underline{U-V})} \xmapsto{R} U - \boxed{\mathsf{s}(\underline{V})} \tag{10}$$

By wave-rule (9) and wave-rule (10) the extended matching procedure generates the following rippling sequence:

$$x - \boxed{\mathsf{s}(\underline{y})} < \boxed{\mathsf{p}(\underline{k})} \quad \xmapsto[{(10)}]{R} \quad \boxed{\mathsf{p}(\underline{x-y})} < \boxed{\mathsf{p}(\underline{k})} \quad \xmapsto[{(9)}]{R} \quad x - y < k$$

---

[4] This induction scheme will result in a more efficient program, namely a while loop. In each iteration y is incremented until $(y+1)^2 > x$.

This example illustrates that the conventional rippling approach [7] to instantiate universally quantified variables in the induction hypothesis by rippling-in is not expressive enough. Moreover, it supports the strength of our approach. The combination of logical proof search and rippling gives us the flexibility to deal with complex logical formulas.

8 Related Work

One of the first approaches to automate the instantiation of existentially quantified variables has been by Biundo [5]. Existentially quantified variables are replaced by Skolem functions which describe the program which is to be synthesized. After induction the formula in the step case is put into clausal form. The synthesis proceeds by *clause-set translations* (e.g. rewriting and case splitting) which induce an AND/OR search space. The work of Kraan *et al.*[10] builds upon the idea to replace existentially quantified variables by skolem functions in order to synthesize logical programs. In order to control better the search space within the inductive step, rippling and middle-out reasoning [9] are used to construct predicate definitions from specifications in classical logic. However, both approaches do not guarantee that the synthesized program is correct, it has to be verified after the synthesis. We believe that constructive type theory provides a firmer mathematical foundation than is found in these systems.

In Smaill & Green [16], an approach for the synthesis of functional programs within the framework of constructive type theory is suggested. This approach builds on higher-order embeddings and higher-order rippling. Middle-out reasoning and higher order embeddings have the disadvantage of a big search space, as rippling in the presence of higher-order function variables does not terminate.

The rippling approaches rely exclusively on this technique and encode logical inference rules as wave-rules. The whole induction conclusion is rippled and these systems aim for a match of the whole induction conclusion with the entire induction hypothesis. The underlying logical calculus is not used to decompose the step case during proof search. This causes major problems when we deal with specifications of more complex formulas, as we illustrated in section 7.

9 Conclusion and Future Work

We have presented an approach for the instantiation of existentially quantified variables which provides a significant degree of automation to proofs in constructive type theory. The key idea is to use first-order meta-variables in place of the existential witness during proof search and rippling and instantiate this meta-variable by an extended matching procedure. Because we reason backwards from the induction hypothesis towards the rippled conclusion by reverse rippling, our approach is highly goal directed and we are able to synthesize lemmata during reverse rippling. By combining logical proof search methods with rippling techniques, we gain flexibility and are able to synthesize case splits which cannot be derived by other comparable systems.

We see our work in a more general framework of matching: two terms t_1 and t_2 match, if the meta-variables in t_1 can be instantiated in such a way that rippling rewrites t_1 towards t_2. This approach allows us to treat meta-variables uniformly. We plan to extend and refine our method in this direction.

Moreover, we plan to explore the use of specially tailored logical proof search methods such as connection method [14] or resolution [17] instead of direct proof search in the sequent calculus. These proof methods are more goal directed. For future research we aim to combine these techniques with a matching procedure which uses rippling and reverse rippling techniques.

References

1. A. Armando, A. Smaill, and I. Green. Automatic synthesis of recursive programs: The proof-planning paradigm. In *Proceedings of the 12th IEEE International Automated Software Engineering Conference*, p 2–9. IEEE Computer Society, 1997.
2. D. Basin and T. Walsh. A calculus for and termination of rippling. *Journal of Automated Reasoning*, 16(2):147–180, 1996.
3. J. L. Bates and R. L. Constable. Proofs as programs. *ACM Transactions on Programming Languages and Systems*, 7(1):113–136, January 1985.
4. W. Bibel, D. Korn, C. Kreitz, F. Kurucz et al.. A multi-level approach to program synthesis. In *Logic Program Synthesis and Transformation*,Springer, 1998.
5. S. Biundo. Automated synthesis of recursive algorithms as a theorem proving tool. In *Proceedings of the 8th ECAI*, 1988.
6. A. Bundy, A. Stevens, F. van Harmelen et al.. Rippling: A heuristic for guiding inductive proofs. *Artificial Intelligence*, 62(2):185–253, August 1993.
7. A. Bundy, F. van Harmelen, A. Smaill et al.. Extensions to the rippling–out tactic for guiding inductive proofs. In *Proceedings of the 10th International CADE*, p 132–146. LNAI, 1990.
8. R. L. Constable, S. F. Allen, H. M. Bromley, and et al. *Implementing Meta-Mathematics with the NuPRLProof Development System*. Prentice-Hall, 1086.
9. Jane T. Hesketh. *Using Middle-Out Reasoning to Guide Inductive Theorem Proving*. PhD thesis, Dept. of Artificial Intelligence, University of Edinburgh, 1991.
10. I. Kraan, D. Basin, and A. Bundy. Logic program synthesis via proof planning. In *Logic Program Synthesis and Transformation*, p 1–14. Springer, 1993.
11. Ferenc Kurucz. Realisierung verschiedender Induktionsstrategien basierend auf dem Rippling-Kalkül. Master's thesis, Technical University Darmstadt, 1997.
12. Per Martin-Löf. Constructive mathematics and computer programming. In *6-th International Congress for Logic, Methodology and Philosophy of Science, 1979*, p 153–175. North-Holland, 1982.
13. B. Nordström, K. Petersson, and J. M. Smith. *Programming in Martin-Löfs Type Theory. An introduction*. Clarendon Press, Oxford, 1990.
14. J. Otten and C. Kreitz. A Uniform Proof Procedure for Classical and Non-classical Logics. *KI-96: Advances in Artificial Intelligence, LNAI* 1137, p 307–319. Springer.
15. B. Pientka. Automating the instantiation of existentially quantified variables. technical report, Dept. of Computer Science, Cornell University,1998.
16. A. Smaill and I .Green. Automating the synthesis of functional programs. Research paper 777, Dept. of Artificial Intelligence, University of Edinburgh, 1995.
17. T. Tammet. A resolution theorem prover for intuitionistic logic. In *Proceedings of the 13th International CADE, LNAI* 1104, p 2–16, 1996.

Knowledge Discovery Objects and Queries in Distributed Knowledge Systems

Zbigniew W. Raś and Jiyun Zheng

University of North Carolina, Dept. of Comp. Science, Charlotte, N.C. 28223, USA
`ras@uncc.edu or jzheng@uncc.edu`

Abstract. The development of many knowledge discovery methods (see [14], [7], [16]) provided us with good foundations to build a kd-Query Answering System ($kdQAS$) for Distributed Knowledge Systems (DKS). By DKS we mean a number of autonomous processing elements (called knowledge systems) that are interconnected by a computer network and that cooperate in their assigned tasks. A knowledge-system we see as a relational database coupled with a discovery layer which is simplified in this paper to a set of rules.

Queries handled by $kdQAS$ are more general than SQL. Also, the queried objects are far more complex than tuples in a relational database. To distinguish them from objects and queries in DBMS, we introduce kd-objects and kd-queries respectively. In general, by $kd - object$ we mean any set of tuples and rules. By $kd - query$ we mean a predicate which queries kd-object in DKS and returns another kd-object for an answer. Our kd-objects may not exist a priori, thus querying them at one site of DKS may require generation, at run time, of new kd-objects either at the same site or at other sites of DKS. So, querying has to major roles: generation of new kd-objects and retrieval of the ones which were generated before.

In relational databases, the result of a query is a relation that can be queried further. This is typically referred to as a closure principle, and it should be one of the most important design principles for $kdQAS$. Our kd-queries satisfy such a closure principle.

Key Words: incomplete information system, cooperative query answering, rough sets, multi-agent system, knowledge discovery.

1 Introduction

In many research fields, such as military, medical, manufacturing, and educational, similar databases are kept at many sites. Each database stores information about local events and the information is expressed in attributes compatible among databases. When similar databases are designed, their events are

Jacques Calmet and Jan Plaza (Eds.): AISC'98, LNAI 1476, pp. 259–269, 1998.

described in terms of the same attributes with some minor or major exceptions. Values of the same attribute may have different generality among databases. The procedures used to collect the data do not have to be the same among databases which means their operational semantics can be different. Also, some attributes might be missing in one database but they occur in many others. Missing attributes lead to problems. Medical doctor may query a database in his hospital to find all patients having certain symptoms, only to realize that one component of that description is a result of a medical test which is missing in the database schema and the same the query cannot be answered. The same query would work on many other databases but the doctor is interested in identifying the patients in his hospital. In this paper we develop a theory for intelligent answering these "locally unreachable" queries.

The task of integrating independently built databases is complicated not only by the differences between the data contents but also by the differences in structure and semantics of the information they contain. The problem is exacerbated when one needs to provide access to such a system for the end-users. For more than 10 years research has been devoted to the question of information retrieval from heterogonous distributed databases. This research has sought to provide integrated access to such databases and has focused on distributed databases, multidatabases, federated databases and their interoperability. The main purpose of integrated access is to enable a number of heterogeneous distributed databases to be queried as if they were a single homogeneous database. Common practice in integrating database systems involves manual integration of each database schema into a global schema [1]. This approach does not work when the number of database systems is large. Navathe and Donahoo [9] propose to allow the database designers to develop a metadata description of their database schema. The collection of metadata descriptions can be automatically processed by a schema builder to create a partially integrated global schema. The heterogeneity problem can be eliminated (see [8]) by using an intermediate model that controls the knowledge translation from a source database or knowledgebase. The intermediate model they developed is based on the concept of abstract knowledge representation. It has two components: a modeling behavior which separates the knowledge from its implementation, and a performative behavior which establishes context abstraction rules over the knowledge. In this paper we propose to handle the heterogeneity problem among independently built databases through the use of discovery layers.

First, we introduce the notion of a Distributed Information System (DIS) which is the main vehicle for development of a Distributed Knowledge System (DKS). Basically to transform DIS to DKS we need to add a discovery layer to every site of DIS. Discovery layers are homogeneous and for simplicity reason they are simplified in this paper to sets of rules. The content of discovery layers may constantly change because by querying one site of DKS, its other sites can be asked for help to answer the query. Rules in this paper are seen as operational definitions providing a commonly sharable information among independently built information systems. So, the transfer of rules from one site of DKS to another site does not cause much problems.

All distributed DBMSs support deduction of information. Our goal is to build a knowledge discovery based Query Answering System ($kdQAS$) for each site of DKS which satisfies the closure principle (the response for a query can be queried further). The access to $kdQAS$ is through WWW.

Predicate logic is the vehicle chosen to represent knowledge in DKS and queries in $kdQAS$. Many other representations are, of course, possible. We have chosen predicate logic because of the need to manipulate queries and rules syntactically without changing their semantical meaning. This syntactical manipulation of queries will be handled by $kdQAS$. By designing an axiomatic system which is sound and complete we are certain that queries we manipulate will not change their semantical meaning. Clearly, this property is very much needed. Without it, we may be looking for an answer to queries which are semantically different from the queries asked by the user. Such a situation has to be avoided.

2   Distributed Information Systems

In this section, we introduce the notion of a Distributed Information System (DIS) which is the main vehicle for development of a Distributed Knowledge System (DKS). Basically to transform DIS to DKS we need to add a discovery layer to every site of DIS. A discovery layer is simplified in our paper to the set of rules. Its content may constantly change because by querying DKS we may discover rules at one site and store them at other sites of DKS.

In this paper, we consider two types of queries called local and global (locally unreachable). Global queries are queries which can be resolved only through the interaction of sites (exchanging knowledge between them) in DKS. Local queries are resolved entirely by a single site of DKS.

So, let us start with basic definitions.

By an information system S we mean a sequence (X, A, V, h), where X is a finite set of objects, A is a finite set of attributes, V is the set-theoretical union of domains of attributes from A, and h is a classification function which describes objects in terms of their attribute values (see [12], [13]). We assume that:

- $V = \bigcup \{V_a : a \in A\}$ is finite,
- $V_a \cap V_b = \emptyset$ for any $a, b \in A$ such that $a \neq b$,
- $h : X \times A \longrightarrow V \cup 2^V$ where $h(x, a) \in V_a$ or $h(x, a) = V_a$ for any $x \in X$ and $a \in A$.

Attribute a is called incomplete in S if there is $x \in X$ such that $h(x, a) = V_a$. By $In(X, a)$ we mean the set $\{x \in X : h(x, a) = V_a\}$. We will be referring to this set when we give the definition of $kdQAS$.

Let $S_1 = (X_1, A_1, V_1, h_1)$, $S_2 = (X_2, A_2, V_2, h_2)$ be information systems. We say that S_2 is a subsystem of S_1 if $X_2 \subseteq X_1$, $A_2 \subseteq A_1$, $V_2 \subseteq V_1$ and $h_2 \subseteq h_1$. By $h_2 \subseteq h_1$ we mean that either $h_2(x, a) = h_1(x, a)$ or [$h_2(x, a) \in V_a$ and $h_1(x, a) = V_a$].

We use a table-representation of a classification function h which is naturally identified with an information system $S = (X, A, V, h)$. For simplicity reason, instead of a set V_a we place the blank symbol which is interpreted here as *all values in V_a are possible*. For example, let $S = (X, A, V, h)$ is an information system where $X = \{a_1, a_6, a_8, a_9, a_{10}, a_{11}, a_{12}\}$, $A = \{C, D, E, F, G\}$ and $V = \{e_1, e_2, e_3, f_1, f_2, g_1, g_2, g_3, c_1, c_2, d_1, d_2\}$. Additionally, we assume here that $V_E = \{e_1, e_2, e_3\}$, $V_F = \{f_1, f_2\}$, $V_G = \{g_1, g_2, g_3\}$, $V_C = \{c_1, c_2\}$, and $V_D = \{d_1, d_2\}$. Then, the function h defined by Table 1 is identified with information system S.

X_2	F	C	D	E	G
$a1$	$f1$	$c1$	$d2$	$e2$	$g1$
$a6$	$f2$		$d2$	$e3$	$g2$
$a8$	$f1$	$c2$			$g1$
$a9$	$f2$	$c1$			$g1$
$a10$	$f2$	$c2$	$d2$	$e3$	$g1$
$a11$	$f1$		$d1$	$e3$	$g2$
$a12$	$f1$	$c1$	$d1$	$e3$	$g1$

Table 1. Information System S

By a Distributed Information System [13] (*DIS*) we mean a pair $DS = (\{S_i\}_{i \in I}, L)$ where:

- $S_i = (X_i, A_i, V_i, h_i)$ is an information system for any $i \in I$,
- L is a symmetric, binary relation on the set I,
- I is a set of sites.

Systems S_{i1}, S_{i2} (or sites $i1, i2$) are called neighbors in a distributed information system DS if $(i1, i2) \in L$. The transitive closure of L in I is denoted by L^+.

A distributed information system $DS = (\{S_i\}_{i \in I}, L)$ is consistent if:

- $(\forall i)(\forall j)(\forall x \in X_i \cap X_j)(\forall a \in A_i \cap A_j)[(x, a) \in Dom(h_i) \cap Dom(h_j) \longrightarrow h_i(x, a) = h_j(x, a)]$.

We assume here that any site of DIS can be queried either for objects or for knowledge. Knowledge in this paper is simplified to a set of rules. Syntactically, a query is built from values of attributes belonging to $V = \bigcup\{V_i : i \in I\}$. A query is called local for a site i, if it is built from values in V_i. Otherwise, it is called global (locally unreachable) for i. Both, local and global queries will be handled by kd-Query Answering System ($kdQAS$). In order to resolve a global query at site i, a transfer of newly discovered knowledge at other sites of DIS to a site i will be needed. This knowledge is stored in discovery layer of site i. If a queried information system S in DIS is incomplete, then a new query has to be invoked and answered first. To be more precise, system S is queried first for certain consistent rules to be discovered locally at S and at its remote sites. Next, this newly discovered set of consistent rules is treated as a new local query which, when applied to the system S, transforms S to a more complete system. In the final step, this new system is queried by the original query. If this original query is global and it is submitted to site i, a transfer of newly discovered knowledge from other sites of DIS to a site i is needed.

In relational databases the result of a query is a relation which can be queried further. Clearly, our $kdQAS$ should have a similar property. To achieve this, we will extend DIS to a Distributed Knowledge System (DKS) where each site is defined as an information system coupled with a discovery layer simplified in this paper to a set of rules. Before we proceed any further, let us give an example of a kd-query.

For instance, SQL-type query

select * from $Flights$
where $airline = "Delta"$
and $departure\_time = "morning"$
and $departure\_airport = "Charlotte"$
and $aircraft = "Boeing"$

is global (locally unreachable) for a database
$Flights(airline, departure\_time, arrival\_time, departure\_airport, arrival\_airport)$
because of the attribute $aircraft$. In order to resolve it, a transfer of newly discovered definitions of $aircraft = "Boeing"$ from other sites of DIS to a site i is needed. So, this query can be called a knowledge discovery query (kd-query).

We begin with a definition of $s(i)$-terms and their standard interpretation M_i in a distributed information system $DS = (\{S_j\}_{j \in I}, L)$, where $S_j = (X_j, A_j, V_j, h_j)$ and $V_j = \bigcup\{V_{ja} : a \in A_j\}$, for any $j \in I$.

By a set of $s(i)$-terms we mean a least set T_i such that:

- $0, 1 \in T_i$,
- $(a, w) \in T_i$ for any $a \in A_i$ and $w \in V_{ia}$,
- $\sim (a, w) \in T_i$ for any $a \in A_i$ and $w \in V_{ia}$,
- if $t_1, t_2 \in T_i$, then $(t_1 + t_2), (t_1 * t_2) \in T_i$.

We say that:

- $s(i)$-term t is *atomic* if $t \in \{(a, w), \sim (a, w), \mathbf{0}, \mathbf{1}\}$ where $a \in B_i \subseteq A_i$ and $w \in V_{ia}$
- $s(i)$-term t is *positive* if it is of the form $\prod\{(a, w) : a \in B_i \subseteq A_i$ and $w \in V_{ia}\}$
- $s(i)$-term t is *primitive* if it is of the form $\prod\{t_j : t_j$ is atomic $\}$
- $s(i)$-term is in *disjunctive normal form* (DNF) if $t = \sum\{t_j : j \in J\}$ where each t_j is primitive.

Standard interpretation M_i of $s(i)$-terms in a distributed information system $DS = (\{S_j\}_{j \in I}, L)$ is defined as follows:

- $M_i(\mathbf{0}) = \emptyset$, $M_i(\mathbf{1}) = X_i$,
- $M_i((a, w)) = \{x \in X_i : w = h_i(x, a)\}$ for any $w \in V_i$,
- $M_i(\sim (a, w)) = \{x \in X_i : \sim (w \in h_i(x, a))\}$ for any $w \in V_i$,
- if t_1, t_2 are s(i)-terms, then
$$M_i(t_1 + t_2) = M_i(t_1) \cup M_i(t_2),$$
$$M_i(t_1 * t_2) = M_i(t_1) \cap M_i(t_2).$$

3   Distributed Knowledge Systems

In this section we introduce the notion of i-rules, kd-objects and, kd-queries. We define a Distributed Knowledge System (DKS) and introduce the notion of its consistency. We also provide a basic architecture of DKS. Finally, we describe the process of querying kd-objects at site i of DKS.

The definition of $s(I)$-terms is similar to the definition of $s(i)$-terms with only one difference. Namely, the set V_i in the definition of $s(i)$-terms is replaced by the set $V = \bigcup\{V_j : j \in I\}$. The meaning of $s(I)$-terms, which forms the foundations for $kdQAS$, is clarified after kd-objects and kd-queries are defined. It depends on the site of DKS it is interpreted in.

By (k, i)-rule in $DS = (\{S_j\}_{j \in I}, L)$, $k, i \in I$, we mean a triple (c, t, s) such that:

- $c \in V_k - V_i$,
- t, s are $s(k)$-terms in DNF and they both belong to $T_k \cap T_i$,
- $M_k(t) \subseteq M_k(c) \subseteq M_k(t + s)$.

By (i, i)-rule in $DS = (\{S_j\}_{j \in I}, L)$, $i \in I$, we mean a triple (c, t, s) such that:

- $c \in V_{ia}$,
- t, s are $s(i)$-terms in DNF built from values of attributes belonging to $V_i - V_{ia}$,
- $M_i(t) \subseteq M_i(c) \subseteq M_i(t + s)$.

For simplicity reason both (i, i)-rules and (k, i)-rules are called i-rules.

System $DS = (\{(S_i, D_i, kdQAS_i, Agent_i)\}_{i \in I}, L)$, where (for any $i \in I$):

- D_i is a discovery layer simplified to a consistent set of i-rules,
- S_i is an information system (a database),

- $kdQAS_i$ is a query answering system for a site i,
- $Agent_i$ is a set of knowledge discovery based client/server protocols.

is called a Distributed Knowledge System (DKS).

If there is $i \in I$ such that S_i is incomplete, then DKS is called incomplete Distributed Knowledge System. Figure 1 shows its basic architecture (WWW interface is added).

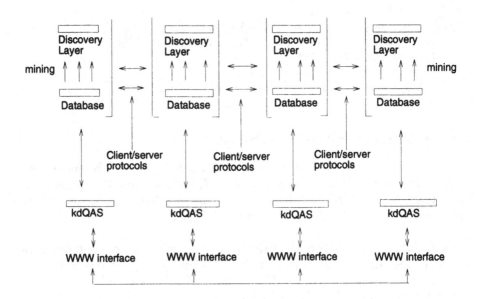

Fig. 1. Distributed Knowledge System

By kd-object at site i of $DS = (\{(S_i, D_i, kdQAS_i, Agent_i)\}_{i \in I}, L)$, we mean any subsystem of (S_i, D_i) or saying another words any subsystem of S_i coupled with a consistent set of i-rules.

By kd-query at site i, $i \in I$, we mean either any $s(I)$-term or any consistent set of (i, i)-rules.

Now, we describe the process of querying a kd-object at site i of DKS. In this paper we consider four options for a kd-query q:

- q is a *primitive* $s(i)$-term,
- q is an $s(i)$-term in DNF,
- q is a *primitive* $s(I)$-term,
- q is an $s(I)$-term in DNF.

Let us assume that q_i is a *primitive* $s(i)$-term. First, $kdQAS_i$ identifies all incomplete attributes among attributes used in q_i. Let us say that $a_{i1}, a_{i2}, ..., a_{ik}$ is the list of all these attributes. In the second step, $kdQAS_i$ finds all certain rules at $S_i = (X_i - In(X_i, a_{ij}), A_i, V_i, h_i)$ describing attribute a_{ij} in terms of attributes from $A_i - \{a_{ij}\}$. Let's denote these rules by R_{ij}. This process is repeated for every $j \in \{1, 2, 3, ..., k\}$. In the third step, $kdQAS_i$ applies the rules in R_{ij} to find the value of a_{ij} for a maximal number of objects from $(In(X_i, a_{ij}), A_i, V_i, h_i)$. These values are stored in a temporary matrix M_{ij}, for every $j \in \{1, 2, 3, ..., k\}$. In the fourth step, all values stored in temporary matrices M_{ij} are moved to corresponding locations in a system $S_i = (X_i, A_i, V_i, h_i)$ to replace some of its null values. Let us denote the resulting information system by S_i^1. At this point, $kdQAS_i$ goes back again to the first step and the process continues to iterate until all newly generated temporary matrices are empty. Let us assume that after m iterations, the process will stop and denote the resulting information system by S_i^m. In the final step, $kdQAS_i$ finds all objects in S_i^m satisfying the description q_i.

If q_i is an $s(i)$-term in DNF, then the strategy described for primitive $s(i)$-terms is repeated for every disjunct of q_i.

Assume now that q_i is a *primitive* $s(I)$-term which is global for S_i. First, $kdQAS_i$ identifies all incomplete attributes in A_i among attributes used in q_i. Let us say that $a_{i1}, a_{i2}, ..., a_{ik}$ is the list of all these attributes. In the second step, $kdQAS_i$ finds all certain rules at $S_i = (X_i - In(X_i, a_{ij}), A_i, V_i, h_i)$ describing attribute a_{ij} in terms of attributes from $A_i - \{a_{ij}\}$. Let's denote these rules by R_{ij}. This process is repeated for every $j \in \{1, 2, 3, ..., k\}$. In the third step, $kdQAS_i$ applies the rules in R_{ij} to find the value of a_{ij} for a maximal number of objects from $(In(X_i, a_{ij}), A_i, V_i, h_i)$. These values are stored in a temporary matrix M_{ij}, for every $j \in \{1, 2, 3, ..., k\}$. In the fourth step, all values stored in temporary matrices M_{ij} are moved to corresponding locations in a system $S_i = (X_i, A_i, V_i, h_i)$ to replace some of its null values. Let us denote the resulting information system by S_i^1. At this point, $kdQAS_i$ goes back again to the first step and the process continues to iterate until all newly generated temporary matrices are empty. Let us assume that after m iterations, the process will stop and denote the resulting information system by S_i^m. Now, $kdQAS_i$ identifies all attributes used in q_i which do not belong to A_i (we call them concepts for the site i). $Agent_i$ sends request to other sites of DKS to find rules describing all these concepts in terms of attributes from A_i. These newly discovered rules are used to approximate query q_i by a new local query p_i for S_i. Also, these rules are stored in the discovery layer D_i which can be used (for new global queries) by $kdQAS_i$ before any help from $Agent_i$ is requested. In the final step, $kdQAS_i$ finds all objects in S_i^m satisfying the description p_i.

If q_i is an $s(I)$-term in DNF, then the strategy described for primitive $s(I)$-terms is repeated for every disjunct of q_i.

4   Interpretation of Primitive *kd*-Queries

In this section we propose a class of i-based operational semantics for a *kd*-query q, assuming that q is an $s(I)$-term. Next, for this class of operational semantics we give a complete and sound set of axioms and rules.

Standard interpretation M_i, introduced in a previous section, shows how to interpret i-queries in DIS. Now, we propose how to interpret non-local queries (called global) bounded in this paper to the class of *primitive* $s(I)$-terms. We call them primitive *kd*-queries.

Parentheses will be used, if necessary, in the obvious way. As will turn out later, the order of a sum or product is immaterial. So, we will abbreviate finite sums and products as $\sum\{t_j : j \in J\}$ and $\prod\{t_j : j \in J\}$, respectively. Intentionally, $s(I)$-terms are names of certain features of parts being processed by *kdQAS*, more complex than those expressed by constants.

Let M_i be a standard interpretation of $s(i)$-terms in $DS = (\{S_j\}_{j \in I}, L)$.

Let $C_i = \bigcup\{V_k : k \in I - \{i\}\} - V_i$ is a set of concepts for S_i. By i-based operational semantics for $s(I)$-terms in $DS = (\{(S_i, D_i)\}_{i \in I}, L)$, $S_i = (X_i, A_i, V_i, h_i)$ and $V_i = \bigcup\{V_{ia} : a \in A_i\}$, we mean the interpretation M_{i,K_i} such that:

- $M_{i,K_i}(0) = \emptyset$, $M_{i,K_i}(1) = X_i$
- for any $w \in V_i$,
  $$M_{i,K_i}(w) = M_i(w),$$
  $$M_{i,K_i}(\sim w) = X_i - M_{i,K_i}(w)$$

- for any $w \in C_i$,
  $$M_i(w) = \{x \in X_i : (\exists t, s)((w, t, s) \in D_i \wedge x \in M_i(t))\}$$
  $$M_i(\sim w) = \{x \in X_i : (\exists t, s)((w, t, s) \in D_i \wedge x \notin M_i(s))\}$$

- for any $s(I)$-terms t_1, t_2
  $$M_{i,K_i}(t_1 + t_2) = M_{i,K_i}(t_1) \cup M_{i,K_i}(t_2),$$
  $$M_{i,K_i}(t_1 * t_2) = M_{i,K_i}(t_1) \cap M_{i,K_i}(t_2),$$
  $$M_{i,K_i}(\sim (t_1 + t_2)) = (\sim M_{i,K_i}(t_1)) \cap (\sim M_{i,K_i}(t_2)),$$
  $$M_{i,K_i}(\sim (t_1 * t_2)) = (\sim M_{i,K_i}(t_1)) \cup (\sim M_{i,K_i}(t_2)),$$
  $$M_{i,K_i}(\sim\sim t) = M_{i,K_i}(t).$$

- for any $s(I)$-terms t_1, t_2
  $$M_{i,K_i}(t_1 = t_2) = (\text{ if } M_{i,K_i}(t_1) = M_{i,K_i}(t_2) \text{ then } T \text{ else } F)$$
  where T stands for $True$ and F for $False$

From the point of view of S_i the interpretation M_{i,K_i} represents a pessimistic approach to query evaluation. It means that $M_{i,K_i}(t)$ is interpreted as a set of objects in X_i which have the property t for sure. We are not retrieving here objects which might have property t.

Let us adopt the following set A of Axiom Schemata:

A1. Substitutions of the axioms of distributive lattices for $s(I)$-terms and the axioms of equality

A2. $\sim w * w = 0$ for any constant w

A3. $\sim w + w = 1$ for any $w \in V_i$

A4. for each $w \in V_i$ there is a subset $w_1, w_2, ..., w_n$ of V_i such that
$\sim w = w_1 + w_2 + ... + w_n$

A5. $v_1 * v_2 = 0$
if $v_1, v_2 \in V_{ia}$ for some $a \in A_i$

A6. for any $s(I)$-term t,
$\sim 0 = 1, \sim 1 = 0, 1 + t = 1, 1 * t = t, 0 * t = 0, 0 + t = t, \sim\sim t = t$

A7. for any $w \notin V_i$
$w = \sum\{t : [w, t, s] \in D_i\}$

A8. for any $w \notin V_i$
$\sim w = \sum\{t : [\sim w, t, s] \in D_i\}$

A9. $\sim (t_1 + t_2) = (\sim t_1) * (\sim t_2)$

A10. $\sim (t_1 * t_2) = (\sim t_1) + (\sim t_2)$

A11. Substitutions of the propositional calculus axioms

The rules of inference for our formal system are the following:

R1. from $(\alpha \Rightarrow \beta)$ and α we can deduce β for any formulas α, β

R2. from $t_1 = t_2$ we can deduce $t(t_1) = t(t_2)$,
where $t(t_1)$ is a $s(I)$-term containing t_1 as a subterm and $t(t_2)$ comes from $t(t_1)$ by replacing some of the occurrences of t_1 with t_2.

We write $A \vdash (t_1 = t_2)$ if there exists a derivation from a set A of formulas as premises to the formula $(t_1 = t_2)$ as the conclusion.

We write $A \models (t_1 = t_2)$ to denote the fact that A semantically implies $(t_1 = t_2)$, that is, for any i-standard interpretation M_{i,K_i} of $s(I)$-terms in DKS we have $M_{i,K_i}(t_1 = t_2) = T$.

Theorem 1. (Completeness). For any $s(I) - terms\ t_1, t_2$, if $A \models (t_1 = t_2)$ then $A \vdash (t_1 = t_2)$.

The above completeness theorem gives us the set of axioms which is sound and sufficient to transform any global $s(I)$-term to its equivalent DNF. So, the set of kd-queries does not have to be bounded to primitive terms.

5   Conclusion

This paper presents a methodology and theoretical foundations of a kd-Query Answering System for DKS which is partially implemented at UNC-Charlotte on a cluster of SPARC 20 workstations.

References

1. Batini, C., Lenzerini, M., Navathe, S., "A comparative analysis of methodologies for database schema integration", in *ACM Computing Surveys*, Vol 18, No. 4, 1986, 325-364

2. Bosc, P., Pivert, O., "Some approaches for relational databases flexible querying", in *Journal of Intelligent Information Systems*, Kluwer Academic Publishers, Vol. 1, 1992, 355-382

3. Chu, W.W., "Neighborhood and associative query answering", in *Journal of Intelligent Information Systems*, Kluwer Academic Publishers, Vol. 1, 1992, 355-382

4. Chu, W.W., Chen, Q., Lee, R., "Cooperative query answering via type abstraction hierarchy", in *Cooperating Knowledge-based Systems* (ed. S.M. Deen), North Holland, 1991, 271-292

5. Cuppers, F., Demolombe, R., "Cooperative answering: a methodology to provide intelligent access to databases", in *Proceedings 2nd International Conference on Expert Database Systems*, Virginia, USA, 1988

6. Gaasterland, T., Godfrey, P., Minker, J., "An overview of cooperative answering", *Journal of Intelligent Information Systems*, Kluwer Academic Publishers, Vol. 1, 1992, 123-158

7. Grzymala-Busse, J., *Managing uncertainty in expert systems*, Kluwer Academic Publishers, 1991

8. Maluf, D., Wiederhold, G., "Abstraction of representation for interoperation", in *Proceedings of Tenth International Symposium on Methodologies for Intelligent Systems*, LNCS/LNAI, Springer-Verlag, No. 1325, 1997, 441-455

9. Navathe, S., Donahoo, M., "Towards intelligent integration of heterogeneous information sources", in *Proceedings of the Sixth International Workshop on Database Re-engineering and Interoperability*, 1995

10. Pawlak, Z., "Rough Sets - theoretical aspects of reasoning about data", Kluwer Academic Publishers, 1991

11. Pawlak, Z., "Rough sets and decision tables", in *Proceedings of the Fifth Symposium on Computation Theory*, Springer Verlag, Lecture Notes in Computer Science, Vol. 208, 1985, 118-127

12. Ras, Z., "Resolving queries through cooperation in multi-agent systems", in *Rough Sets and Data Mining*, (Eds. T.Y. Lin, N. Cercone), Kluwer Academic Publishers, 1997, pp. 239-258

13. Ras, Z., Joshi, S., "Query approximate answering system for an incomplete DKBS", in *Fundamenta Informaticae Journal*, IOS Press, Vol. 30, No. 3/4, 1997, 313-324

14. U.M. Fayyad, G. Piatetsky-Shapiro, P. Smyth, R. Uthurusamy, "Advances in Knowledge Discovery and Data Mining", AAAI Press/MIT Press, 1996

15. Ras, Z., "Collaboration control in distributed knowledge-based systems", in *Information Sciences Journal*, Elsevier, Vol. 96, No. 3/4, 1997, pp. 193-205

16. Skowron, A., "Boolean reasoning for decision rules generation", in *Methodologies for Intelligent Systems, Proceedings of the 7th International Symposium on Methodologies for Intelligent Systems*, (eds. J. Komorowski, Z. Ras), Lecture Notes in Artificial Intelligence, Springer Verlag, No. 689, 1993, 295-305

ALLTYPES:
An ALgebraic Language and TYPE System

Fritz Schwarz

GMD, Institute SCAI, 53754 Sankt Augustin, Germany
fritz.schwarz@gmd.de,
http://www.gmd.de/SCAI/people/schwarz.html

Abstract. The software system ALLTYPES provides an environment
that is particularly designed for developing computer algebra software
in the realm of differential equations. Its most important features may
be described as follows: A set of about thirty parametrized algebraic
types is defined. Data objects represented by these types may be manipulated by more than one hundred polymorphic functions. Reusability of
code is achieved by genericity and inheritance. The user may extend the
system by defining new types and polymorphic functions. A language
comprising seven basic language constructs is defined for implementing
mathematical algorithms. The easy manipulation of types is particularly
supported by ALLTYPES. To this end a special portion of the language
that is enclosed by a pair of absolute bars is dedicated to manipulating
typed objects, i. e. user-defined or automatic type coercions. Type in-
quiries are also included in the language. A small amount of parallelism
is supported in terms of two language constructs **pand** and **por** where
the letter **p** indicates a parallel version of the respective logical function.
Currently ALLTYPES is implemented in Reduce and Macsyma (to be
completed soon). Software implemented on top of ALLTYPES should
work independent of the underlying computer algebra language.

1 Organization of Computer Algebra Software

The origin of the software described in this article is the desire to provide an environment for implementing high quality computer algebra software for working
with differential equations. Areas of its application are for example the symmetry analysis of ordinary and partial differential equations, finding closed form
solutions of ordinary differential equations and computations in $\mathcal{D}$-modules, for
example Janet base algorithms.

Before entering into the details of ALLTYPES, the following question will
be dealt with: Why not work with one of the popular systems like for example
Axiom, Macsyma, Maple or Reduce directly? Apart from the first system, the
answer is obvious. In our terminology, Macsyma, Maple or Reduce are computer
algebra *languages* because they neither allow the *structuring* of large pieces of
software in terms of an algebraic type system, nor do they allow the reuse of
code by some kind of inheritance mechanism. ALLTYPES provides a collection of

Jacques Calmet and Jan Plaza (Eds.): AISC'98, LNAI 1476, pp. 270–283, 1998.
© Springer-Verlag Berlin Heidelberg 1998

types especially designed for working with differential equations and differential algebra in general. This fine-structured type system requires special tools for manipulating objects of these various types easily; they are made available by ALLTYPES in terms of a specialized portion of the language. Furthermore this language defines a small number of powerful control constructs that are especially well suited for writing computer algebra code. Software developed on t op of ALLTYPES has the following characteristic properties.

▷ Due to the powerful language constructs, code may be written such that each line may be executed individually, its action may be specified in mathematical terms, and the result may be checked against this specification;

▷ Whenever a bug occurs, in almost all cases the type of this bug in a newly implemented piece of software is such that its repair does not create new bugs somewhere else.

These features make ALLTYPES into an environment that has turned out to be better suited for implementing differential equations software than other systems. Nevertheless the current implementation is considered only as a first version of a more advanced system. These questions will be discussed further in the summary at the end of this article.

Principles from Software Engineering. This section contains a short review of those concepts from software engineering that are relevant for the design of computer algebra software. It turns out that many aspects that arise during the design and the implementation of computer algebra software may be better understood if they are considered in this more general context. References for this part are the books by Booch [1], Coad and Yourdon [2] and Meyer [3]. A recent review by Taivalsaari [6] is also useful.

A little consideration leads to the conclusion that the ultimate reason for most of the problems which occur during the development and the maintenance of large pieces of computer algebra software is the fact that too many lines of code have to be considered at a single time. As a consequence, the two most important principles of software design follow rather naturally.

▷ Decompose large pieces of software into smaller ones through *modularity*;

▷ Limit the growth of software through *reusability*.

The principles according to which modularization is achieved is the most important part of the design process. In a computer algebra context appropriate modules are represented by *algebraic types*. Each type should represent a concept from the underlying mathematical field. In order to achieve a high degree of reusability, the flexibility of most of the types is increased by *parametrization*.

Another important mechanism for increasing reusability is *inheritance*. Its basic idea is to allow new types to be based on already existing ones. As a consequence, often the functionality of software can be extended by adding a small amount of new code instead of modifying existing code, or unnecessarily reimplementing features that are already there. This mechanism is established by the *Is_a* relationship between two types. If A and B are types, the statement

Abstraction Level	Description
Application Software	*Main Objective:* Generation of results in application field
	Operations: Provides user interface
Logic Verification System	*Main Objective:* Correctness of Implementation
	Operations: Theorem Provers
	Example: Theorema
Algebraic Type System	*Main Objective:* Structure
	Data & Operations: Parametrized algebraic types
	Polymorphic functions
	Examples: Axiom, Magma
Algebraic Language	*Main Objective:* Efficiency
	Data & Operations: Polynomials and rational functions
	Arithmetic, gcd's, factorization
	Examples: Macsyma, Maple, Mathematica, Reduce, Derive
Implementation Language	*Objective:* Connection to hardware
	Examples: C, LISP

Fig. 1. Hierarchical levels of computer algebra software

A *Is_a* B means that A has all the features of B and in addition those that are defined particularly for the new type A. This is expressed by saying A *inherits* from B or A is a *descendant* or *successor* of B, or B is an *ancestor* or *predecessor* of A. In ALLTYPES any type may inherit from any number of predecessors, i. e. *multiple inheritance* is supported. In this way the *Is_a* relation generates a hierarchical structure between the various members of the type system.

In Figure 1 the organization of computer algebra software according to these principles is shown. The various levels of abstraction necessary for closing the huge gap between the mathematics on the top and the hardware at the bottom are obvious. Their distinguishing feature is the functionality they provide. The distinction between an *algebraic language* and an *algebraic type system* is especially emphazised. The former provides above all *efficiency*, the latter *structure*. The logic verification system *Theorema* is currently under development by Buchberger [7].

In principle ALLTYPES is an independent piece of software. It is connected to the underlying computer algebra language by a well defined interface that is described in a separate article by Scheller [11]. In this document type names are always written in capitals, for example POLY, PARF or LIEALG. Usually they abbreviate the full name of a mathematical object like polynomial, partial fraction or Lie algebra in these examples. Polymorphic functions are written in italics and begin always with a capital letter, examples are *Degree*, *Substitute* or *CoefficientVariables*.

2   Design of ALLTYPES

There are basically two different constituent parts of ALLTYPES. On the one hand there is the type system comprising about 30 parametrized algebraic types for modeling mathematical objects supplemented by about 100 polymorphic functions for working with them. Their action is achieved by a multiple dispatching mechanism that selects the proper methods for the data objects at hand, either at compile time or at run time. A good comparison of single dispatching vs. multiple dispatching may be found in the article by Taivalsaari [6], section 3.9. Secondly there is a language provided that is especially designed for implementing mathematical algorithms efficiently and safely. These components are described in detail in this section. Theoretical questions on the design of an algebraic type system are discussed in [12], see also [8].

Informally a type describes the features that are common among a collection of mathematical objects. The user interface of any type is made up of its name, its parameters and perhaps certain attributes that are collectively called the *category* of the type, they may be considered as the *type of a type* or a *second order type*. The totality of all types decomposes naturally into the following three kinds.

▷ *Basic types* are data objects supported by the underlying language, they are essentially various kinds of numbers, variables and collections of any kind of objects represented as lists. They are introduced for efficiency reasons;

▷ *Algebraic types* are the main components of the type system, they serve to represent structured objects like e. g. univariate polynomials, partial fractions or linear differential equations. Most of them have one or more parameters and belong to one or more categories;

▷ *Set types* are applied for representing homogeneous collections of objects of the same type like e. g. polynomial ideals, rational function fields or Lie algebras. In the object-oriented terminology they correspond to *container classes*.

Basic Types. For efficiency reasons, the representation of numbers and variables is essentially taken over from the underlying computer algebra language.

Number NUM. Subtypes *Integer: IN, Rational Number: RN, Algebraic Number: AN(k), Modular Number: MN(k)* are the number types currently supported by the system. The single argument of MN defines the modulus. The argument of AN has the following meaning. Algebraic numbers are not considered as individual elements but as members of an appropriate algebraic number field that is determined by a primitive element. This number field determines the type. The totality of algebraic number fields required in a problem is organized globally, the individual fields are enumerated by the parameter k. Any change of these number types is handled by user defined or automatic type coercions. In general the system tries to avoid the introduction of unnecessary field extensions.

Variable: VAR and *Derivative: DER.* Next to the numbers, these are the most elementary building blocks of all symbolic expressions. A derivative may have a number of dependencies attached to it so that it behaves differently within a type of category *DIFFERENTIAL*.

Algebraic Types. They form the backbone of the type system of ALLTYPES. Each individual type is characterized by its parameters and the categories which are attached to it. The parameter values are either a type denoted by *TYPE*, or a data object denoted by letters followed by a colon and its type. Examples of this notation are $v : VAR$, $\{v_1, v_2 \ldots\} : SET\,VAR$ or $n : IN$. In some cases a predicate name is also allowed. It is not always necessary that all type parameters are specified. Beginning with the leftmost parameter, a sufficient set has to be specified such that the required action may be executed with a meaningful result. For example, if an object of type *POLY* is to be coerced into a distributive polynomial, calling *DPOLY* without any parameter yields a *DPOLY* w.r.t. to *all* variables or derivatives with coefficients of some number type. If the first parameter is *POLY* and the second a set of variables $\{v_1, v_2, \ldots\}$, the result is a distributive polynomial in these variables with polynomial co efficients in the remainig ones that may be considered as parameters. A similar systematics applies to the other types of the system. If the required coercion is mathematically not meaningful an error will occur, for example if it is requested that a genuine rational function of type *RATF* be coerced to a *DPOLY* over all its variables.

There are four categories known to the system at the moment, they are *RING*, *FIELD*, *DIFFERENTIAL* and *CONTAINER*. They are applied for controling various operations like for example how to divide out the content of a polynomial or how to apply derivatives.

Subsequently a selection of types is listed, supplemented by some remarks on the meaning of the type parameters. The usage of these type notations for manipulating the types of data objects will be explained in the next section.

Polynomial: POLY NUM, CATEGORY RING. The main purpose of this type is to act as a mediator between the representation of polynomials in the underlying computer algebra language. Within the type system they often represent those parts of a more structured object for which at a certain point additional structure is not available or not relevant, for example the coefficients of a univariate polynomial that depend on some additional parameters. The ordering of these objects is determined globally by the underlying computer algebra language. The coefficient type of any *POLY* is some number type *NUM*. It is carried along with the individual objects.

Rational Function: RATF NUM, CATEGORY FIELD. The same remarks apply as for the preceding type.

Univariate Polynomial: UPOLY(TYPE, v : VAR), CATEGORY RING. The first parameter determines the coefficient type, the second parameter is a variable name with respect to which the data object must have a polynomial dependency.

Partial Fraction: PARF(TYPE, v : VAR), CATEGORY FIELD. The parameters have the same meaning as for *UPOLY*. Univariate partial fractions may be considered as the natural extension of univariate polynomials.

Factored Polynomial: FPOLY(TYPE). The single parameter determines the type of the irreducible components of an object of this type.

Linear Form: LFORM(TYPE, {$v_1, v_2, \ldots$} : SET VAR| fn : predicate, Ordering). The first parameter determines the coefficient type. Objects of this type must be linear and homogeneous w.r.t. the variables on the list {$v_1, v_2, \ldots$} or w.r.t. to those variables for which the predicate returns true. If the *Ordering* parameter is provided, it is applied for establishing the respective term ordering, if not, a system dependent default ordering is applied.

Differential Polynomial: DFPOLY(TYPE, {$d_1, d_2, \ldots$} : SET DER, {$v_1, v_2, \ldots$} : SET VAR, Ordering), CATEGORY {DIFFERENTIAL, RING}.

Distributive Polynomial: DPOLY(TYPE, {$v_1, v_2, \ldots$} : SET VAR, Ordering), CATEGORY RING. The first argument determines the coefficient type, objects of this type must be polynomial in the variables $v_1, v_2, \ldots$.

Function Field: FFIELD(TYPE, {$e_1, e_2, \ldots$} : SET VAR), CATEGORY FIELD. Objects of this type must be rational in the variables $e_1, e_2, \ldots$ that may be elementary or algebraic functions.

Linear Differential Form: LDF(TYPE, {$d_1, d_2, \ldots$} : SET DER, {$v_1, v_2, \ldots$} : SET VAR, Ordering). Objetcs of this type must be linear and homogeneous in the derivatives of a set of functions $d_1, d_2, \ldots$. With respect to these derivatives they are *LFORM*'s with coefficient type determined by the first parameter.

Linear Differential Operator: LDO(TYPE, {$v_1, v_2, \ldots$} : SET VAR). The first argument determines the coefficient type, the first order partial derivatives are taken w.r.t. to the variables {$v_1, v_2, \ldots$}.

Linear Ordinary Differential Equation: LODE(TYPE, y : DER, x : VAR). Objects of this type are linear and homogeneous in y and its derivatives w.r.t. x with coefficient type determined by the first parameter.

Ordinary Differential Equation: ODE(TYPE, y : DER, x : VAR)). Objects of this type must be polynomial in y and its derivatives w.r.t. x with coefficient type determined by the first parameter.

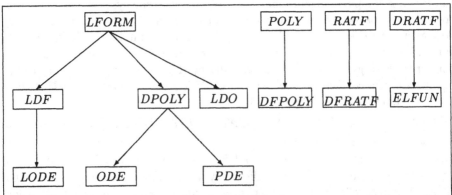

Fig. 2. The inheritance hierarchy of the algebraic type system is shown. The full names of the types and its parameters are given in the text.

Polymorphic Functions and Methods. The operations that may be performed on the various data objects are specified in mathematical terminology for the *polymorphic functions*. They are applied by the user for writing the code that realizes the desired algorithms. A polymorphic function cannot come into action by itself, only after it is replaced by the appropriate *method* the code may be executed. In general many methods corresponding to different types, possibly with varying number of arguments, may be defined for a single polymorphic function name. The proper method is determined by the types of the data objects to be operated upon. It is the responsibility of the user to assure that the required method is available at execution time. A method may either be defined in the code of the respective type, or it may be inherited from some other type.

There are several advantages of organizing software in this manner. On the one hand, the degree of reusablitiy is increased by writing code in terms of polymorphic functions instead of the methods itself. Secondly introducing a new type means going through the collection of polymorphic functions and implementing those methods for the newly defined type that apply to it, and possibly additional methods that are valid exclusively for this newly defined type and its descendants. This is a unique and well-defined procedure including the guarantee that *all* necessary changes have been performed *without* explicit changes of already existing code. Finally for the user it is much more efficient to become acquainted with a single set of polymorphic functions providing essentially the same functionality for the complete type system, instead of repeating this effort for the individual types. In other words, the *learning curve* increases faster.

The mechanism applied in ALLTYPES that gives rise to a polymorphic function call being replaced by the proper method will be explained now in more detail. The methods corresponding to a polymorphic function are denoted by extending the polymorphic function name by the type names, separated by underscores; they are called *type extensions*. There may be as many type extensions as there are parameters, however a single extension is also possible. Providing fewer extensions than parameters is used as another mechanism for increasing reusability. The significance of the various type extensions is not symmetric however. The first type extension of a method determines the type where it is defined. For a *type homogeneous* polymorphic function the first extension of any of its methods determines also the type of the returned result. As an example the type homogeneous polymorphic function *Multiply* may be called as $Multiply(u, v)$ by a user. Valid examples of method calls that may be generated from it are

$$Multiply\_POLY\_POLY(u: POLY, v: POLY): POLY$$
$$Multiply\_POLY\_BASIC(u: POLY, v: BASIC): POLY$$
$$Multiply\_LFORM(u: LFORM, v: TYPE): LFORM$$

The first two methods define the multiplication of an object of type *POLY* by another *POLY* or an object of type *BASIC* which may be some number or a variable respectively. The last method defines the multiplication of an object of type *LFORM* by an object of any type that is consistent with the type homogeneity of *Multiply*. Essentially this means that the coefficients of the first argument u are multiplied by v. In accordance with the type homogeneity of the polymorphic function $Multiply$ the type of the returned object is as indicated after the closing bracket of the method calls. A method

$$Multiply\_LFORM\_LFORM(u : LFORM, v : LFORM)$$

is not allowed by type homogeneity because the result cannot be coerced to an object of type $LFORM$. If u and v for example both have the same type $LFORM(POLY\,NUM, \{a, b, c\})$, an appropriate call for multiplying them would be

$$Multiply(u|QFORM(POLY\,NUM, \{a, b, c\})|, v)$$

where QFORM abbreviates Quadratic Form. It generates the method call

$$Multiply\_QFORM\_LFORM(u|QFORM(POLY\,NUM, \{a, b, c\})|, v)$$

is generated and the result of type $QFORM(POLY\,NUM, \{a, b, c\})$ is returned. Type homogeneity entails also the different answer for the two polymorphic function calls

$$Multiply(u : LFORM, v : RATF) : LFORM$$

and

$$Multiply(u : RATF, v : LFORM) : RATF$$

Another example showing the flexibility gained by this software organization is the polymorphic function $Coefficient$. Some of its method calls are

$$Coefficient\_POLY\_POLY(u : POLY, v : POLY)$$
$$Coefficient\_POLY\_BASIC(u : POLY, v : VAR)$$
$$Coefficient\_UPOLY\_BASIC(u : UPOLY, n : IN)$$
$$Coefficient\_LFORM\_BASIC(u : LFORM, v : VAR)$$

In the first call, the second argument v must be a monomial, its coefficient in u is returned. In the second call, u must be linear in the variable v, otherwise a runtime error occurs. In third call n must be a positive integer, the coefficient of the $n - th$ power of the univariate polynomial is returned. In the last call, v must be a variable w.r.t. which u is linear and homogeneous.

There are more than 1000 methods defined in ALLTYPES. All the information necessary for working with them may be obtained by applying a few online utility functions provided for this purpose.

The Language Constructs. The complete language is composed of two fundamentally different parts. On the one hand, there is the *algebraic language* which is applied for manipulating algebraic objects according to the mathematical rules that are valid for them, for example arithmetic, differentiation, integration and many others. Secondly, there is the *type manipulation part* that is applied for manipulating the types of the mathematical objects such that they satisfy the constraints imposed by the type system. This latter part of the language is syntactically separated from the former by the occurrence of a pair of |, i. e. any piece of input code enclosed between two bars specifies a type manipulation. These two parts of the language will be described now one after the other. The fundamental syntactic units are the *FunctionConstructor* and the *IterationConstructor*.

They are partially described by the following set of productions where keywords are given in boldface letters.

$$FunctionConstructor = FunctionName(Expression|Predicate,$$
$$Iterator[, Iterator])$$

$$FunctionName = \textbf{applies}|\textbf{select}|\textbf{exists}|\textbf{satisfies}|\textbf{separate}$$

$$Iterator = Variable \textbf{ from}[..\textbf{from}]Set|Variable$$
$$\textbf{from } Integer ..[Integer ..]Integer$$

$$IterationConstructor = \textbf{iterate}(\textbf{true}?Predicate,$$
$$Expression[, Expression])|\textbf{iterate}($$
$$Expression[, Expression], \textbf{true}? Predicate)$$

$$OrderingConstructor = \textbf{orders}(BooleanExpression,$$
$$Variable \textbf{ and } Variable \textbf{ from } Set[, Set])$$

The remaining part of the syntax and the semantics will be explained informally. An *Expression* is any piece of code that is syntactically and semantically accepted by the interpreter or compiler and returns a result. In *Axiom* it is essentially an expression, in a LISP system it is very similar to a *form*. The free variables occurring in an *Expression* or a *Predicate* are evaluated in the environment in which the *FunctionConstructor* is called. All local variables are attached to certain sets by the subsequent *Iterator*.

The *Variable* in the first production alternative for the *Iterator* determines a variable name which may occur in the *Expression*, it specifies an element of the *Set*. Each occurrence of the keyword **from** determines a level of nesting of *Set*. If there is a single occurrence, the iteration occurs over elements of a set, if there are two, elements of a set of sets are taken, and so forth. If there is more than a single iterator, the iteration takes place over several sets, set of sets etc. in parallel. Both the number of iterators and the depth of nestings of sets are arbitrary up to the following constraints. The level of nestings in each iterator must be the same, the same is true for the cardinality of sets at each level. The second production for the *Iterator* allows it to iterate over intervals of integers, the stepsize being determined by the optional middle *Integer*.

The **applies** constructor causes the *Expression* to be evaluated for each selection of variables as determined by the *Iterator*. It returns an object with the same structure as it occurs in the latter with the result of the evaluations taken at the respective positions. The objects may be sets, sets of sets etc.

In the **select** and the **exists** constructors the *Expression* must be a predicate. The returned object of the **select** constructor has the same structure as those occurring in the iterators, it contains those elements of the *Set* occurring in the last *Iterator* for which the *Expression* evaluates to true. The **exists** constructor returns only the first element occurring in this *Set* for which the *Expression* is true. If the *Expression* begins with the keywords **maximal** or **minimal**, the remaining part of it must evaluate to an integer. In this case the **select** constructor returns all elements from the *Set* occurring in the last *Iterator* for which the *Expression* returns the maximal or minimal value respectively. The same applies to the **exists** constructor except that only the first element with this property is returned. A meaningful application will presume the proper number of occurrences in either case.

The **satisfies** constructor is actually a predicate. It returns true if all evaluations of its first argument which must be a predicate also evaluate to true.

The **separate** constructor generates new sets of objects from the *Set* which occurs in the last *Iterator*. The newly created sets are formed such that the value of the *Expression* is the same for all elements of its member sets.

The two versions of the *IterationConstructor* are the basic syntactic unit for an unspecified number of repetitions of an action. In the first version, the *Predicate* is evaluated before any other action is taken. If the result is true, the *Expression*'s are evaluated from left to right one after the other. This proceeding is repeated until the *Predicate* evaluates to something that is not true. In this case the result of the last evaluation in the *Expression* is returned. If the *Predicate* is not true at the first evaluation, the value of the complete *IterationConstructor* is false. In the second version, a similar evaluation scheme is applied except that the *Predicate* is evaluated at the end of each round. The returned value of the constructor is always that of the last evaluation in the *Expression*.

The *BooleanExpression* in the **orders** constructor contains the two arguments *Variable* of the subsequent **and** clause. It causes the *Set* to be rearranged such that the evaluation of the *BooleanExpression* with any pair of elements such that the first member is located to the left of the second member evaluates to *true*. If the last optional argument of type *Set* is provided, its elements are rearranged in the same way as the first argument and this latter set is returned.

Parallel Jobs with pand and por. The language constructs **pand** and **por** which abbreviates *parallel and* and *parallel or* respectively allow a limited amount of parallelism in all situations where the hardware on which ALLTYPES is running provides at least two processors. In computer algebra applications it occurs frequenly that there are various alternatives available at a certain point of an algorithm for obtaining the desired result, it is not known however in advance which alternative will produce the answer more efficiently. Due to the overhead for generating a copy of a process and the communication between both of them, they should be applied only for a coarse parallelization. Examples meeting this criterion are Gröbner- or Janet- base calculations w.r.t. different term orderings, or heuristic solutions as opposed to decision procedures, e.g. for determining the symmetries of an ordinary differential equation. If the code for these two alternatives is denoted by $\mathcal{A}_1$ or $\mathcal{A}_2$ respectively, the call

$$z := \mathcal{A}_1 \textbf{ por } \mathcal{A}_2$$

may be applied with the following semantics. At first an exact copy of the original process is generated. Then the two processes are started executing version $\mathcal{A}_1$ or $\mathcal{A}_2$ respectively. The process that terminates first cancels the other one and returns the produced result as the value of z. The syntax for calling **pand** is similar.

$$z := \mathcal{A}_1 \textbf{ pand } \mathcal{A}_2$$

After a copy of the first process is generated, the two processes execute the code for A_1 and A_1 respectively. If either process terminates with a value nil, the other process is cancelled and the value nil is assigned to z. If both processes terminate with a value different from nil, the result A_2 is assigned to z.

3 Working with ALLTYPES

Typically working with ALLTYPES means implementing a piece of software in terms of the type system and the language described above. Before it may be applied for generating results, it has to be compiled. This proceeding is explained first.

Writing and Compiling ALLTYPES Code. The compilation process comprises two passes. In the first pass, the compiler of the underlying computer algebra language generates essentially the same kind of output as is generated from code that does not use the features of ALLTYPES. Upon completion of the first pass, all methods defined in the system are globally known. Whenever possible, this information is applied in the second pass for attaching the proper types to the polymorphic function calls of the user code, i. e. to replace them by the methods. This is true in those cases where the types of the arguments of a polymorphic function call are known at compile time. In the remaining cases the dispatching takes place at runtime. There is a continuous transition between those two alternatives, the only difference is the efficiency of the runtime system provided all type declarations are correct. Some details of the underlying *type inference* and *inheritance mechanism* will be discussed later on.

Methods are implemented as procedures of the underlying language. The most distinctive feature of ALLTYPES code is the declaration of types and how they are handled. This process is governed by the following rules.

▷ The user calls *exclusively* the polymorphic functions and *not* the methods for achieving a certain action;

▷ In all cases where the type of a variable is known at compile time it should be declared in order for enabling the compiler to insert the proper methods. This is true also for the parameters of the polymorphic function calls.

Defining New Methods. The details of this proceeding are explained informally for a polymorphic function $f00$ and a method $f00\_TYPE_1\_TYPE_2$ corresponding to it. The general structure of such a definition may be seen from the following example.

```
procedure f00_TYPE₁_TYPE₂(u,v);
  begin TYPE₁ u; TYPE₂ v; TYPE₃ w,z;
  BASIC x; SET UPOLY p; scalar y;
  w:=f01(u,x);
  w:=f02(u,y);
  z:=f03(v,w);
    ⋮
```

```
    return applies(Lcoef x,x from p);
    end;
```
The body of any method consists of a *BEGIN - END* block, f01, f02 and f03 are polymorphic function names. The declaration **scalar** is applied for those variables the type of which cannot be determined at compile time. Consequently for a polymorphic function with any argument declared as **scalar** the method cannot be determined at compile time. If a homogeneous collection of objects of the same type T is assigned to a variable, the proper declaration is SET T. In the above example, during the second pass of the compiler the code

```
    w:=f01_TYPE_{i_1}_BASIC(u,x);
    w:=f02(u,y);
    z:=f03_TYPE_{i_2}_TYPE_{i_3}(v,w);
    return applies(Lcoef_UPOLY x,x from p);
```

is generated. Whenever the compiler succeeded in substituting the proper method for a polymorphic function call, the compiled code for the corresponding method is inserted. In the above example this applies to the polymorphic functions $f01$ and $f03$ where the type extensions $TYPE_{i_k}$ have been attached to the respective polymorphic functions. Type $TYPE_{i_k}$ may or may not be identical to type $TYPE_k$ depending on whether the respective method is obtained by inheritance or not. If a method cannot be determined at compile time, a piece of code is inserted that performs this substitution at runtime as it is true for the polymorphic function $f02$.

Creating New Types. The language construct *DefineAlgebraicType* is provided for extending the type system. Its syntax is formally defined by the following production rules.

DefineAlgebraicType *Type* [**CATEGORY** *Category*] **REP**(*Field*[, *Field*] *)
DefineAlgebraicType *Type* Is_a *Parent*
    [**CATEGORY** *Category*][**with**(*Field*[, *Field*]*)]
Type ::= *TypeName*|*TypeName*(*ParameterName*[, *ParameterName*] *)
Parent ::= *ParentName*|{*ParentName*[, *ParentName*] * }
Category ::= *CategoryName*|{*CategoryName*[, *CategoryName*] * }
Field ::= *FieldName* := *Expression*.

The first version of *DefineAlgebraicType* is applied if a new type is defined from scratch. Whenever it is based on types already existing in the system, the second option with the inheritance operator *Is_a* is applied. Any type A may inherit from one or more types $B, C, \ldots$ in the system. Valid type definitions are

$$DefineAlgebraicType\ A\ Is\_a\ B \quad or$$
$$DefineAlgebraicType\ A\ Is\_a\ \{B, C, D, \ldots\} \tag{1}$$

defining a single or multiple inheritance relation respectively. In detail, any inheritance relation implies the following features for the descendant.

▷ All methods of the ancestors are inherited to the descendant. Details of how repeated inheritance and multiple inheritance are handled are explained below;
▷ All categories of the ancestors are inherited to the descendant;
▷ In the internal representation, the fields of the descendant are the union of the fields of its anchestors.

Inheritance relations create a hierarchical structure between the individual types of the system. For the current implementation of ALLTYPES these relations are shown in Figure 2. If new types are defined by the user, these hierarchies have to be extended appropriately. The inheritance relation (1) may be represented by the following tree.

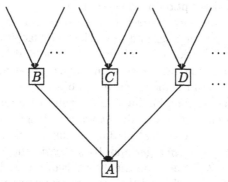

It indicates that type A has parents B, C, D and possibly more. Each of these parent types may inherit from some other types, and so forth. If A is the type of the single argument of a polymorphic function call $f00$, this tree is traversed in a depth-first-search from right to left until a method for $f00$ is found. If upon completion of this traversal no method has been found, an error occurs. This rule resolves uniquely the ambiguities that may occur due to repeated inheritance. If a polymorphic function call has $n > 1$ arguments with types $A_1, A_2, \ldots, A_n$, for each argument type the inheritance tree is constructed similar as for type A above. The first $n - 1$ types are fixed to its root values and A_n is treated like the single argument A above. If this search fails, a single step in the traversal of the tree for A_{n-1} is performed and the tree for A_n is traversed again. This procedure is extended in an obvious way until a method has been found or until all trees are traversed.

4   Summary and Outlook

The system ALLTYPES described in this article is very well suited for implementing sophisticated software for working with differential equations. Its organization by means of a fine structured type system of well adjusted types has turned out to be a fundamental prerequisite for developing high quality software efficiently. In particular the following features of ALLTYPES should prove to be valid design principles for any advanced computer algebra software.

▷ The separation of the computer algebra *language* level responsible for efficiency and the computer algebra *type system* on top of it responsible for structure;
▷ A small set of powerful language constructs bringing about a programming style such that each single line of code may be specified in mathematical terms, executed and the result may be tested against this specification;
▷ The syntax and semantics of the type manipulation facilities implemented as a specialized part of the language.

The implementation of ALLTYPES on top of Reduce will be supplemented soon by an implementation on top of Macsyma [13]. Upon completion, the software for working with differential equations should run without change on both software platforms. Similar implementations based on other systems like for example *MathEdge* of Maple or *MathLink* of Mathematica are highly desirable as well as a detailed description of the system [14].

The most important shortcomings of the current implementation are the lack of type checking by the compiler and the lack of any facilities for encapsulation like for example the definition of local functions in a module. The reason lies in the fact that the underlying implementation language does not support it, all the less does it support any of the more advanced object-oriented principles like e. g. inheritance and dynamic type binding. These will be important considerations for a future implementation in a modern computer language.

Acknowledgments. A critical reading of this article and numerous helpful comments by Bruno Buchberger and Tony Hearn, the continuous advice of Winfried Neun of the ZIB Berlin on the internals of the Reduce system and the careful implementation by Dietlind Scheller are gratefully acknowledged.

References

1. Booch, G, Object Oriented Design, Benjamin/ Cummings Publishing, 1991.
2. Coad, P., Yourdon, E., Object-Oriented Analysis and Object-Oriented Design, both published by Yourdon Press, Englewood Cliffs, 1991.
3. Meyer, B., Object-oriented Software Construction, Prentice Hall, 1988.
4. Budd, T., Object-Oriented Programming, Addison-Wesley, 1991.
5. Cardelli, L., Wegner, P., On Understanding Types, Data Abstraction, and Polymorphism, Computing Surveys **17**, 471-522 (1985).
6. Taivalsaari, A., On the Notion of Inheritance, ACM Computing Surveys **28**, 438-479 (1996).
7. Buchberger, B., Symbolic Computation: Computer Algebra and Logic, Proceedings of the *Frontiers in Combining System* Conference, Munich, F. Baader and K. U. Schulz, eds, Applied Logic Series, Kluwer Academic Publishers, 1996.
8. Fedoraro, J. F., The Design of a Language for Algebraic Computation Systems, Thesis, Berkeley, 1983.
9. Hearn, A. C., Reduce User's Manual, part of Reduce.
10. Jenks, R. D., Sutor, B., Axiom, Springer, 1992.
11. Scheller, D., ALLTYPES: The Reduce Implementation, GMD Report, 1998.
12. Weber, A., Structuring the Type System of a Computer Algebra System, Dissertation, Universität Tübingen, 1992.
13. Golden, J., private communication.
14. Schwarz, F., ALLTYPES: The User Manual, GMD Report, to appear.

Real Parametrization of Algebraic Curves *

J. Rafael Sendra[1] and Franz Winkler[2]

[1] Dpto de Matemáticas, Universidad de Alcalá de Henares, E-28871 Madrid, Spain
mtsendra@alcala.es
[2] RISC-Linz, J. Kepler Universität Linz, A-4040 Linz, Austria
Franz.Winkler@risc.uni-linz.ac.at

Abstract. There are various algorithms known for deciding the parametrizability (rationality) of a plane algebraic curve, and if the curve is rational, actually computing a parametrization. Optimality criteria such as low degrees in the parametrization or low degree field extensions are met by some parametrization algorithms. In this paper we investigate real curves. Given a parametrizable plane curve over the complex numbers, we decide whether it is in fact real. Furthermore, we discuss methods for actually computing a real parametrization for a parametrizable real curve.

1 Introduction

In [SW91], [SW97] we have described a symbolic algorithm for computing a rational parametrization $(x(t), y(t))$ of a plane algebraic curve $\mathcal{C}$ of genus 0 (only these curves have a rational parametrization). This algorithm is implemented in the system CASA [MW96]. Approaches to the parametrization problem for algebraic curves are also described in [AB88] and [vH97].

Definition: Let $\mathbb{K}$ be a field of characteristic 0, $\mathcal{K}$ the algebraic closure of $\mathbb{K}$. Let the irreducible affine curve $\mathcal{C}$ be defined as the set of solutions in the affine plane $\mathbb{A}(\mathcal{K})^2$ of the polynomial equation

$$f(x, y) = 0$$

over $\mathbb{K}$, i.e. $f \in \mathbb{K}[x, y]$. Then $x(t), y(t)$ in $\mathcal{K}(t)$, the field of rational functions over $\mathcal{K}$, constitute a *rational parametrization* of $\mathcal{C}$, if and only if, except for finitely many exceptions, every evaluation $(x(t_0), y(t_0))$ at $t_0 \in \mathcal{K}$ is a point on $\mathcal{C}$, and, conversely, almost every point on $\mathcal{C}$ is the result of evaluating the parametrization at some element of $\mathcal{K}$.

In this case $\mathcal{C}$ is called *parametrizable* or *rational*.

Equivalently, $\mathcal{P}(t) = (x(t), y(t))$ is a rational parametrization of $\mathcal{C}$ if $\mathcal{P}$: $\mathbb{K} \longrightarrow \mathcal{C}$ is rational and not both $x(t)$ and $y(t)$ are constant. Furthermore, if $\mathcal{P}$ is birational we say that $\mathcal{P}(t)$ is a *proper parametrization*.   □

---

* The first author was supported by DGICYT PB 95/0563 and UAH-Proj. E010/97. The second author was supported by the Austrian *Fonds zur Förderung der wissenschaftlichen Forschung* under Proj. HySaX, P11160-TEC. Both authors were supported by the Austrian-Spanish exchange program Acción Integrada 30/97.

The parametrization problem for algebraic curves consists in first deciding whether the given curve C has such a rational parametrization, and if so finding one.

In our geometrical approach to parametrization we basically determine the singularities of the curve C, decide the genus of C (C can be parametrized if and only if genus(C) = 0), find a couple of simple points on C in a low degree (i.e. degree 1 or 2) algebraic extension $\mathbb{K}(\gamma)$ of $\mathbb{K}$, and from the singularities and these simple points derive a parametrization of C. The parametrization will have coefficients in $\mathbb{K}(\gamma)$. So starting with a parametrizable curve defined by a polynomial $f(x, y) \in \mathbb{Z}[x, y]$ we will either get a parametrization with rational coefficients or with coefficients in $\mathbb{Q}(\gamma)$, γ algebraic over $\mathbb{Q}$ of degree 2. If the curve C is a real curve, i.e. if C has infinity many real points, then for obvious reasons we will not be satisfied with a parametrization with complex coefficients.

For practical applications, such as in computer aided geometric design, we need to be able to parametrize real curves with real coefficients, if possible. In this paper we demonstrate that if C is a real curve, then our algorithm actually computes a parametrization of C with *real* coefficients. Alternatively, one could take a possibly complex parametrization and, if possible, transform it to a real one. This approach is developed in [RS95] and [RS97a].

2 Real Curves

Let $f(x, y) \in \mathbb{C}[x, y]$ be a non-constant polynomial. f defines a plane affine curve C over the complex numbers, i.e. $C \subset \mathbb{A}^2(\mathbb{C})$, the affine plane over $\mathbb{C}$. Whenever useful or necessary, we will also consider the curve C in the projective plane, i.e. $\mathbb{P}^2(\mathbb{C})$. Points in the projective plane are written as $(a : b : c)$.

Definition: The curve C is a called a *real* curve, if and only if C has infinitely many points in $\mathbb{A}^2(\mathbb{R})$. □

A real curve always has a defining polynomial over the reals. A proof of the following Lemma is given in [RS97a].

Lemma 1: If the curve C is real, then it can be defined by a real polynomial. □

Not only can every real curve be defined over the reals, also the irreducibility of the curve is independent from whether we view it in $\mathbb{A}^2(\mathbb{R})$ or in $\mathbb{A}^2(\mathbb{C})$. A proof of the following Lemma is given in [Wi96], Theorem 5.5.3.

Lemma 2: Let C be a real curve. C is irreducible over $\mathbb{R}$ if and only if it is irreducible over $\mathbb{C}$. □

The algorithm presented in [SW91] implies that every parametrizable plane curve over an algebraically closed field $\mathcal{K}$ of characteristic zero can be parametrized over any subfield of $\mathcal{K}$ that contains the coefficients of the irreducible polynomial defining the curve, and the coordinates of one simple point of the curve. Thus, as a consequence of Lemma 1, one deduces that every real parametrizable plane curve can be parametrized over the reals (this result is also known as the algebraic version of the real Lüroth's theorem [RS97b]), and that a parametrizable plane curve is real if and only if it has at least one real simple point.

Example 1: Let the curve C_1 ([Wi96]) be defined by

$$f_1(x, y) = (x^2 + 4y + y^2)^2 - 16(x^2 + y^2) = 0.$$

C_1 is a real curve. In fact C_1 is a parametrizable real curve, and a parametrization over the reals is

$$x(t) = \frac{-1024t^3}{256t^4 + 32t^2 + 1}, \qquad y(t) = \frac{-2048t^4 + 128t^2}{256t^4 + 32t^2 + 1}.$$

On the other hand, the curve C_2 ([RS97a]) defined by

$$f_2(x, y) = 2y^2 + x^2 + 2x^2y^2 = 0$$

is not a real curve. The only point of C_2 in the affine plane over the reals is the double point $(0, 0)$. The complex curve C_2 is parametrizable, and a parametrization is

$$x(t) = \frac{-t^2 - 2t + 1}{t^2 - 2t - 1}, \qquad y(t) = \frac{it^2 + 2it - i}{2t^2 + 2}.$$

$\square$

3   A Real Parametrization Algorithm

Let $\mathbb{L}$ be a computable subfield of $\mathbb{C}$, and let the irreducible affine curve C be defined by the polynomial $f(x, y) \in \mathbb{L}[x, y]$. We may assume that f is a real polynomial (indeed associated with a real polynomial), for otherwise by Lemma 1 one knows that C is not a real curve. Thus, we may suppose that $\mathbb{L}$ is a subfield of $\mathbb{R}$. In the process of parametrization it is necessary to view C as a projective curve in the projective plane $\mathbb{P}^2(\mathbb{C})$. This projective curve, also denoted by C, is defined by the homogeneous polynomial $F(x, y, z) \in \mathbb{L}[x, y, z]$, where F is the homogenization of f.

The property of parametrizability is independent of whether we view C in the affine or the projective plane, and parametrizations can be easily converted [SW91].

Definition: Let the irreducible curve C of degree d be defined by the irreducible polynomial $f(x, y) \in \mathbb{L}[x, y]$ of degree d. The singular point $P \in \mathbb{P}^2(\mathbb{C})$ of multiplicity m on C is an *ordinary* singular point, if and only if there are m different tangents to C at P.

If C has only ordinary singularities $P_1, \ldots, P_n$ of multiplicities $r_1, \ldots, r_n$, respectively, then the genus of C is defined as

$$\text{genus}(C) = \frac{1}{2}[(d-1)(d-2) - \sum_{i=1}^{n} r_i(r_i - 1)].$$

This definition, and also the method described in this paper, can be extended to curves with non-ordinary singularities. For the sake of simplicity we do not consider this situation here.

The linear system of *adjoint curves* of degree d' to C consists of all the curves of degree d' having every point $P_i, 1 \leq i \leq n$, as a point of multiplicity at least $r_i - 1$. □

Since the adjoint curves of any degree d' to a rational plane curve C have defining polynomial over $\mathbb{L}$, and can be computed in a finite number of ground field operations [SW91], the problem of parametrizing is reduced to the problem of determining a simple point on the curve. We will do that by transforming C birationally to a conic $\mathcal{D}$. The simple real points on C and on $\mathcal{D}$ correspond uniquely to each other, except for finitely many exceptions. So there is a simple real point on C if and only if there is a simple real point on $\mathcal{D}$. This question can be decided. If the answer is yes, a real point on $\mathcal{D}$ can be computed, transformed to a real point on C, and from this point we can derive a parametrization of C over $\mathbb{R}$.

In [SW97] we prove the following generalization of a theorem by Hilbert and Hurwitz [HH90].

Theorem 1: *Let C be a rational plane curve of degree d defined by a polynomial over $\mathbb{L}$, $\mathcal{H}_a$ the linear system of adjoint curves to C of degree $a \in \{d, d-1, d-2\}$, and $\tilde{\mathcal{H}}_a^s$ a linear subsystem of $\mathcal{H}_a$ of dimension s with all its base points on C. Then we have the following:*

(i) *If $\Phi_1, \Phi_2, \Phi_3 \in \tilde{\mathcal{H}}_a^s$ are such that the common intersections of the three curves Φ_i and C are the set of base points of $\tilde{\mathcal{H}}_a^s$, and such that*

$$\mathcal{T} = \{y_1 : y_2 : y_3 = \Phi_1 : \Phi_2 : \Phi_3\}$$

is a birational transformation, then the birationally equivalent curve to C, obtained by $\mathcal{T}$, is irreducible of degree s.

(ii) *Those values of the parameters for which the rational transformation $\mathcal{T}$ is not birational satisfy some algebraic conditions.* □

We will use Theorem 1 to transform the curve C to either a line or a conic. For the transformed curves it will be easy to decide the existence of real points and if so to determine a real point. So we need to select a linear subsystem of low dimension in the system of adjoint curves, e.g. by fixing additional base points. These additional base points will introduce algebraic coefficients into the system, unless we can find rational ones or whole conjugate families of such points.

Definition: Let $F \in \mathbb{L}[x_1, x_2, x_3]$ be a homogeneous polynomial defining a parametrizable projective curve C. Let $p_1, p_2, p_3, m \in \mathbb{L}[t]$. The set of projective points $\mathcal{F} = \{(p_1(\alpha) : p_2(\alpha) : p_3(\alpha)) \mid m(\alpha) = 0\} \subset \mathbb{P}^2(\mathbb{C})$ is a *family of s conjugate simple points on C* if and only if the following conditions are satisfied: m is squarefree, $\deg(m) = s$, $\deg(p_i) < \deg(m)$ for $i = 1, 2, 3$, $\gcd(p_1, p_2, p_3) = 1$,

$\mathcal{F}$ contains exactly s points of $\mathbb{P}^2(\mathbb{C})$, $F(p_1(t), p_2(t), p_3(t)) = 0 \bmod m(t)$, and there exists $i \in \{1, 2, 3\}$ such that $\frac{\partial F}{\partial x_i}(p_1(t), p_2(t), p_3(t)) \bmod m(t) \neq 0$.   □

If we choose all the points in a family of conjugate points as additional base points in the system of adjoint curves, then the corresponding subsystem will again have coefficients over the ground field $\mathbb{L}$.

Definition: Let $\mathcal{C}$ be a plane curve defined by a polynomial over $\mathbb{L}$, $\mathcal{H}$ a linear system of curves in which all the elements are of the same degree, $\tilde{H}$ the defining polynomial of a linear subsystem $\tilde{\mathcal{H}}$ of $\mathcal{H}$, and let $\tilde{\mathcal{S}}$ be the set of base points of $\tilde{\mathcal{H}}$ that are not base points of $\mathcal{H}$. Then, we say that $\tilde{\mathcal{H}}$ is a *rational subsystem* of $\mathcal{H}$ if the following conditions are satified:

(1) $\tilde{H}$ has coefficients in $\mathbb{L}$.
(2) For almost every curve $\Phi \in \mathcal{H}$, and $\tilde{\Phi} \in \tilde{\mathcal{H}}$ it holds that

$$\dim(\mathcal{H}) - \dim(\tilde{\mathcal{H}}) = \sum_{P \in \tilde{\mathcal{S}}} (\text{mult}_P(\tilde{\Phi}, \mathcal{C}) - \text{mult}_P(\Phi, \mathcal{C})),$$

where $\text{mult}_P(\mathcal{C}_1, \mathcal{C}_2)$ denotes the multiplicity of intersection of the curves $\mathcal{C}_1$, $\mathcal{C}_2$ at the point P.   □

Essentially, this notion requires that when a point or a family of points on $\mathcal{C}$ are used to generate a subsystem $\tilde{\mathcal{H}}$ of $\mathcal{H}$ (by introducing some points on $\mathcal{C}$ as new base point on $\mathcal{H}$ with specific multiplicities) the linear system of equations containing the contraints is over $\mathbb{L}$, and its rank equals the number of new known intersection points between $\mathcal{C}$ and a generic representative of the subsystem. In the next proposition some special cases of rational linear subsystem are analyzed. The following Proposition 1 and Theorem 2 are proved in [SW97].

Proposition 1: *Let $\mathcal{C}$ be a rational plane curve of degree d defined by a polynomial over $\mathbb{L}$, $\mathcal{H}_a$ the linear system of adjoint curves to $\mathcal{C}$ of degree $a \in \{d, d-1, d-2\}$, and $\mathcal{F} = \{(p_1(t) : p_2(t) : p_3(t))\}_{A(t)}$ a family of k conjugate points on $\mathcal{C}$ over $\mathbb{L}$. Then we have the following:*

(i) *If $\mathcal{F}$ is a family of simple points, $k \leq \dim(\mathcal{H}_a)$, and $\tilde{\mathcal{H}}_a$ is the subsystem of $\mathcal{H}_a$ obtained by forcing every point in $\mathcal{F}$ to be a simple base point of $\tilde{\mathcal{H}}_a$, then $\tilde{\mathcal{H}}_a$ is rational, and $\dim(\tilde{\mathcal{H}}_a) = \dim(\mathcal{H}_a) - k$.*
(ii) *If $\mathcal{F}$ is a family of r-fold points, $r \cdot k \leq \dim(\mathcal{H}_a)$, and $\tilde{\mathcal{H}}_a$ is the subsystem of $\mathcal{H}_a$ obtained by forcing every point in $\mathcal{F}$ to be a base point of multiplicity r, then $\tilde{\mathcal{H}}_a$ is rational, and $\dim(\tilde{\mathcal{H}}_a) = \dim(\mathcal{H}_a) - r\,k$.*   □

Theorem 2: *Let $\mathcal{C}$ be a rational plane curve of degree d defined by a polynomial over $\mathbb{L}$, and $\mathcal{H}_a$ the linear system of adjoint curves to $\mathcal{C}$ of degree $a \in \{d, d-1, d-2\}$. Then every rational linear subsystem of $\mathcal{H}_a$ of dimension s with all its base points on $\mathcal{C}$ provides curves that generate families of s conjugate simple points over $\mathbb{L}$ by intersection with $\mathcal{C}$.*   □

As a consequence of Proposition 1 and Theorem 2 we get the following algorithmically important facts.

Theorem 3: *Let C be a rational plane curve of degree d, defined by a polynomial $f(x,y) \in \mathbb{L}[x,y]$.*

(i) C has families of $d-2, 2d-2$, and $3d-2$ conjugate simple points over $\mathbb{L}$.
(ii) C has families of 2 conjugate simple points over $\mathbb{L}$.
(iii) If d is odd, then C has a simple point over $\mathbb{L}$.
(iv) If d is even, then C has simple points over an algebraic extension of $\mathbb{L}$ of degree 2.

Proof: (i) Let $P_1, \ldots, P_n$ be the singular points on C, having multiplicities $r_1, \ldots, r_n$, respectively. Since we assume that all singularities are ordinary and C is rational, we have

$$(d-1)(d-2) = \sum_{i=1}^{n} r_i(r_i - 1).$$

By application of Proposition 1 we see that the dimension of the system of adjoint curves of degree $d-2$, $\mathcal{H}_{d-2}$, to C is

$$\frac{(d-1)d}{2} - 1 - \sum_{i=1}^{n} \frac{(r_i - 1)r_i}{2} = d - 2.$$

Now we can apply Theorem 2 for $s = d-2$ (i.e. choosing the whole system) and we get that C has families of $d-2$ conjugate simple points. Similarly, by using systems of adjoint curves of degrees $d-1$ and d, respectively, we get that C has families of $2d-2$ and $3d-2$ conjugate simple points.
(ii) We first apply statement (i) to obtain two different families of $(d-2)$ simple points. Let $\mathcal{H}_{d-1}$ be the system of adjoint curves of degree $(d-1)$. Applying Proposition 1 one has that the linear subsystem $\tilde{\mathcal{H}}_{d-1}$ obtained by forcing all the points in these two families to be simple base points of $\mathcal{H}_{d-1}$ is rational of dimension 2. Thus, applying Theorem 2 to $\tilde{\mathcal{H}}_{d-1}$ one obtains families of two simple points.
(iii) Applying statement (ii) one can determine $\frac{d-3}{2}$ different families of two simple points on C. Let $\mathcal{H}_{d-2}$ be the system of adjoint curves of degree $(d-2)$. Applying Proposition 1 one has that the linear subsystem $\tilde{\mathcal{H}}_{d-2}$ obtained by forcing all the points in these families to be simple base points of $\mathcal{H}_{d-2}$ is rational of dimension one. Thus, applying Theorem 2 one concludes that C has simple points over $\mathbb{L}$.
(iv) This is an inmediate consequence of statement (ii). □

Summarizing we get the following algorithm for deciding the parametrizability over $\mathbb{R}$ and, in the positive case, computing such a parametrization.

Algorithm REAL-PARAM(f)

 – **Input:** $F(x_1, x_2, x_3) \in \mathbb{L}[x_1, x_2, x_3]$ is an irreducible homogeneous polynomial of degree d, that defines a rational plane curve C.

- **Output:** a real parametrization of C, or
"`no-real-parametrization`" if no real parametrization exists.

(1) Compute the linear system H of adjoint curves to C of degree $(d-2)$.
(2) If d is odd, apply Theorem 3 (iii) to find $(d-3)$ simple points of F over $\mathbb{L}$.
(3) If d is even, apply Theorem 3 (ii) to find $\frac{d-3}{2}$ families of two simple points of F over $\mathbb{L}$.
(4) Determine the linear rational subsystem $\tilde{H}$ obtained by forcing the points computed in steps (2) and (3) to be simple base points on H.
(5) Take $\tilde{\Phi}_1, \tilde{\Phi}_2, \tilde{\Phi}_3 \in \tilde{H}$ such that the common intersections of the three curves $\tilde{\Phi}_i$ and F are the set of base points of $\tilde{H}$, and such that

$$\mathcal{T} = \{y_1 : y_2 : y_3 = \tilde{\Phi}_1 : \tilde{\Phi}_2 : \tilde{\Phi}_3\}$$

is a birational transformation (Theorem 1).
(6) Determine the transformed curve $\mathcal{D}$ to C obtained by $\mathcal{T}$. Note that applying Theorem 1 one has that $\mathcal{D}$ is either a conic or a line depending on whether d is even or odd, respectively. $\mathcal{D}$ can be easily determined by sending a few points from C to $\mathcal{D}$ and then interpolating.
(7) If d is odd, parametrize the line $\mathcal{D}$ over $\mathbb{L}$. Apply the inverse transformation $\mathcal{T}^{-1}$ to find a parametrization of C over $\mathbb{L}$, and therefore over $\mathbb{R}$. (Or, alternatively, determine as many points on $\mathcal{D}$ over $\mathbb{L}$ as necessary, transfer them back to C by $\mathcal{T}^{-1}$, and use them for computing a parametrization of C over $\mathbb{L}$.)
(8) If d is even, decide whether the conic $\mathcal{D}$ can be parametrized over $\mathbb{R}$. If so, parametrize $\mathcal{D}$ over $\mathbb{R}$. Apply the inverse transformation $\mathcal{T}^{-1}$ to find a real parametrization of C over $\mathbb{R}$. (Or, alternatively, determine as many points on $\mathcal{D}$ over $\mathbb{R}$ as necessary, transfer them back to C by $\mathcal{T}^{-1}$, and use them for computing a parametrization of C over $\mathbb{R}$.)
If not, report "`no-real-parametrization`". $\square$

In step (8) we have to decide whether an irreducible conic $\mathcal{D}$, defined by a homogeneous polynomial $G(y_1, y_2, y_3) \in \mathbb{L}[y_1, y_2, y_3]$, contains a real point P. If so, then we can obviously parametrize $\mathcal{D}$ by intersecting it by lines through P.

In fact we can decide whether $\mathcal{D}$ contains a rational point. For details see [IR82] and [HW98]. If this is not the case, we can transform $\mathcal{D}$ to an equivalent conic $\mathcal{D}'$ by a birational mapping over $\mathbb{R}$, such that the defining equation of $\mathcal{D}'$ has the form

$$y_1^2 \pm y_2^2 \pm y_3^2 = 0$$

(see, e.g., the Law of Inertia in [Ga59]). Then $\mathcal{D}'$, and hence $\mathcal{D}$, has a real point if and only if the defining equation of $\mathcal{D}'$ is not equivalent to $y_1^2 + y_2^2 + y_3^2 = 0$. In fact, if $\mathcal{D}$ contains a real point, then we can compute one. Thus, the question in step (8) can be completely decided.

An alternative approach is to decide the reality of C by computing the signature of the corresponding quadratic form, and once the reality is decided, to find a family of two conjugate points on C (Theorem 3 (ii)) whose quadratic defining polynomial has real roots.

4   Real Reparametrization

If a complex rational parametization $\mathcal{P}(t)$ of an irreducible affine plane curve $\mathcal{C}$ over $\mathbb{C}$ is given, or computed by any parametrization algorithm, the alternative approach presented in [RS97a] may be considered. In this situation, the reality of $\mathcal{C}$ is decided by computing a gcd of two real bivariate polynomonials, and if the curve is real, a linear parameter change is determined to transform the original parametrization into a real one.

The main idea of the algorithm presented in [RS97a] is to associate with the original parametrization a plane curve that contains as points the complex values (taking the real and imaginary parts) of the parameter that generates, via the parametrization, the real points on the original curve. Then the reality of the original curve is characterized by means of the reality of the associated curve, that is proved to be either a line or a circle. More precisely, let $\mathcal{P}(t)$ be the proper complex parametrization of $\mathcal{C}$:

$$x(t) = \frac{q_1(t)}{h(t)}, \quad y(t) = \frac{q_2(t)}{h(t)},$$

where $q_1, q_2, h \in \mathbb{C}[t]$ and $\gcd(q_1, q_2, h) = 1$. Then, we apply the formal change of variable $t = t_1 + it_2$ to $\mathcal{P}(t)$ to obtain:

$$x(t_1 + it_2) = \frac{u_1(t_1, t_2) + iv_1(t_1, t_2)}{h_1(t_1, t_2)^2 + h_2(t_1, t_2)^2},$$

$$y(t_1 + it_2) = \frac{u_2(t_1, t_2) + iv_2(t_1, t_2)}{h_1(t_1, t_2)^2 + h_2(t_1, t_2)^2}$$

where $h_1, h_2 \in \mathbb{R}[t_1, t_2]$, $u_1, v_1 \in \mathbb{R}[t_1, t_2]$ and $u_2, v_2 \in \mathbb{R}[t_1, t_2]$ are the real and imaginary parts of $h(t_1 + it_2)$, $q_1(t_1 + it_2) \cdot \bar{h}(t_1 - it_2)$ and $q_2(t_1 + it_2) \cdot \bar{h}(t_1 - it_2)$, respectively ($\bar{h}$ denotes here the conjugate of h). Then, it is proved in [RS97a] that the plane curve $\mathcal{C}$ is real if and only if $\gcd(v_1, v_2)$ is either a real line or a real circle. Furthermore, if the plane curve $\mathcal{C}$ is real, and $(m_1(t), m_2(t))$ is a real proper rational parametrization of $\gcd(v_1, v_2)$, then $\mathcal{P}(m_1(t) + i\, m_2(t))$ is a real proper rational parametrization of $\mathcal{C}$.

Clearly, these two results provide an algorithm for deciding the reality of curves, and in the affirmative case, computing the linear change of parameter that reparametrizes the original complex proper parametrization into a real proper parametrization.

5   Examples

Example 2: We consider the curve $\mathcal{C}_1$ of Example 1. $\mathcal{C}_1$ is defined by

$$f_1(x, y) = (x^2 + 4y + y^2)^2 - 16(x^2 + y^2) = 0.$$

Let us first apply the algorithm REAL-PARAM to $\mathcal{C}_1$ to see whether it is parametrizable.

C_1 has 3 double points in the projective plane, namely

$$(0:0:1) \quad \text{and} \quad (1:\pm i:0).$$

So genus(C_1) = 0, which means that C_1 is rational and must have a parametrization over $\mathbb{C}$ (the picture actually suggests that it is a real curve, and therefore
must have a parametrization over $\mathbb{R}$).

The system $\tilde{\mathcal{H}}$ of conics (curve of degree 2) passing through all three of these
double points is defined by

$$h(x,y,z,s,t) = x^2 + sxz + y^2 + tyz,$$

so it is a system of dimension 2. Let the birational transformation $\mathcal{T}$ be

$$\mathcal{T} = (\Phi_1 : \Phi_2 : \Phi_3) = (h(x,y,z,0,1) : h(x,y,z,1,0) : h(x,y,z,1,1)),$$

i.e.

$$\Phi_1 = x^2 + y^2 + yz, \quad \Phi_2 = x^2 + xz + y^2, \quad \Phi_3 = x^2 + xz + y^2 + yz.$$

We determine the birationally equivalent conic $\mathcal{D}_1$ to C_1 by sending the 6 points
in the families

$$\mathcal{F}_1 = \{(t:-t+2:1) \mid 4t^4 - 32t^3 + 80t^2 - 128t + 80\},$$
$$\mathcal{F}_2 = \{(t:1-2t:1) \mid t^2 - 4t + 1\}$$

onto $\mathcal{D}_1$ by $\mathcal{T}$. This gives us the conic defined by

$$15x^2 + 7y^2 + 6xy - 38x - 14y + 23.$$

$\mathcal{D}_1$ has the real (in fact, rational) point $(1, 8/7)$, which (by $\mathcal{T}^{-1}$) corresponds to
the point $P = (0, -8)$ on C_1.

Now we restrict $\tilde{\mathcal{H}}$ to the curves through P. This restricted linear system is
defined by (after renaming of the free parameter)

$$h^*(x,y,z,t) = x^2 + txz + y^2 + 8yz.$$

Computing the resultants of $f_1(x,y)$ and $h^*(x,y,1,t)$, with respect to x and y,
respectively, and taking the primitive parts with respect to the parameter t, one
gets two polynomials $R_1 \in \mathbb{C}[y,t]$ and $R_2 \in \mathbb{C}[x,t]$, such that the degrees of
R_1 and R_2, with respect to y and x, respectively, are one. Hence, solving the
system $\{R_1 = 0, R_2 = 0\}$ in the variables $\{x,y\}$, one gets the following real
parametrization $\mathcal{P}(t)$ of C_1:

$$x(t) = \frac{-1024t^3}{256t^4 + 32t^2 + 1}, \qquad y(t) = \frac{-2048t^4 + 128t^2}{256t^4 + 32t^2 + 1}.$$

Let us now apply the real reparametrization approach of section 4 to C_1.
The idea is, therefore, to apply any basic parametrization algorithm to C_1 without taking care of the field extensions, and afterwards to analyze the possible

reparametrization of the achieved parametrization over the ground field. Thus, the first steps are the same. We consider the linear system $\tilde{\mathcal{H}}$ of adjoint curves of degree 2, and then we force $\tilde{\mathcal{H}}$ to pass through any simple point on the curve. We take, for instance, $Q = (-\frac{128}{9}i : -\frac{160}{9} : 1)$. The obtained linear subsystem of $\tilde{\mathcal{H}}$ has dimension 1 and is defined by the polynomial

$$h^*(x, y, z, t) = x^2 + txz + y^2 + (\frac{34}{5} - \frac{4}{5}it)\, yz.$$

Now, proceeding as above, we get the following parametrization $Q(t)$ over $\mathbb{C}$ of C_1:

$$x(t) = -32\,\frac{-1024\, i + 128\, t - 144\, i\, t^2 + i\, t^4 - 22\, t^3}{2304 - 3072\, i\, t - 736\, t^2 + 9\, t^4 - 192\, i\, t^3}$$

$$y(t) = -40\,\frac{1024 - 256\, i\, t - 80\, t^2 + t^4 + 16\, i\, t^3}{2304 - 3072\, i\, t - 736\, t^2 + 9\, t^4 - 192\, i\, t^3}.$$

Now, we execute the formal change of parameter $t = t_1 + i\, t_2$ in $Q(t)$, and we compute the gcd of the imaginary parts, $v_1(t_1, t_2)$ and $v_2(t_1, t_2)$, of the normalized (i.e. with denominators in $\mathbb{R}[t_1, t_2]$) rational functions $x(t_1 + i\, t_2)$ and $y(t_1 + i\, t_2)$, respectively:

$$D(t_1, t_2) = \gcd(v_1, v_2) = t_1^2 + t_2^2 + 6t_2 - 16.$$

In this situation, since $D(t_1, t_2)$ defines a real circle, it follows that the original curve C_1 is real and, therefore, parametrizing over the reals the curve defined by $D(t_1, t_2)$ one gets the linear change of parameter to transform $Q(t)$ into a real parametrization. More precisely, one takes the real parametrization of $D(t_1, t_2)$:

$$t_1(t) = \frac{-10t}{t^2 + 1}, \quad t_2(t) = \frac{-10t}{t^2 + 1}.$$

Therefore, $\mathcal{L}(t) = Q(\frac{-10t}{t^2+1} + i\,\frac{-10t}{t^2+1})$ is a real parametrization of C_1. In fact, $\mathcal{L}(t)$ is the parametrization:

$$x(t) = \frac{-32t}{16t^4 + 8t^2 + 1}, \quad y(t) = 8\,\frac{4t^4 - 1}{16t^4 + 8t^2 + 1}.$$

$\square$

Example 3: We consider the curve C_2 of Example 1. C_2 is defined by

$$f_2(x, y) = 2y^2 + x^2 + 2x^2y^2 = 0.$$

Let us apply the algorithm REAL-PARAM to C_2 to see whether it can be parametrized. The singularities of C_2 in the projective plane are

$$(1 : 0 : 0), \quad (0 : 1 : 0), \quad (0 : 0 : 1),$$

each of which is a double point. So genus$(C_2) = 0$, which means that C_2 can be parametrized over $\mathbb{C}$.

The system $\tilde{\mathcal{H}}$ of conics passing through all three of these double points is defined by

$$h(x, y, z, s, t) = xz + tyz + sxy,$$

so it is a system of dimension 2. Let the birational transformation $\mathcal{T}$ be

$$\mathcal{T} = (\Phi_1 : \Phi_2 : \Phi_3) = (h(x, y, z, 1, 0) : h(x, y, z, 0, 1) : h(x, y, z, 1, 1)),$$

i.e.

$$\Phi_1 = xz + xy, \quad \Phi_2 = xz + yz, \quad \Phi_3 = xz + yz + xy.$$

We determine the birationally equivalent conic $\mathcal{D}_2$ to $\mathcal{C}_2$ by sending the 8 points in the families

$$\mathcal{F}_1 = \{(t : -2t + 1 : 1) \mid 8t^4 - 8t^3 + 11t^2 - 8t + 2\},$$
$$\mathcal{F}_2 = \{(t : -t + 2 : 1) \mid 2t^4 - 8t^3 + 11t^2 - 8t + 8\}$$

onto $\mathcal{D}_2$ by $\mathcal{T}$. This gives us the conic defined by

$$5z^2 - 6xz - 6yz + 3x^2 + 2xy + 3y^2.$$

$\mathcal{D}_2$ has no real point. So also $\mathcal{C}_2$ can have no real point, i.e. it is NOT parametrizable over $\mathbb{R}$.

But we can parametrize $\mathcal{C}_2$ over $\mathbb{C}$ by passing the system of adjoint curves through the point

$$P = (-\alpha, \alpha), \quad \text{where } 2\alpha^2 + 3 = 0,$$

getting

$$h^*(x, y, z, t) = xz + tyz + \frac{2}{3}(\alpha - \alpha t)xy.$$

Now, proceeding as in Example 2, we get the following parametrization $\mathcal{P}(t)$ of $\mathcal{C}_2$ over $\mathbb{C}$:

$$x(t) = \frac{-\alpha t^2 - 2\alpha}{t^2 - 2t - 2}, \quad y(t) = \frac{-\alpha t^2 - 2\alpha}{t^2 + 4t - 2}.$$

We leave the application of the reparametrization algorithm to the reader. $\square$

Conclusion

So, as we have seen above, we can parametrize any parametrizable real curve by a real parametrization $\mathcal{P}(t) = (x(t), y(t))$, i.e. $x(t), y(t) \in \mathbb{R}(t)$. This is what we usually need in applications, such as in computer aided geometric design. The algorithms described in this paper allow us, for the first time, to decide the possibility of a real parametrization and, if it exists, to actually compute one.

References

[AB88] Abhyankar S.S, Bajaj C.L., (1988), *Automatic parametrization of rational curves and surfaces III: Algebraic plane curves.* Computer Aided Geometric Design **5**, 309-321.

[Ga59] Gantmacher F.R., (1959), *The Theory of Matrices.* Chelsea, New York.

[HH90] Hilbert D., Hurwitz A. (1890), *Über die Diophantischen Gleichungen vom Geschlecht Null.* Acta math. **14**, 217-224.

[HW98] Hillgarter E., Winkler F. (1998), *Points on algebraic curves and the parametrization problem.* In: D. Wang (ed.), Automated Deduction in Geometry, 185-203, Lecture Notes in Artif. Intell. 1360, Springer Verlag Berlin Heidelberg.

[vH97] van Hoeij M. (1997), *Rational parametrizations of algebraic curves using a canonical divisor.* J. Symbolic Computation **23/2&3**, 209-227.

[IR82] Ireland K., Rosen R. (1982), *A classical introduction to modern number theory.* Springer Verlag, Graduate Texts in Mathematics, New York.

[MSW96] Mňuk M., Sendra J.R., Winkler F. (1996), *On the complexity of parametrizing curves.* Beiträge zur Algebra und Geometrie **37/2**, 309-328.

[MW96] Mňuk M., Winkler F. (1996), *CASA — A System for Computer Aided Constructive Algebraic Geometry.* In: J. Calmet and C. Limongelli (eds.), Proc. Internat. Symp. on Design and Implementation of Symbolic Computation Systems (DISCO'96), 297-307, LNCS 1128, Springer Verlag Berlin Heidelberg New York.

[RS95] Recio T., Sendra J.R. (1995), *Reparametrización real de curvas reales paramétricas.* Proc. EACA'95, 159-168, Univ. de Cantabria, Santander, Spain.

[RS97a] Recio T., Sendra J.R. (1997), *Real parametrizations of real curves.* J. Symbolic Computation **23/2&3**, 241-254.

[RS97b] Recio T., Sendra J.R. (1997), *A really elementary proof of real Lüroth's theorem.* Revista Matemática de la Universidad Complutense de Madrid **10**, 283-291.

[SW91] Sendra J.R., Winkler F. (1991), *Symbolic parametrization of curves.* J. Symbolic Computation **12/6**, 607-631.

[SW97] Sendra J.R., Winkler F. (1997), *Parametrization of algebraic curves over optimal field extensions.* J. Symbolic Computation **23/2&3**, 191-207.

[Wa50] Walker R.J. (1950), *Algebraic curves.* Princeton Unversity Press.

[Wi96] Winkler F. (1996), *Polynomial algorithms in computer algebra.* Springer-Verlag Wien New York.

Non-clausal Reasoning with Propositional Definite Theories

Zbigniew Stachniak*

York University, Toronto, Canada
zbigniew@cs.yorku.ca

Abstract. In this paper we propose a non-clausal representational formalism (of *definite formulas*) that retains the syntactic flavor and algorithmic advantages of Horn clauses. The notion of a definite formula is generic in the sense that it is available to any logical calculus. We argue that efficient automated reasoning techniques which utilize definite formula representation of knowledge (such as SLD-resolution) can be developed for classical and a variety of non-classical logics.

1 Introduction

Among the most effective methods by which the efficiency of automated reasoning in classical logic can be achieved is the restriction of the reasoning process to formulas in a specific syntactic form, most notably to clauses, and the use of inference rules and proof techniques that are tailored to the selected normal form. Resolution (Robinson, [7]) and Clausal Boolean Constrain Propagation (McAllester, [5]) for clauses, and SLD-resolution for Horn clauses (cf. [2]) can serve as examples.

The construction of efficient clause-based automated reasoning methods for non-classical logics presents two major obstacles. First, the syntactic analogue of a clause that would provide a syntactic base for an efficient reasoning could be difficult to find. The notion of a clause in classical logic relies on the standard interpretation of logical connectives. In a non-standard logic classical connectives can be either absent, supplemented with other connectives (modal, temporal, etc), or interpreted in a non-standard way. Second, some applications require that the logical equivalence, rather than satisfiability, is to be preserved through the transformation of formulas of an object language into clauses. Even for classical logic such a transformation can result in an exponentially larger theory. For these reasons non-clausal proof techniques have been developed for AI applications that do not (or cannot) relay on clause manipulation (e.g, some truth-maintenance systems or general purpose knowledge representation systems). Some of the recent examples include Restricted Fact Propagation (Roy-Chowdhury and Dalal, [8]), linear resolution for theories in negation normal form [1], resolution proof systems for Resolution Logics (Stachniak, [10]).

* Research supported by a grant from the Natural Sciences and Engineering Research Council of Canada.

In this paper we propose a representational formalism (of *definite formulas*) that is intrinsically non-clausal, but at the same time it retains the syntactic flavor and algorithmic advantages of Horn clauses. It does not relay on a rigid structure of a normal form but, instead, it captures the logical features of literals in a clause. The proposed non-clausal analogue of a Horn clause has decided advantages. Due to its free-form nature, "we avoid the proliferation of sentences and the disintegration of intuition that accompany the translation to clausal form", [4]. Another is that it allows a universal Horn close-like representation of knowledge across the class of logical systems. For classical logic, definite formula and Horn clause fragments are of equal expressive power. However, representing knowledge using definite formulas rather than Horn clauses may result in an exponentially smaller theory. And finally, efficient automated reasoning techniques that utilize definite formula representation of knowledge can be developed for classical and a variety of non-classical logics. In this paper we adopt SLD-resolution to definite formula fragments of logical systems for which non-clausal resolution proof systems are available.

The rest of the paper is divided into three parts. In the first part, Section 2, we use the example of classical logic for a brief and informal introduction to definite formulas. In Section 3 we discuss ways of defining definite formulas in the context of non-classical logics. Our extension relies on the notion of polarity which we discuss in the first place. We then briefly review the non-clausal resolution framework for non-classical logics and define its SLD refinement. The last part, Section 4, comprises the mathematical results on non-clausal SLD-resolution. We give a list of conditions sufficient for the completeness of SLD-resolution. We use classical propositional logic and Łukasiewicz three-valued propositional logic to illustrate concepts and theoretical results reported in this paper. Only propositional logics are discussed; the extension of the proposed representational framework to first-order logic is the subject of a forthcoming paper.

2   From Polarity to Definite Formulas

The basic idea behind the notion of a definite formula of classical logic is quite simple: atoms in a Horn clause (literals) satisfy two polarity constraints:

(p1) every atom has a polarity value '+' (positive, e.g. p is '+' in $p \leftarrow q\,r$),
     '−' (negative, e.g. atoms q and r are both '−' in $p \leftarrow q\,r$), or
     '0' (no polarity, e.g. p in $p \leftarrow p\,q$);
(p2) at most one of the atoms is non-negative (i.e., it is either '+' or '0').

Since polarity values of atoms can be computed for an arbitrary formula (cf. [4] and [6], see also Example B in Section 3.1) these two constraints determine a special class of formulas whose atoms have the same polarity features as literals in a Horn clause. And this is exactly the class of formulas we are interested in. As an example, consider

$$\alpha = (p_1 \wedge \neg q_1) \vee (\neg p_2 \wedge \neg q_2) \vee \ldots \vee (\neg p_n \wedge \neg q_n).$$

If we accept the position that polarity values of variables of α reflect the monotonic behavior of the Boolean function $f_\alpha(p_1, \ldots, p_n, q_1, \ldots, q_n)$ defined by α, then we can assign '+' to p_1 to indicate that f_α is nondecreasing in p_1, and '−' to all the other variables of α, to reflect the fact that f_α is nonincreasing in any of these variable arguments. In short, the polarity status of the variables in α is the same as in the Horn clause $\alpha_H = p_1 \vee \neg q_1 \vee \neg p_2 \vee \neg q_2 \vee \ldots \vee \neg p_n \vee \neg q_n$. Indeed, Boolean function defined by α_H is nondecreasing in p_1 and nonincreasing in all the other variable arguments. Let us stress however, that α and α_H are not equivalent in classical logic. Moreover, the transformation of α into conjunctive normal form using the distributivity law yields 2^n Horn clauses.

To sum things up, by adopting a certain notion of polarity that effectively labels atoms in formulas as positive, negative, or of no polarity, we can select a class of *definite formulas* (in short, *d-formulas*), i.e., formulas of an object language whose atoms satisfy (p1) and (p2). The formulaic analogues of the familiar notions such as 'goal clause', 'axiom clause', or 'rule clause', can be easily reconstructed (e.g., a 'goal formula' is a d-formula with only negative atoms).

Having the notion of a d-formula introduced (still informally at this point), we can now ask the question if efficient automated reasoning methods for Horn clauses can be adopted for efficient reasoning with d-formulas. To answer this question we select a well-known method – SLD-resolution. Any other method could have been chosen (e.g., Clausal Boolean Constraint Propagation, cf. [5]).

SLD-resolution procedure (which we identify in this paper with the input refinement of resolution with a goal as the top clause) restricts deductions to sequences

$$G_0, C_0, \ldots, G_{s-1}, C_{s-1}, G_s,$$

of Horn clauses, where: G_0 is a goal clause (the *top goal*) and $C_0, \ldots, C_{s-1}$ are non-goal clauses which, together with G_0, represent a given reasoning task. Every $G_{i+1}, i < s$, is a goal clause obtained by resolving G_i with C_i. The well-known fact for ground Horn clauses states that if $X = \{C_0, \ldots, C_{s-1}\}$ consists of non-goal clauses, then $X \cup \{G_0\}$ is inconsistent *iff* there is SLD-refutation of this set (i.e., SLD-deduction of the empty clause) with G_0 as the top goal. First-order logic analogue of this fact is the basis of *SLD*-refutation procedure employed in logic programming systems for definite clauses (cf. [2]).

In order to obtain a similar result for d-formulas of classical logic, we must replace the clausal form of the resolution principle by its non-clausal analogue (cf. [4], [6], [10]):

$$Res: \quad \frac{\alpha(p), \beta(p)}{\alpha(p/F) \vee \beta(p/T)}$$

where F is a contradictory formula (say $p \wedge \neg p$) while T is $\neg F$. If we view F and T as constants (i.e., if we ignore the fact that they contain variables), then the resolvent $\alpha(p/F) \vee G(p/T)$ of a goal formula $G(p)$ with a d-formula $\alpha(p)$ in which p is non-negative, is also a goal formula. This allows an almost direct adaptation of the SLD-resolution procedure to d-formulas. We informally describe this in the following example (for the detailed discussion of SLD-resolution for d-formulas we have to wait until Section 3.2).

EXAMPLE A: Let the set X consist of the following four formulas of classical propositional logic:

(a) r, p, $(p \wedge q) \rightarrow \neg r$, $p \rightarrow q$.

Every formula of X is a d-formula under the standard notion of polarity (which we recall in Example B of the next section). The variables of $(p \wedge q) \rightarrow \neg r$ have negative polarities which, for simplicity, we record by superscripting them by '$-$' (the superscript '$+$' will be used to indicate the positive occurrence of a variable). Hence, $(p^- \wedge q^-) \rightarrow \neg r^-$ is a goal d-formula. The remaining formulas of X can be rewritten as: r^+, p^+, and $p^- \rightarrow q^+$. To refute X we use the non-clausal resolution rule to deduce F in the following way:

$$G_0 : (p^- \wedge q^-) \rightarrow \neg r^- \qquad \text{[top goal]}$$
$$C_0 : r^+ \qquad \text{[formula of } X\text{]}$$

Resolving C_0 with G_0 upon r gives us $F \vee ((p^- \wedge q^-) \rightarrow \neg T)$ which can be simplified to

$$G_1 : \quad F \vee ((p^- \wedge q^-) \rightarrow F)$$

by replacing $\neg T$ with syntactically 'simpler' but equivalent F. We continue the refutation by selecting

$$C_1 : p^- \rightarrow q^+ \qquad \text{[formula of } X\text{]}$$
$$G_2 : (p^- \rightarrow F) \vee F \vee ((p^- \wedge T) \rightarrow F) \qquad \text{[resolvent of } C_1 \text{ and } G_1 \text{ upon } q\text{]}$$
$$C_2 : p^+ \qquad \text{[formula of } X\text{]}$$

By resolving C_2 and G_2 upon p we get $F \vee (T \rightarrow F) \vee F \vee ((T \wedge T) \rightarrow F)$ which can be simplified to

$$G_3 : \quad F$$

by applying Boolean-like reduction rules (e.g., substituting $T \rightarrow F$ by F, or $F \vee F$ by F). This successfully terminates the deductive process. The structure of this refutation resembles clausal SLD-refutation; we always resolve the current goal with a formula of X. The simplified forms of goals are obtained by applying Boolean-like reduction rules. $\qquad \Box$

3   Reasoning with d-Formulas

Among interesting features of d-formulas there is almost universal applicability of this representational framework to a variety of non-classical logics; all that is needed is a suitable generalization of the notion of polarity and, for resolution-based reasoning with such formulas, an extension of the resolution principle beyond classical logic. We derive both by adopting or reusing ideas from various sources, primarily from ([4], [6], [10]). We begin with the formal definition of a definite formula.

3.1   d-Formulas Defined

Let $\mathcal{P} = \langle L, \vdash \rangle$ be an arbitrary propositional logic, where L denotes the language and $\vdash$ the inference operation of $\mathcal{P}$ (we identify L with the set of all well-formed formulas of $\mathcal{P}$). Our task is to select a set $\mathcal{D}_L \subseteq L$ of *definite formulas* according to the criterion spelled out in the preceding section:

DEFINITION: $\alpha \in \mathcal{D}_L$ *(is a d-formula) iff atoms of α satisfy* (p1) *and* (p2).

This criterion is well-defined only when a suitable method of assigning polarity values '+', '−', and '0' to variables in formulas of L is selected. Clearly, not every assignment of polarity values to variables would be useful. Indeed, our intention is to model the notion of a d-formula after the role of positive and negative literals in a Horn clause of classical logic: polarity values of these literals indicate whether or not the Boolean function defined by a given clause is nonincreasing or nondecreasing in its arguments (under the assumption that the truth-value F is smaller than T). Although this interpretation of polarity values is semantic and our definition of a logic $\mathcal{P}$ does not explicitly involve any semantic concepts ($\mathcal{P}$ is just a pair $\langle L, \vdash \rangle$), we can still formulate an analogous but syntactic interpretation of polarity values. The clue is to view formulas of L as reference points, or truth-values, that are ordered using some binary preorder relation $\prec$ on L. We read '$\alpha \prec \beta$' as 'β is logically at least as large as α' and require only that $\alpha \prec \beta$ implies $\alpha \vdash \beta$. Although $\prec$ can be chosen in a number of ways, the relation that is most frequently used in the literature is, in fact, the inference operation $\vdash$ itself, i.e., we let $\alpha \prec \beta$ iff $\alpha \vdash \beta$. This relation is implicitly used in [6]; it is the '*if-then*' relation in [4] and the $\leq_{Rs}$ *polarity relation* in [10]. Other choices for the relation 'logically larger than' can be found in the works just quoted.

Having chosen a 'logically larger than' relation $\prec$, we can require that a polarity assignment algorithm should label variables of formulas of L with '+', '−', or '0' in such a way that:

(p3) if a variable p is labeled '+' in a formula α and if $\beta \prec \gamma$, then
     $\alpha(p/\beta) \prec \alpha(p/\gamma)$.

Informally speaking, (p3) says that a polarity assignment algorithm may label a variable p with '+' in α, only when α is nondecreasing in p with respect to $\prec$. Similarly,

(p4) if a variable p is labeled '−' in a formula α and if $\beta \prec \gamma$, then
     $\alpha(p/\gamma) \prec \alpha(p/\beta)$.

In conclusion, for every logic $\mathcal{P}$ a class of d-formulas can be defined by selecting a polarity assignment algorithm that satisfies (p3) and (p4) with respect to a preselected 'logically larger than' relation. Clearly, in practical applications we would like the assignment of polarity values to be a computationally feasible process. In the examples that follow we discuss two such algorithms that define classes of d-formulas for classical logic and the three-valued logic of Łukasiewicz.

EXAMPLE B: We describe the class of d-formulas for classical logic determined by polarity assignment algorithm whose variant is called the *relational polarity* in [4] and the *operator polarity* in [10]. Let L be the language of classical logic with the connectives $\neg$, $\vee$, $\wedge$, and $\rightarrow$. Let $\alpha(p) \in L$ be a formula containing a variable p. To assign a polarity value to p in α, we first compute the polarity value of every occurrence of p in this formula. This is done by induction on the complexity of α.

- If $\alpha(p)$ is the variable p, then the only occurrence of p (in itself) is positive.
- Having the polarity values of occurrences of p in a formula β assigned, we proceed as follows. Select an occurrence of p in β. This occurrence retains its polarity value in: $\gamma \vee \beta, \beta \vee \gamma, \gamma \wedge \beta, \beta \wedge \gamma$, and in $\gamma \rightarrow \beta$; it changes to the opposite polarity value in: $\neg\beta$ and $\beta \rightarrow \gamma$.

Finally, we declare p positive (negative) in α if every occurrence of p in α is positive (negative). If p is neither positive nor negative, then the polarity value '0' is assigned to it. The reader is asked to verify that this polarity assignment method satisfies (p3) and (p4) with respect to $\vdash$.

If we run this algorithm on the formulas of the set X of Example A, then the polarity values generated for the variables occurring in these formulas will be as indicated in Example A.

Under this polarity assignment algorithm, Horn clause and d-formula fragments of classical logic have the same expressive power: every Horn clause is a d-formula and, conversely, every d-formula can be transformed into an equivalent set of Horn clauses. However, representing knowledge using d-formulas rather than Horn clauses may result in an exponentially smaller theory. Indeed, the transformation of the formula α discussed at the beginning of Section 2 into conjunctive normal form using the distributivity law yields 2^n Horn clauses.

We finally note that the membership in $\mathcal{D}_L$ can be decided in polynomial time.   □

EXAMPLE C: We now turn to the three-valued logic $\mathbf{L}_3$ of Łukasiewicz. We have selected this particular logic since it is a well-known non-classical calculus (see [3] for its definition) and since a non-clausal resolution proof system for $\mathbf{L}_3$ is available (cf. [10], Appendix B).

Although $\mathbf{L}_3$ and classical logic have the same language, the polarity assignment algorithm described in Example B cannot be fully applied to formulas of $\mathbf{L}_3$ and $\vdash$ without violating (p3) and (p4). While assigning '+' to p in, say, $p \vee q$ (q being a variable) would not violate (p3), making p negative in, e.g., $\neg p$ would certainly be wrong. Indeed, if we let α be $q \vee \neg q$ and β be $\neg(\alpha \rightarrow \neg\alpha)$, then in $\mathbf{L}_3$, $\alpha \vdash \beta$ but not $\neg\beta \vdash \neg\alpha$, violating (p4). Even if we restrict the class of d-formulas to only those formulas whose variables satisfy (p3) and (p4) with respect to $\vdash$, the resulting class will not be a very impressive collection.

In principle, switching to another polarity assignment algorithm and/or a different 'logically larger than' relation the selection of d-formulas could be more favorable. This is indeed the case for $\mathbf{L}_3$; under vo-polarity assignment algorithm (for its definition see [10]) the class of d-formulas is the same as for classical logic defined in Example B (cf. [10], Section 5.8).   □

Automated reasoning literature contains a number of polarity assignment algorithms for classical and non-classical logics, and we refer the interested reader to [4], [9], [10]. As we have indicated in Example C, different polarity assignment algorithms typically define different classes of d-formulas, and it is most likely the trade-of between the efficiency of labeling and the expressiveness of the resulting class of d-formulas that will guide the selection of a polarity assignment algorithm for practical applications.

3.2   Non-clausal Resolution

Having the notion of a definite formula at our disposal, we can now move to the next task – efficient resolution-based proof procedures that manipulate d-formulas. As indicated in Section 2, our target is the generalization of SLD-resolution. In this section we provide the necessary definitions; in Section 4 we investigate the scope of applicability of SLD-resolution.

Non-clausal form of the resolution principle comes up in a natural way. The resolution rule Res of classical logic (see Section 2) is a case analysis on the truth of the common atom p: if $\alpha(p)$ and $\beta(p)$ are simultaneously satisfiable, then $\alpha(p/F)$ or $\beta(p/T)$ is satisfiable, and hence, so is the resolvent $\alpha(p/F) \vee \beta(p/T)$. Under the truth-functional semantics (such as logical matrix semantics, cf. [3]) a non-classical logic may require more than two truth-values (and some of these truth-values might not be definable in the object language). Hence, in general, the analysis of satisfiability may consist of more than two cases. In [10] the generalized non-clausal resolution rule is defined by the expression

$$Res : \frac{\alpha_0(p), \ldots, \alpha_n(p)}{\alpha_0(p/v_0) \vee \ldots \vee \alpha_n(p/v_n)},$$

where $\alpha_0(p), \ldots, \alpha_n(p)$ are arbitrary formulas of the object language with a common variable p and $v_0, \ldots, v_n$ are preselected formulas called *verifiers*. The idea behind this generalization of the resolution rule is straightforward: verifiers $v_0, \ldots, v_n$ play a similar role to that of the formulas F and T in the classical case; they realize the case analysis of simultaneous satisfiability of $\alpha_0, \ldots, \alpha_n$. (We assume here that the disjunction connective is available; this assumption is not necessary but it simplifies things mildly.)

During the refutation process new information is generated by applying the resolution rule. Additional rules, called the *reduction rules*, are also used to keep the deduced facts in a syntactically simple form (see [6] and the transformation rules in [4] and [10]). The reduction rules are instructions of the form

replace a subformula $f(w_1, \ldots, w_k)$ *with a simpler* w

(in symbols, $f(w_1, \ldots, w_k) \Rightarrow w$), where f is a k-ary connective of a logic in question and $w, w_1, \ldots, w_k$ are verifiers. In Example A we have already seen such rules in action: $\neg T \Rightarrow F$ has been used to obtain formula G_1 while the rules $(T \rightarrow F) \Rightarrow F, (T \wedge T) \Rightarrow T$, and $(F \vee F) \Rightarrow F$ have been used to deduce G_3.

To be able to determine at what point a refutation process should be terminated we select a special subset $\mathcal{F}$ of the set of verifiers. These are the *terminal verifiers*; the deduction of any of these verifiers from a set X of formulas successfully terminates the deductive process and we declare X *refutable*. In Example A, we had implicitly assumed that $\mathcal{F} = \{F\}$; with the deduction of F (formula G_3) we had successfully terminated the refutation of X. The resolution rule, the reduction rules, and the set of terminal verifiers constitute a deductive proof system called a *resolution proof system*.

3.3   SLD-Resolution

Let L be a propositional language, let $\mathcal{D}_L$ be a given class of d-formulas, and let Rs be a resolution proof system on L whose rule is $n + 1$-argument.

DEFINITION: *Let $X \cup \{G_0\} \subseteq \mathcal{D}_L$ be finite, where G_0 is a goal. SLD-deduction from X with G_0 as the top goal is a sequence*

$$G_0, \; C_0, \dots, G_{s-1}, \; C_{s-1}, \; G_s.$$

Every C_i is a set of $\leq n$ formulas of X. Every $G_{i+1}, i < s$, is a goal obtained by first, resolving G_i with the formulas of C_i and then by simplifying the resolvent using the reduction rules of Rs. This deduction is called a refutation, if G_s is a terminal verifier of Rs.

In the definition of a SLD-deduction every goal G_{i+1} is obtained by resolving the goal formula G_i with the formulas of C_i We should not, however, interpret this to mean that exactly one copy of G_i and exactly one copy of each formula of C_i is used for the resolution. Indeed, G_i may occupy several argument positions of an instance of the resolution rule and so may the formulas of C_i.

EXAMPLE D: Let us return to the three-valued logics L_3 of Łukasiewicz. In [10], Appendix B, one can find the description of the non-clausal resolution proof system, called Rs_3, for this logic. It can be briefly described in the following way.

Rs_3 is defined in terms of six verifiers $v_0, \dots, v_5$. While all these verifiers are required by the reduction rules, only three of them (v_3, v_4, v_5) are needed by the resolution rule

$$\frac{\alpha(p), \beta(p), \gamma(p)}{\alpha(p/v_3) \vee \beta(p/v_4) \vee \gamma(p/v_5)}.$$

The terminal verifiers are: v_0, v_1, v_3, v_4. Some of the reduction rules are listed in the following refutation of the set X defined in Example A:

$G_0 : \; (p^- \wedge q^-) \rightarrow \neg r^-$ [top goal]

$C_0 : \; \{r^+\}$ [subset of X]

Resolving r, r and G_0 upon r gives us $v_3 \vee v_4 \vee ((p^- \wedge q^-) \to \neg v_5)$ which reduces to the goal

$$G_1 : \quad v_4 \vee ((p^- \wedge q^-) \to v_0)$$

using the rules $(v_3 \vee v_4) \Rightarrow v_4$ and $\neg v_5 \Rightarrow v_0$. We continue the refutation with

$C_1 : \{p^- \to q^+\}$ [subset of X]

$G_2 : \quad (p^- \to v_3) \vee (p^- \to v_4) \vee v_4 \vee ((p^- \wedge v_5) \to v_0)$
                     [resolvent of $p \to q, p \to q$, and G_1 upon q]

$C_2 : \{p^+\}$ [subset of X]

Resolving p, p, and G_2 upon p gives us $v_3 \vee v_4 \vee (v_5 \to v_3) \vee (v_5 \to v_4) \vee v_4 \vee ((v_5 \wedge v_5) \to v_0)$. Using the reduction rules of Rs_3 (such as $(v_3 \vee v_4) \Rightarrow v_4$, $(v_5 \to v_3) \Rightarrow v_3$, and $(v_5 \to v_4) \Rightarrow v_4$) we simplify this resolvent to

$$G_3 : \quad v_4$$

which is a terminal verifier of Rs_3.   □

There are logics $\mathcal{P}$ for which no resolution proof system Rs (as described in Section 3.2) can be constructed so that the notions of inconsistency in $\mathcal{P}$ and refutability in Rs coincide (this problem is throughly investigated in [10]). Hence, SLD-resolution method could be unavailable to some logics even when a rich class of d-formulas can be found. In such cases, other reasoning methods for d-formulas should be sought.

4   Technical Results

Having the principles of SLD-resolution spelled out, it is now time to answer a question of a technical nature: *for which resolution proof systems the notions of a refutable set of d-formulas and of SLD-refutable set of d-formulas coincide?*

To answer this question let us select a propositional language L and let $\mathcal{D}_L \subseteq L$ be the set of d-formulas determined by some 'logically larger than' relation $\prec$ and some polarity assignment algorithm. Moreover, let Rs be a resolution proof system on L. Henceforth, $L, \mathcal{D}_L, \prec$, and Rs are fixed. So far in our discussion we have not required that the choice of $\mathcal{D}_L$ and of the relation $\prec$ is to be linked in any way to the properties of Rs. This has to change if we want SLD-resolution to be complete for d-theories (i.e., if we want every refutable set of d-formulas to be SLD-refutable). Below we state four conditions that imply completeness of SLD-resolution (cf. Theorem 1 below).

(s1) If v and w are verifiers such that $v \prec w$ and w is a terminal verifier of Rs, then so is v.

(s1) states that terminal verifiers should not be logically larger than non-terminal verifiers. In classical logic, (s1) prohibits $T \prec F$ while $F \prec T$ may or may not hold. To force $F \prec T$ to be true, we add:

(s2) If v and w are two different verifiers that appear in the resolution rule of Rs, then either $v \prec w$ or $w \prec v$.

Next, we want proper 'refutational' behavior of disjunction, i.e., if the disjunction $v \lor w$ of two verifiers is refutable, then both v and w should be refutable:

(s3) If v and w are verifiers of Rs, then $v \lor w$ can be reduced to a terminal verifier iff both v and w are terminal verifiers.

The role of the reduction rules is not to generate new information during the deductive process but only to rewrite resolvents into a simpler form. Therefore, this process should preserve $\prec$:

(s4) If α^* is obtained from α by an application of a reduction rule and if β^* is obtained from β in the same way, then $\alpha \prec \beta$ implies $\alpha^* \prec \beta^*$.

THEOREM 1: *Let $S \cup \{G\}$ be a finite minimal refutable set of d-formulas, where G is a goal. If Rs satisfies (s1)–(s4), then $S \cup \{G\}$ has SLD-refutation with G as the top goal.*

The non-clausal resolution proof system for classical logic discussed informally in Section 2, and the resolution system Rs_3 for the three-valued logic of Łukasiewicz satisfy (s1)–(s4) with respect to the class of d-formulas defined in Examples B and C. By Theorem 1, SLD-resolution is available to these systems.

The rest of this section is devoted to the proof of Theorem 1. This is accomplished using the semantic tree argument.

Let $v_0, \ldots, v_n$ be the verifiers used in the resolution rule of Rs. To make the presentation reasonably simple, we shall treat the verifiers of Rs as logical constants, i.e., we shall be ignoring the fact that they may contain variables. Let X be a finite set of formulas of L and let $p_1, \ldots, p_k$ be all the variables that occur in formulas of X. A *semantic tree* of X is a finite tree T_X that represents all possible assignments of verifiers $v_0, \ldots, v_n$ to the variables $p_1, \ldots, p_k$. The root node of T_X (on level 0) represents the empty assignment. The path from the root of T_X to a node N at level l determines a partial assignment h_N which assigns verifiers to the first l variables $p_1, \ldots, p_l$ and which is undefined for the remaining variables. A non-leaf node N has $n+1$ children; the i-th child extends the assignment h_N by assigning the verifier v_i to p_{l+1}.

A node N of T_X is a *failure node* if and only if there is $\alpha \in X$ which can be reduced to a terminal verifier when all its variables are replaced by verifiers as indicated by the assignment h_N; in such a case, we shall say that N falsifies α. We do not expand failure nodes further, hence they are leaves in T_X. A node N is an *inference node* if it is a non-failure node whose every child is a failure node. A semantic tree T_X of X is said to be *closed* provided that every leaf of T_X is a failure node.

Given a closed semantic tree T_X of X, we label all the nodes of T_X in the following way. The label of a leaf N is any formula of X that is falsified by N. If N is an internal node on level l and if $\alpha_0, \ldots, \alpha_n$ are the labels of the children of N, then the label of N is the resolvent $\alpha_0(p_l/v_0) \lor \ldots \lor \alpha_n(p_l/v_n)$.

LEMMA 2: *Let X be a finite set of formulas refutable in Rs. Then every semantic tree T_X of X is closed and the label of the root of T_X can be transformed into a terminal verifier using the reduction rules of Rs.*

Proof: Let X be as stated, let T_X be a semantic tree of X, and let $p_1, \ldots, p_k$ be all the variables that occur in formulas of X. Since X is refutable, the resolution rule of Rs can be used to deduce a formula $\alpha = \beta_0 \vee \ldots \vee \beta_s$, where:

(a) for every choice $v_{i_1}, \ldots, v_{i_k}$ of k verifiers there exists $\beta^* \in X$ such that $\beta^*(p_1/v_{i_1}, \ldots, p_k/v_{i_k})$ is one of the disjuncts of α (cf. [10], Lemma 4.17).

Moreover, α can be reduced to a terminal verifier. By (s3), every $\beta_j, j \leq s$, can be reduced to a terminal verifier. So, for every leaf N we can select $\beta^* \in X$ (as in (a)) that is falsified by N. This means that T_X is closed. Finally, we can label the root of T_X with the disjunction of some (possibly all) disjuncts of α and, by (s3), this label is reducible to a terminal verifier.   □

LEMMA 3: *If a finite set X of formulas has a closed semantic tree, then X is refutable in Rs.*

Proof: The proof (by induction on the number k of inference nodes of T_X) is left to the reader.   □

PROOF OF THEOREM 1: By Lemma 2, there is a closed semantic tree T of $S \cup \{G\}$. Since G is a goal, at least one variable occurs in it and, hence, T has at least one non-leaf node. We prove this theorem by induction on the number k of non-leaf nodes.

Let $k = 1$. By Lemma 2, the label of the root N of T, i.e., the resolvent of labels of the children of N, can be reduced to a terminal verifier G^*. Since $S \cup \{G\}$ is minimal refutable, at least one of the children of N is falsified by G while the remaining children are falsified by the formulas of S. Hence, the sequence G, S, G^* is a required SLD-refutation.

Next, assume that if a minimal refutable set $S_1 \cup \{G_1\}$ has a closed semantic tree of $\leq k$ internal nodes, then it has SLD-refutation with G_1 as the top goal. Suppose that our tree T has $k + 1$ internal nodes. First, we have to do some leaf relabeling. Select an inference node N, on some level l, and consider the labels $\alpha_0(p_l), \ldots, \alpha_n(p_l)$ of the children of N. Suppose that p_l is negative in some $\alpha_i(p_l)$ and that $\alpha_j(p_l) = G$. If $v_i \prec v_j$, then, by (p4), $\alpha_i(p_l^-/v_j) \prec \alpha_i(p_l^-/v_i)$ and, by (s1) and (s4), the j-th child of N also falsifies α_i. So, we can replace the label of the j-th child by α_i. By repeating this process of substitution enough times we can guarantee that:

(a) if a child N_j of N is labeled with G, then for every child $N_i \neq G$ such that $v_i \prec v_j$, p_l is positive or of no polarity in α_i.

We repeat this procedure for the remaining inference nodes. Next, let us select and fix an inference node N of T, say on level l, such that at least one of its children falsifies G (such an N exists since $S \cup \{G\}$ is minimal refutable). Assuming (a) and performing the 'renaming' operation on children of N similar to that described above, we use (s2) to conclude that:

(b) if $\alpha_i(p_l^-)$ is a label of a child of N, then $\alpha_i = G$.

By (b), p_l is negative only in G. So, the label G_1 of N must be a goal formula.

Finally, we form S_1 by removing from S all the formulas that label the children of N and which are not the labels of some other leaves of T. The tree T_1 obtained from T by removing the children of N is a closed semantic tree of $S_1 \cup \{G_1\}$ with k internal nodes. By Lemma 3, $S_1 \cup \{G_1\}$ is refutable (without any loss of generality we can also assume that $S_1 \cup \{G_1\}$ is minimal refutable). By the inductive hypothesis, there is a SLD-refutation $\mathcal{R}$ of $S_1 \cup \{G_1\}$ with G_1 as the top goal. Hence, the sequence $G, X, \mathcal{R}$ is a SLD-refutation of $S \cup \{G\}$ with G as the top goal, where X is the set of all the labels of children of N different from G. $\qquad \square$

5   References

1. Hähnle, R., Murray, N. and Rosenthal, E.: Completeness for Linear Regular Negation Normal Form Inference Systems. *State University of New York, Albany, Technical Report 97-2* (1997).
2. Lloyd, J. W.: *Foundations of Logic Programming*, 2nd ed. Springer-Verlag (1987).
3. Malinowski, G.: *Many-Valued Logics*. Oxford University Press (1993).
4. Manna, Z. and Waldinger, R.: Special Relations in Automated Deduction. *J. ACM* 33 (1986) 1–59.
5. McAllester, D.: Truth Maintenance. *Proc. AAAI-90* (1990) 1109–1116.
6. Murray, N.: Completely Non-Clausal Theorem Proving. *Artificial Intelligence* 18 (1982) 67–85.
7. Robinson, J.A.: A Machine-Oriented Logic Based on the Resolution Principle. *J. ACM* 12 (1965) 23–41.
8. Roy-Chowdhury, R. and Dalal, M.: Model Theoretic Semantics and Tractable Algorithm for CNF-BCP. *Proc. AAAI-97* (1997) 227–232.
9. Shankar, S. and Slage, J.: Connection Based Strategies for Deciding Propositional Temporal Logic. *Proc. AAAI-97* (1997) 172–177.
10. Stachniak, Z.: *Resolution Proof Systems: An Algebraic Theory*. Kluwer Academic Publishers (1996).

Author Index

Lecture Notes in Artificial Intelligence (LNAI)

Lecture Notes in Computer Science